CHILTON'S

TIMING BELT
SERVICE MANUAL

**COVERS ALL 1990-99 U. S. AND CANADIAN VEHICLES
DOMESTIC AND IMPORTED**

President
Dean F. Morgantini, S.A.E.

Vice President–Finance　　　**Vice President–Sales**
Barry L. Beck　　　　　　　　　　Glenn D. Potere

Executive Editor
Kevin M. G. Maher, A.S.E.

Production Manager　　　**Production Assistant**
Ben Greisler, S.A.E.　　　　　Melinda Possinger

Project Managers
George B. Heinrich III, A.S.E., S.A.E., Will Kessler, A.S.E., S.A.E., James R. Marotta, A.S.E.,
S.T.S., Richard Schwartz, A.S.E., Todd W. Stidham, A.S.E., Ron Webb

Schematics Editor
Christopher G. Ritchie

Editors
Christopher Bishop, Leonard Davis, A.S.E., S.T.S., Matthew E. Frederick, A.S.E., S.A.E., Frank
Keytanjian, A.S.E., S.A.E., Thomas A. Mellon, A.S.E., S.A.E., Eric Michael Mihalyi, A.S.E., S.A.E.,
S.T.S., Christine Nuckowski, S.A.E.

CHILTON™ Automotive Books
PUBLISHED BY **W. G. NICHOLS, INC.**

Manufactured in USA, © 1999 W. G. Nichols, 1020 Andrew Drive, West Chester, PA 19380
ISBN 0-8019-9125-0
Library of Congress Catalog Card No. 98-74941
1234567890 8765432109

™Chilton is a registered trademark of the Chilton Company and is licensed to W. G. Nichols, Inc.

Table of Contents

...RMATION 1
...e Engines 1
...ON 1
...AL & INSTALLATION 2
...cura 2
  1.8L (B18B1 & B18C1) ENGINES 2
  2.2L (F22B1) ENGINE 3
  2.5L (G25A4) ENGINE 5
  3.2L & 3.5L ENGINES 6
Audi 11
  4-CYLINDER ENGINES 11
  5-CYLINDER ENGINES 12
  8-CYLINDER ENGINES 14
BMW 15
  2.5L (M20B25) ENGINE 15
Chrysler Corporation 17
  DOMESTIC MODELS 17
    1.6L and 2.0L (VIN V) Engines 17
    2.0L (VIN C) SOHC Engine 18
    2.0L (VIN F) Engine 19
    2.0L (VIN Y) Engine 20
    2.2L SOHC & 2.5L Engines 22
    2.2L DOHC Engine 23
    2.4L (VIN B) Engine 24
    2.4L (VIN X and S) Engines 25
    2.5L (VIN G) Engine 26
    2.5L (VIN H) Engine 27
    2.5L (VIN K) Engine 28
    2.5L (VIN N) Engine 30
    3.0L (VIN 3) Engine 31
    3.0L (VIN J) Engine..............................32
    3.2L (VIN J) and 3.5L (VIN F)
      Engines 34
  IMPORT MODELS 35
    1.5L and 1.8L Engines 35
    1.6L DOHC Engine 37
    2.4L Engine 37
Ford Motor Company 41
  1.3L (VIN H) ENGINE 41
  1.6L (VIN Z & 6) ENGINES 42
  1.8L (VIN 8) ENGINE 43
  1.9L (VIN J) ENGINE 44
  2.0L (VIN 3) ENGINE 47
  2.0L (VIN A) ENGINE 49

Ford Motor Company (cont.)
  2.2L (VIN C) ENGINE 50
  2.3L & 2.5L (VIN C) ENGINES 51
  2.5L (VIN B) ENGINE 54
  3.0L (VIN W) ENGINE 55
  3.0L (VIN Y) SHO ENGINE 57
  3.2L (VIN P) SHO ENGINE 59
Geo/Chevrolet 62
  1.0L (VIN 6) & 1.3 (VIN 9) ENGINES 62
  1.6L (VIN 6) & 1.8L (VIN 8)
    ENGINES 62
  1.6L (VIN U) ENGINE 67
General Motors Corporation 68
  2.0L ENGINE 68
  3.0L (VIN R) ENGINE 70
  3.4L (VIN X) ENGINE 73
Honda 76
  1.5L & 1.6L ENGINES 76
  2.0L (B20A5), 2.1L (B21A1), 2.2L
    (F22B4) & 2.3L (H23A1) ENGINES 79
  2.0L (B20B4) ENGINE 84
  2.2L (F22A1) ENGINE 85
  2.2L (F22B1 & F22B2) ENGINES 86
  2.2L (F22B6) & 2.3L (F23A7)
    ENGINES 88
  2.2L (H22A1) ENGINE 89
  2.6L ENGINE 91
  2.7L ENGINE 92
  3.2L ENGINE 93
Hyundai 95
  1.5L ENGINE 95
  1.6L & 2.0L (VIN P) ENGINES 97
  1.8L (VIN M) ENGINE............................100
  2.0L (VIN F) ENGINE.............................104
  3.0L ENGINE 105
Infiniti 107
  3.0L (VG30) ENGINE 107
  3.0L (VG30DE) ENGINE 108
  3.3L (VG33E) ENGINE 110
Isuzu 111
  1.6L (SOHC) ENGINE 111
  1.6L (DOHC) & 1.8L ENGINES 112
  2.2L (F22B6) & 2.3L (F23A7)
    ENGINES 113

Table of Contents

Isuzu (cont.)
  2.6L ENGINE 114
  3.2L & 3.5L ENGINES 115
Kia..118
  SOHC ENGINE 118
  DOHC ENGINE 119
Lexus..119
  2.5L (2VZ-FE) & 3.0L (3VZ-FE)
    ENGINES 119
  3.0L (1MZ-FE) ENGINE 122
  3.0L (2JZ-GE) ENGINE 123
  4.0L (1UZ-FE) ENGINE 125
Mazda..129
  1.5L (Z5D), 1.6L (B6) & 1.8L (BP)
    ENGINES 129
  1.6L (B6E) & 1.8L (BPE)
    ENGINES 131
  1.6L (B6ZE) ENGINE..........................132
  1.8L (K8) ENGINE 133
  1.8L (K8D) ENGINE 134
  1.8L (BPD) ENGINE 135
  2.0L (FS) ENGINE 136
  2.2L (F2) ENGINE 137
  2.3L (KJ) ENGINE 138
  2.3L (VIN A) & 2.5L (VIN C)
    ENGINES 139
  2.5L (KL) ENGINE 140
  3.0L (JE) ENGINE...............................141
Mitsubishi 144
  1.5L ENGINE 144
  1.6L & 2.0L (NON-TURBO) DOHC
    ENGINES 145
  1.8L, 2.0L & 2.4L SOHC
    ENGINES 147
  2.0L (TURBO) DOHC
    ENGINE 148
  2.4L (VIN W) ENGINE 150
  2.4L (VIN G) ENGINE 151
  3.0L (6G72) SOHC ENGINE................152
  3.0L (6G72) DOHC ENGINE 155
  3.5L ENGINE 158
Nissan 158
  3.0L (VG30DE & VG30DETT)
    ENGINES 158

Nissan (cont.)
  3.0L (VG30E) EXCEPT QUEST
    & 3.3L (VG33E) ENGINES 162
Saab...168
  2.5L (B258L) ENGINE 168
  3.0L (B308L) ENGINE 169
Subaru...170
  1.2L ENGINE 170
  1.8L ENGINE 171
  2.2L ENGINE 176
  2.5L ENGINE 177
  2.7L ENGINE 184
  3.3L ENGINE 186
Suzuki...187
  1.3L ENGINE 187
  1.6L 8-VALVE ENGINE 189
  1.6L 16-VALVE ENGINE 191
Toyota...192
  1.5L (3E-E & 3E) ENGINES 192
  1.5L (5E-FE) ENGINE.........................193
  1.6L (4A-FE) & 1.8L (7A-FE)
    ENGINES 196
  1.6L (4A-GE) ENGINE 198
  2.0L (3S-GTE) ENGINE......................200
  2.0L (3S-FE) & 2.2L (5S-FE)
    ENGINES 201
  2.5L (2VZ-FE) ENGINE 205
  3.0L (1MZ-FE) ENGINE 207
  3.0L (2JZ-GTE & 2JZ-GE)
    ENGINES 208
  3.0L (3VZ-FE) ENGINE 209
  3.0L (7M-GE & 7M-GTE)
    ENGINES 213
  3.4L (5VZ-FE) ENGINE 214
Volkswagen..218
  1.6L DIESEL ENGINE 218
  1.8L ENGINE 218
  1.9L DIESEL ENGINE 219
  2.0L ENGINE 220
Volvo..221
  2.3L (B230F & B230FT) ENGINES 221
  2.3L (B234F) ENGINE 222
  2.3L (B5234) & 2.4L ENGINES 224
  2.9L (6304F) ENGINE.........................225

HOW TO USE THIS MANUAL

Locating Information

The Table of Contents, located at the front of this manual, lists all contents in the order in which they appear in the manual.

To find where a particular manufacturer's engine information is located, you need only look in the Table of Contents. Once you have found the proper listing in the Table of Contents turn to the indicated page to find the applicable information.

Safety Notice

Proper service and repair procedures are vital to the safe, reliable operation of all motor vehicles, as well as the personal safety of those performing the repairs. This manual outlines procedures for servicing and repairing vehicles using safe effective methods. The procedures contain many NOTES, WARNINGS and CAUTIONS which should be followed along with standard safety procedures to eliminate the possibility of personal injury or improper service which could damage the vehicle or compromise its safety.

It is important to note that repair procedures and techniques, tools and parts for servicing vehicles, as well as the skill and experience of the individual performing the work vary widely. It is not possible to anticipate all of the conceivable ways or conditions under which vehicles may be serviced, or to provide cautions as to all of the possible hazards that may result. Standard and accepted safety precautions and equipment should be used when handling toxic or flammable fluids, and safety goggles or other protection should be used during cutting, grinding, chiseling, prying, or any other process that can cause material removal or projectiles.

Some procedures require the use of tools specially designed for a specific purpose. Before substituting another tool or procedure, you must be completely satisfied that neither your personal safety, nor the performance of the vehicle will be endangered.

Although information in this manual is based on industry sources and is as complete as possible at the time of publication, the possibility exists that some vehicle manufacturers made later changes which could not be included here. Information on very late models may not be available in some circumstances. While striving for total accuracy, NP/Chilton cannot assume responsibility for any errors, changes, or omissions that may occur in the compilation of this data.

Part Numbers

Part numbers listed in this book are not recommendations by NP/Chilton for any product by brand name. They are references that can be used with interchange manuals and aftermarket supplier catalogs to locate each brand supplier's discrete part number.

Special Tools

Special tools are recommended by the vehicle manufacturer to perform their specific job. Use has been kept to a minimum, but where absolutely necessary, they are referred to in the text by the part number of the tool manufacturer. These tools may be purchased, under the appropriate part number, from your local dealer or regional distributor, or an equivalent tool can be purchased locally from a tool supplier or parts outlet. Before substituting any tool for the one recommended, read the previous Safety Notice.

Acknowledgments

NP/Chilton wishes to express its appreciation to all manufacturers involved in the production of this book. A special thanks goes to Lincoln Automotive Products, 1 Lincoln Way, St. Louis, MO 63120, for providing the professional shop equipment used in our tear-down facility, including jacks, engine stands, fluid and lubrication tools, and shop presses.

Copyright Notice

No part of this publication may be reproduced, transmitted or stored in any form or by any means, electronic or mechanical, including photocopy, recording or by information storage or retrieval system without prior written permission from the publisher.

GENERAL INFORMATION

Timing belts are typically only used on overhead camshaft engines. Timing belts are used to synchronize the crankshaft with the camshaft, similar to a timing chain on a overhead valve (pushrod) engine. Unlike a timing belt, a timing chain will normally last the life of the engine without needing service or replacement. Timing belts use raised teeth to mesh with sprockets to operate the valvetrain of an overhead camshaft engine.

Whenever a vehicle with an unknown service history comes into your repair facility or is recently purchased, here are some points that should be asked to help prevent costly engine damage:
- Does the owner know if, or when the belt was replaced?
- If the vehicle purchased is used, or the condition and mileage of the last timing belt replacement are unknown, it is recommended to inspect, replace, or at least inform the owner that the vehicle is equipped with a timing belt.
- Note the mileage of the vehicle. The average replacement interval for a timing belt is approximately 60,000 miles (96,000 km).

Interference Engines

Engines, chain- or belt-driven, can be classified as either free-running or interference, depending on what would happen if the piston-to-valve timing is disrupted. A free-running engine is designed with enough clearance between the pistons and valves to allow the crankshaft to rotate (pistons still moving) while the camshaft stays in one position (several valves fully open). If this condition occurs normally, no internal engine damage will result. In an interference engine, there is not enough clearance between the pistons and valves to allow the crankshaft to turn without the camshaft being in time.

An interference engine can suffer extensive internal damage if a timing belt fails. The piston design does not allow clearance for the valve to be fully open and the piston to be at the top of its stroke. If the belt fails, the piston will collide with the valve and will bend or break the valve, damage the piston, and/or bend a connecting rod. When this type of failure occurs, the engine will need to be replaced or disassembled for further internal inspection; either choice costing many times that of replacing the timing belt.

INSPECTION

➡ **For manufacturer's recommended service interval, refer to the maintenance interval chart located in this manual.**

The average replacement interval for a timing belt is approximately 60,000 miles (96,000km). If, however, the timing belt is inspected earlier or more frequently than suggested, and shows signs of wear or defects, the belt should be replaced at that time.

✽✽ WARNING

Never allow antifreeze, oil or solvents to come into with a timing belt. If this occurs immediately wash the solution from the timing belt. Also, never excessive bend or twist the timing belt; this can damage the belt so that its lifetime is severely shortened.

Inspect both sides of the timing belt. Replace the belt with a new one if any of the following conditions exist:

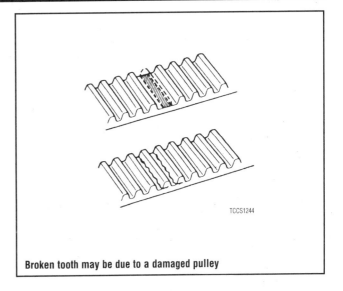

Broken tooth may be due to a damaged pulley

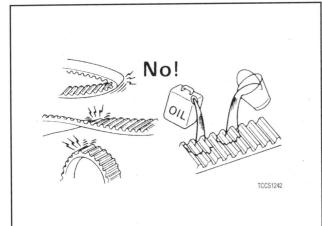

Never bend or twist a timing belt excessively, and do not allow solvents, antifreeze, gasoline, acid or oil to come into contact with the belt

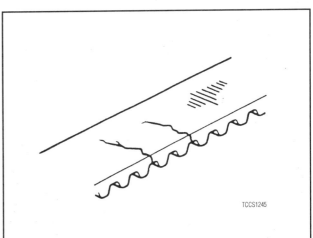

Back surface worn or cracked from a possible overheated engine or interference with the belt cover

2 TIMING BELTS—ACURA

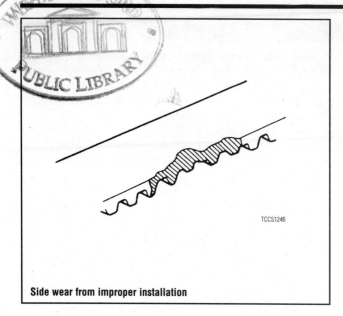

Side wear from improper installation

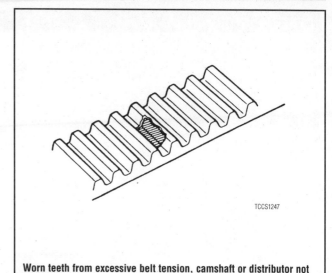

Worn teeth from excessive belt tension, camshaft or distributor not turning properly, or fluid leaking on the belt

- Hardening of the rubber—back side is glossy without resilience and leaves no indentation when pressed with a fingernail
- Cracks on the rubber backing
- Cracks or peeling of the canvas backing
- Cracks on rib root
- Cracks on belt sides
- Missing teeth or chunks of teeth
- Abnormal wear of belt sides—the sides are normal if they are sharp, as if cut by a knife.

If none of these conditions exist, the belt does not need replacement unless it is at the recommended interval. The belt MUST be replaced at the recommended interval.

✱✱ WARNING

On interference engines, it is very important to replace the timing belt at the recommended intervals, otherwise expensive engine damage will likely result if the belt fails.

REMOVAL & INSTALLATION

Acura

1.8L (B18B1 & B18C1) ENGINES

1. Disconnect the negative battery cable.
2. Turn the crankshaft pulley until cylinder No. 1 is set to TDC, compression. The white crankshaft pulley mark should be aligned with the pointer on the lower timing belt cover.
3. Raise and safely support the vehicle. Remove the left front wheel. Remove the wheel well splash guard.
4. Remove the power steering belt and power steering pump. Do not disconnect the power steering fluid lines.
5. Remove the air conditioning belt and the alternator belt.
6. Remove the cruise control actuator.
7. Support the engine.
8. Remove the side engine mount.
9. Tag, then disconnect the spark plug wires from the spark plugs.
10. Remove the cylinder head cover.
11. Remove the crankshaft pulley bolt and the crankshaft pulley. To remove the crankshaft pulley bolt a pulley holder (holder attachment tool part No. 07MAB-PY3010A and holder handle tool part No. 07JAB-001020A, or equivalent) will be needed to keep the crankshaft from turning.

✱✱ WARNING

Do not use the timing belt covers to store small parts. Grease or oil can transfer from the parts to the cover, then to the belt. Clean the covers thoroughly before installation.

12. Recheck that the No. 1 piston is at TDC on its compression stroke. Align the groove on the toothed side of the crankshaft timing belt drive sprocket to the arrow pointer on the oil pump.
13. To set the camshafts to top dead center for the No. 1 cylinder, align the hole in each camshaft with the holes in the No. 1 camshaft holders, then push 5.0mm pin punches into the holes. Be sure that the **UP** arrows are pointing up and that the TDC marks on the intake and exhaust sprockets are aligned.
14. Loosen the tensioner adjusting bolt 180 degrees (½ turn). Push on the tensioner to remove the tension from the timing belt, then retighten the bolt. If the timing belt is to be reinstalled, mark the direction of rotation on the belt with a crayon or white paint.
15. Remove the timing belt from the sprockets.

➡ **Be sure the water pump pulley turns counterclockwise freely. Check for signs of seal leakage; a small amount of weeping from the bleed hole is normal.**

16. If necessary, remove the timing belt tensioner by performing the following:
    a. Remove the timing belt tensioner spring.
    b. Remove the bolt from the timing belt tensioner and remove the tensioner.

To install:

17. If the timing belt tensioner was removed, perform the following:

ACURA—TIMING BELTS

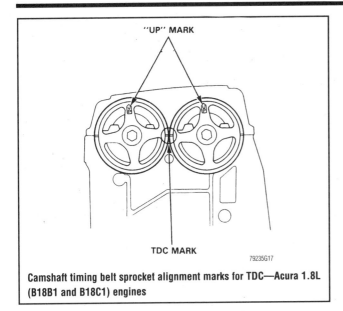

Camshaft timing belt sprocket alignment marks for TDC—Acura 1.8L (B18B1 and B18C1) engines

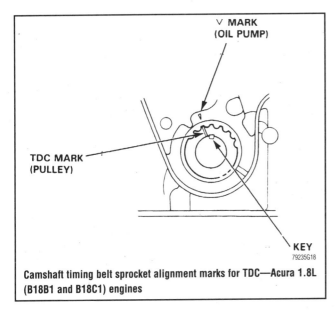

Camshaft timing belt sprocket alignment marks for TDC—Acura 1.8L (B18B1 and B18C1) engines

    a. Position the timing belt tensioner on the engine and install the attaching bolt loosely.
    b. Install the timing belt tensioner spring.
    c. Push the tensioner down, then snug the tensioner bolt to hold this position.

➡**Before reinstallation, check every component for cleanliness. All covers, pulleys, shields, etc. must be completely free of grease and oil.**

➡**Install the timing belt in the correct sequence. Also, if installing the old belt, be sure it is turning the same direction.**

  18. Install the timing belt first to the crankshaft pulley, then to the adjuster, then to the water pump pulley, the exhaust camshaft and finally to the intake camshaft pulley.
  19. Install the lower belt cover. Install the crankshaft pulley, tightening the bolt to 130 ft. lbs. (177 Nm). Lubricate the threads and the flange of the bolt with engine oil before installation.
  20. Loosen the adjusting bolt, allowing the adjuster to tension the belt. Retighten the bolt to 40 ft. lbs. (54 Nm).

  21. Remove the pin punches from the camshafts.
  22. Rotate the crankshaft 4 to 6 turns counterclockwise. This allows the belt to equalize tension across all of the pulleys.
  23. Once again, set the engine to TDC compression for cylinder No. 1. Check that all timing marks for the cam and crankshaft are properly aligned. If any mark is out of alignment, remove the timing belt and reinstall it.
  24. Loosen the adjusting bolt 180 degrees (½ turn). Rotate the crankshaft counterclockwise until the camshaft pulleys have moved 3 teeth. Retighten the adjusting bolt to 40 ft. lbs. (54 Nm).
  25. Check the torque of the crankshaft pulley bolt.
  26. Install the other timing belt covers.
  27. Install the rubber seal in the groove of the cylinder head cover. Be sure that the seal and groove are thoroughly clean first.
  28. Apply liquid gasket to the rubber seal at the eight corners of the recesses. Do not install the parts if 20 minutes or more have elapsed since applying the liquid gasket. Instead, reapply liquid gasket after removing the old residue.
  29. Install the cylinder head cover and all other applicable components. Tighten the cylinder head cover nuts in 2 steps to 88 inch lbs. (10 Nm).

2.2L (F22B1) ENGINES

  1. Disconnect the negative battery cable.
  2. Set the No. 1 piston is at TDC/compression. The white TDC mark on the crankshaft pulley must line up with the pointers on the lower belt cover.
  3. Label and disconnect the engine wiring harness that is routed over the upper belt cover to the alternator and temperature sending unit. Move the harness out of the work area.
  4. Loosen the component adjusting bolts and remove the accessory drive belts.
  5. Remove the dipstick tube.
  6. Mark the sparkplug wires and remove the cylinder head cover.
  7. Remove the timing belt upper cover.
  8. Turn the engine to align the timing marks and set cylinder No.1 to Top Dead Center (TDC). The white mark on the crankshaft sprocket should align with the pointer on the timing belt cover. The words **UP** embossed on the camshaft sprocket should be aligned in the upward position. The marks on the edge of the sprocket should be aligned with the cylinder head or the back cover upper edge. Once in this position, the engine must **NOT** be turned or disturbed.
  9. There are two belts in this system; the belt running to the camshaft sprocket is the timing belt, the other, shorter belt drives the balance shaft and is referred to as the balancer shaft belt, or timing balancer belt. Lock the timing belt adjuster in position, by installing one of the lower timing belt cover bolts in the adjuster arm.
  10. Loosen the timing belt and balancer shaft tensioner adjuster nut, do not loosen the nut more than one revolution. Push the tensioner for the balancer belt away from the belt to relieve the tension. Hold the tensioner and tighten the adjusting nut to hold the tensioner in place.
  11. Carefully remove the balancer belt. Do not crimp or bend the belt; protect it from contact with oil or coolant.
  12. Remove the balancer belt sprocket from the crankshaft.
  13. Loosen the lockbolt, installed in the timing belt adjuster, and the adjusting nut. Push on the timing belt adjuster to remove the tension on the timing belt, then tighten the adjuster nut.
  14. Remove the timing belt by sliding it off the sprockets. Do not crimp or bend the belt; protect it from contact with oil or coolant.

4 TIMING BELTS—ACURA

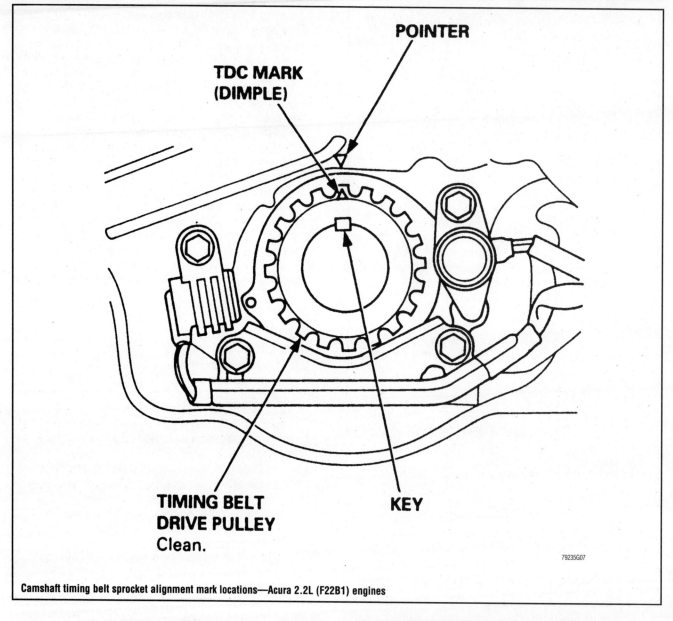

Camshaft timing belt sprocket alignment mark locations—Acura 2.2L (F22B1) engines

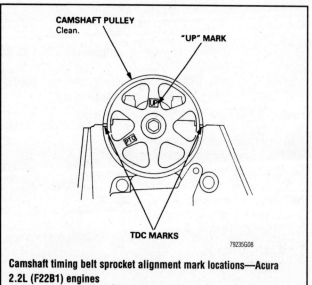

Camshaft timing belt sprocket alignment mark locations—Acura 2.2L (F22B1) engines

15. If necessary, remove the belt tensioners by performing the following:
    a. Remove the springs from the balancer belt and the timing belt tensioners.
    b. Remove the adjusting nut from the belt tensioners.
    c. Remove the bolt from the balancer belt adjuster lever, then remove the lever and the tensioner pulley.
    d. Remove the lockbolt from the timing belt tensioner lever, then remove the tensioner pulley and lever from the engine.
16. This is an excellent time to check or replace the water pump. Even if the timing belt is only being replaced as part of a good maintenance schedule, consider replacing the pump at the same time.

To install:
17. If the water pump is to be replaced, install a new O-ring on the pump and make certain it is properly seated. Install the water pump and tighten the mounting bolts to 106 inch lbs. (12 Nm).
18. If the tensioners were removed, perform the following to install them:
    a. Install the timing belt tensioner lever and the tensioner pulley.

b. Install the balancer belt pulley and adjuster lever.

c. Install the adjusting nut and bolt to the balancer belt adjuster lever.

d. Install the springs to the tensioners.

e. Install the lockbolt to the timing belt tensioner, then move the tensioner it's full deflection and tighten the lockbolt.

f. Move the balancer belt tensioner it's full deflection and tighten the adjusting nut to hold its position.

19. The pointer on the crankshaft sprocket should be aligned with the pointer on the oil pump; camshaft sprocket must be aligned so that the word **UP** is at the top of the sprocket and the marks on the edge of the sprocket are aligned with the surfaces of the head, or the back cover upper edge.

20. Install the timing belt in the following sequence:

    a. Start the belt on the crankshaft sprocket.
    b. Then, around the tensioner sprocket.
    c. On the water pump sprocket.
    d. Finally, around the camshaft sprocket.

21. Check the timing marks to be sure that they did not move.

22. Loosen, then retighten the timing belt adjusting nut.

23. Install the timing/balancer belt drive sprocket and the lower timing belt cover.

24. Install the crankshaft pulley and bolt, tighten the bolt to 181 ft. lbs. (245 Nm). Rotate the crankshaft sprocket five or six revolutions to properly position the timing belt on the sprockets.

25. Set the No. 1 cylinder to TDC and loosen the timing belt adjusting nut one turn. Turn the crankshaft counterclockwise until the cam sprocket has moved 3 teeth; this creates the proper tension on the timing belt.

26. Tighten the timing belt adjusting nut.

27. Set the crankshaft sprocket and the camshaft sprocket to TDC. If the sprockets do not align, remove the belt to realign the marks, then install the belt.

28. Remove the crankshaft pulley and the lower cover.

29. With the timing marks aligned, lock the timing belt adjuster in place with one of the lower cover mounting bolts.

30. Loosen the adjusting nut and ensure the timing balancer belt adjuster moves freely.

31. Align the rear timing balancer sprocket using a 6x100mm bolt or rod. Mark the bolt or rod at a point 2.9 in. (74mm) from the end. Remove the bolt from the maintenance hole on the side of the block; insert the bolt/rod into the hole and align the 2.9 in. (74mm) mark with the face of the hole. This will hold the shaft in place during installation.

32. Align the groove on the front balancer shaft sprocket with the pointer on the oil pump.

33. Install the balancer belt. Once the belts are in place, be sure that all the engine alignment marks are still correct. If not, remove the belts, realign the engine and reinstall the belts. Once the belts are properly installed, slowly loosen the adjusting nut, allowing the tensioner to move against the belt. Remove the bolt from the maintenance hole and reinstall the bolt and washer.

34. Install the crankshaft pulley, then turn the crankshaft sprocket one turn counterclockwise and tighten the timing belt adjusting nut to 33 ft. lbs. (45 Nm).

35. Remove the crankshaft pulley and the bolt locking the timing belt adjuster in place.

36. Install the lower timing belt cover and tighten the bolts to 106 inch lbs. (12 Nm).

37. Install a new seal around the adjusting nut. Do not loosen the adjusting nut.

38. Install the crankshaft pulley. Coat the threads and seating face of the pulley bolt with engine oil, then install and tighten the bolt to 181 ft. lbs. (250 Nm).

39. Install a the dipstick tube, then install the side engine mount. Tighten the bolt and nut attaching the mount to the engine to 40 ft. lbs. (55 Nm). Tighten the through-bolt and nut to 47 ft. lbs. (65 Nm), then remove the jack from under the engine.

40. Install the upper timing belt cover. Tighten the bolt toward the exhaust manifold to 89 inch lbs. (10 Nm) and the bolt toward the intake manifold to 106 inch lbs. (12 Nm).

41. Install the cylinder head cover gasket in the groove of the cylinder head cover. Before installing the gasket, thoroughly clean the seal and the groove. Seat the recesses for the camshaft first, then work it into the groove around the outside edges. Be sure the gasket is seated securely in the corners of the recesses.

42. Apply liquid gasket to the four corners of the recesses in the cylinder head cover gasket. Do not install the parts if five minutes or more have elapsed since applying liquid gasket. After assembly, wait at least 20 minutes before filling the engine with oil.

43. Install the cylinder head (valve) cover and all other applicable components. Tighten the bolts attaching the cylinder head cover in two steps to the proper sequence of 89 inch lbs. (10 Nm).

2.5L (G25A4) ENGINES

1. Disconnect the negative battery cable.
2. Set the No. 1 piston is at TDC/compression. The white TDC mark on the crankshaft pulley must line up with the pointers on the lower belt cover.
3. Label and disconnect the engine wiring harness that is routed over the upper belt cover to the alternator and temperature sending unit. Move the harness out of the work area.
4. Loosen the component adjusting bolts and remove the accessory drive belts.
5. Remove the dipstick tube.
6. Mark the sparkplug wires and remove the cylinder head cover.

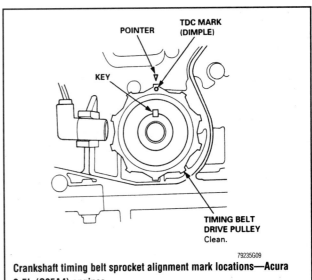

Crankshaft timing belt sprocket alignment mark locations—Acura 2.5L (G25A4) engines

6 TIMING BELTS—ACURA

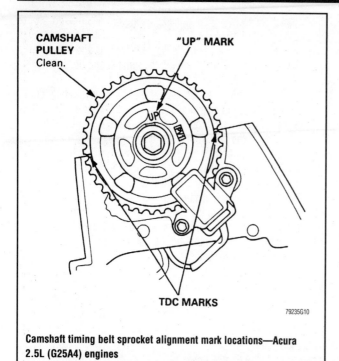

Camshaft timing belt sprocket alignment mark locations—Acura 2.5L (G25A4) engines

7. Remove the timing belt upper cover.
8. Be sure the **UP** mark and the TDC marks on the camshaft sprocket are correctly positioned.
9. Rotate the crankshaft to align the white timing mark on the crankshaft pulley with the pointer on the lower cover. There are similar alignment marks on the crankshaft sprocket and oil pump housing
10. Remove the timing belt lower cover.
11. Loosen the adjusting bolt 180 degrees (½ turn), then push the tensioner down to relieve the belt tension. Retighten the adjusting bolt to 33 ft. lbs. (44 Nm).

➡ **Do not remove the adjusting bolt and tensioner pulley unless they are to be replaced. The bolt is only loosened and tightened in this procedure to tension the timing belt.**

12. Remove the timing belt.
13. Inspect the timing belt tensioner pulley and tension spring for signs of wear. Remove and replace the tensioner assembly as necessary.

To install:

➡ **Replace the timing belt if it shows any signs of wear, damage, or contamination from oil or coolant. The source of any oil or coolant contamination must be determined and corrected before the new timing belt may be installed.**

14. Verify that the crankshaft and camshaft sprocket matchmarks are properly aligned.
15. Install the timing belt in the following order: first onto the crankshaft sprocket, then the tensioner pulley, the water pump sprocket, and finally the camshaft sprocket.
16. Loosen the tensioner adjusting bolt to allow the spring to set the tension. Then, tighten the bolt to 33 ft. lbs. (44 Nm). Rotate the crankshaft six full turns to seat the belt and verify that the timing marks align properly.
17. Install the lower and upper timing belt covers, and tighten the bolts to 106 inch lbs. (12 Nm).

18. Oil only the threads on the crankshaft pulley bolt. Install the crankshaft pulley and tighten the bolt to 181 ft. lbs. (245–250 Nm).
19. Install the dipstick tube.
20. Install the cylinder head cover with new gaskets and washers. Tighten the cap nuts on Vigor models to 84 inch lbs. (9.5 Nm), or to 106 inch lbs. (12 Nm) on 2.5TL models.

3.2L & 3.5L ENGINES

Car Models

LEGEND

1. Disconnect the negative battery cable.
2. Remove the damper, center bracket, and center mount. The center bracket mounting bolts are corrosion resistant and should be replaced during installation.
3. Remove the upper and lower TCS control valve brackets.
4. On GS models, disconnect the TCS throttle sensor connector and TCS throttle actuator connector, then remove the TCS control valve assembly from the throttle body.
5. Remove the engine wiring harness covers and the wiring harness from the front of the engine.
6. Remove the oil pressure switch connector and the engine ground cable.
7. Remove the alternator, power steering, and air conditioning compressor belts.
8. Remove the drive belts for the alternator, air conditioner compressor, and power steering pump.
9. Remove the upper timing belt covers.
10. Rotate the crankshaft to Top Dead Center (TDC) on the compression stroke for the No. 1 piston. The white mark on the crankshaft pulley will be aligned with the pointer on the lower cover, and the camshaft sprocket marks will be aligned with the yellow marks on the rear covers.

❄❄ **WARNING**

Do not rotate the crankshaft or camshafts with the belt removed. The pistons will contact the valves and cause engine damage. If it is necessary to move the camshaft(s)

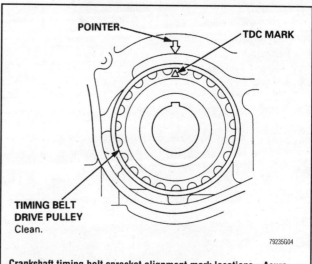

Crankshaft timing belt sprocket alignment mark locations—Acura Legend 3.2L (C32A1) engines

ACURA—TIMING BELTS

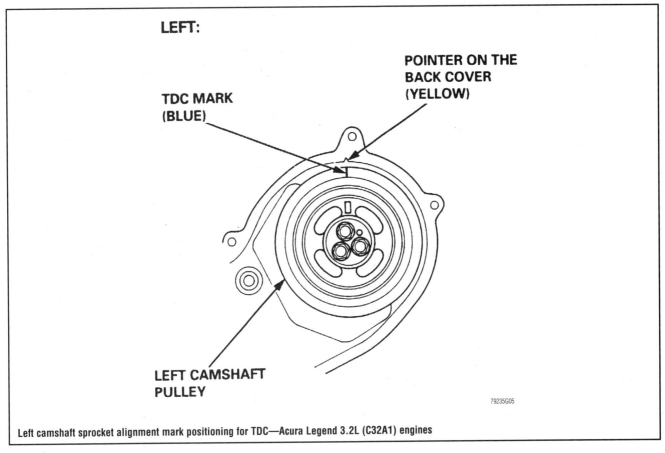

Left camshaft sprocket alignment mark positioning for TDC—Acura Legend 3.2L (C32A1) engines

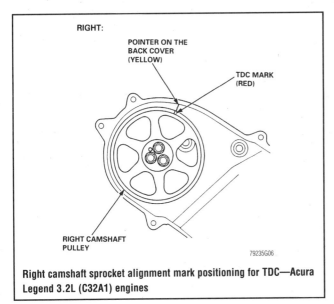

Right camshaft sprocket alignment mark positioning for TDC—Acura Legend 3.2L (C32A1) engines

for proper alignment, first advance the crankshaft by 15 degrees from TDC. Adjust the camshaft position as needed, then return the crankshaft 15 degrees to TDC.

11. Remove the lower timing belt cover.
12. Loosen the timing belt tensioner pulley bolt approximately ½ turn (180 degrees) and push the pulley to slacken the belt tension. Tighten the bolt and remove the belt. If the belt is to be reinstalled, mark the direction of rotation on it with a crayon or with white paint.

To install:

→Replace the belt if it is worn, cracked, or oil soaked. Find and repair the source of the oil leak before installing a new belt. Inspect the water pump. If there is any doubt about the condition of the water pump, it should be replaced now that the timing belt is removed.

13. Install the belt in the following sequence around the sprockets: crankshaft, adjuster pulley, left camshaft, water pump, right camshaft.
14. To adjust the timing belt tension, loosen the tensioner pulley bolt approximately ½ turn (180 degrees). The spring will automatically set the proper tension.
15. Install the lower cover and the crankshaft pulley. Apply oil to the pulley bolt threads and washer, then tighten the pulley bolt to 174 ft. lbs. (240 Nm).
16. Rotate the crankshaft five or six turns clockwise and check that the timing marks on the crankshaft and camshafts align properly. Adjust the timing belt tension again by rotating the crankshaft to align the blue mark on the pulley with the pointer.
17. Loosen the tensioner pulley bolt so that the tensioner can automatically adjust the belt tension, then retighten the bolt to 31 ft. lbs. (43 Nm).
18. Install the dipstick pipe, then the upper covers. Install all applicable components. When installing the center mount, center bracket, and damper, tighten the center mount and bracket bolts to 28 ft. lbs. (39 Nm) and the damper bolts to 16 ft. lbs. (22 Nm).

EXCEPT LEGEND

1. Turn the engine to align the timing marks and set cylinder No.1 to Top Dead Center (TDC) on the compression stroke. The

8 TIMING BELTS—ACURA

white mark on the crankshaft pulley should align with the pointer on the timing belt cover. Remove the inspection caps on the upper timing belt covers to check the alignment of the timing marks. The pointers for the camshafts should align with the marks on the camshaft sprockets.

  2. Remove all necessary component for access to the timing belt covers.

  3. Remove the upper and lower timing belt covers. Clean any dirt, oil or grease from the covers. Do not use the covers for storing removed items.

  4. Loosen the timing belt tensioner adjusting bolt 180 degrees (½ turn). Push on the tensioner to remove tension from the timing belt, then tighten the adjusting bolt.

  5. Remove the timing belt.

  6. If necessary, remove the timing belt tensioner by performing the following:
    a. Remove the spring from the tensioner.
    b. Remove the bolt mounting the tensioner, then remove the tensioner.

To install:

✱✱ CAUTION

Do not rotate the crankshaft pulley or camshaft pulleys with the timing belt removed. The pistons may hit the valves and cause damage.

  7. If necessary, install the timing belt tensioner by performing the following:
    a. Install the tensioner and the attaching bolt.
    b. Move the tensioner its full deflection to the left and tighten the bolt.
    c. Install the spring to the tensioner.

  8. Remove the spark plugs.

  9. Set the timing belt drive (crankshaft) sprocket so that the No. 1 piston is at top dead center (TDC). Align the TDC mark on the tooth of the timing belt drive sprocket with the pointer on the oil pump.

  10. Set the camshaft pulleys so that the No. 1 piston is at TDC. Align the TDC mark on the camshaft pulleys to the pointers on the back covers.

  11. Install the timing belt on the sprockets in the following sequence: drive sprocket (crankshaft), tensioner pulley, left camshaft sprocket, water pump pulley, right camshaft sprocket.

  12. Loosen, then retighten the timing belt adjuster bolt to tension the timing belt.

  13. Install the lower timing belt cover.

  14. Install the crankshaft sprocket and pulley bolt. Tighten the bolt to 174 ft. lbs. (235 Nm) with the aid of the crank pulley holder.

  15. Rotate the crankshaft five or six turns clockwise so that the timing belt positions itself properly on the sprockets.

  16. Set cylinder No. 1 to TDC by aligning the timing marks. If the timing marks do not align, remove the timing belt, then adjust the components and reinstall the timing belt.

  17. Rotate the crankshaft clockwise enough to move the camshaft pulley nine teeth (the blue mark on the crankshaft pulley should line up with the pointer on the lower cover).

  18. Loosen the timing belt adjusting bolt 180 degrees (½ turn), then tighten the bolt to 31 ft. lbs. (42 Nm).

  19. Install the upper timing belt covers, then install all applicable components. When installing the center bracket, tighten the bolts attaching the brackets to 40 ft. lbs. (54 Nm), then the mount through-bolt to 40 ft. lbs. (54 Nm).

Truck and SUV Models

  1. Disconnect the negative battery cable.
  2. Drain the engine coolant into a sealable container.
  3. Remove the air cleaner assembly and intake air duct.
  4. Disconnect the upper radiator hose from the coolant inlet.
  5. Remove the upper fan shroud from the radiator.
  6. Remove the four nuts retaining the cooling fan assembly. Remove the cooling fan from the fan pulley.
  7. Loosen and remove the drive belts.
  8. Remove the upper timing belt covers.
  9. Remove the fan pulley assembly.
  10. Rotate the crankshaft to align the camshaft timing marks with the pointer dots on the back covers. Verify that the pointer on the crankshaft aligns with the mark on the lower timing cover.

➡ **When the timing marks are aligned on 1995 vehicles, no pistons will be at TDC/compression. When the timing marks are aligned on 1995½–99 vehicles, the No. 2 piston is at TDC/compression.**

✱✱ WARNING

Align the camshaft and crankshaft sprockets with their alignment marks before removing the timing belt. Failure to align the belt and sprocket marks may result in valve damage.

  11. Use tool No. J-8614–01, or a suitable pulley holding tool to remove the crankshaft pulley center bolt. Remove the crankshaft pulley.

  12. If present, disconnect the two oil cooler hose bracket bolts on the timing cover. Move the oil cooler hoses and bracket off of the lower timing cover.

  13. Remove the lower timing belt cover.

  14. Remove the pusher assembly (tensioner) from below the belt tensioner pulley. The pusher rod must always face upward to prevent oil leakage. Depress the pusher rod, and insert a wire pin into the hole to keep the pusher rod retracted.

  15. Remove the timing belt.

  16. Inspect the water pump and replace it if there is any doubt about its condition.

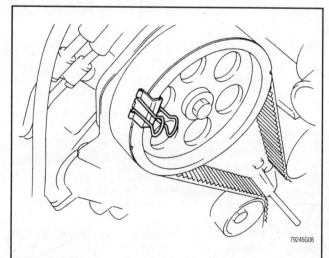

Using a double clip to hold the belt in place—Acura 3.2L and 3.5L truck engines

ACURA—TIMING BELTS

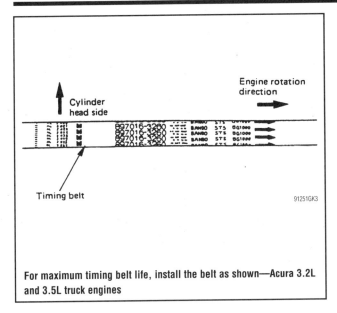

For maximum timing belt life, install the belt as shown—Acura 3.2L and 3.5L truck engines

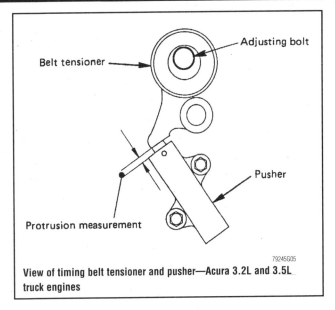

View of timing belt tensioner and pusher—Acura 3.2L and 3.5L truck engines

17. Repair any oil or coolant leaks before installing a new timing belt. If the timing belt has been contaminated with oil or coolant, or is damaged, it must be replaced.

To install:

18. Verify that the sprocket timing marks are still aligned and that the groove and the keyway on the crankshaft timing sprocket align with the mark on the oil pump. The white pointers on the camshaft timing sprockets should align with the dots on the front plate.

19. Install the timing belt. Use clips to secure the belt onto each sprocket until the installation is complete. Align the dotted marks on the timing belt with the timing mark opposite the groove on the crankshaft sprocket.

➡ The arrows on the timing belt must follow the belt's direction of rotation. The manufacturer's trademark on the belt's spine should be readable left-to-right when the belt is installed.

20. Align the white line on the timing belt with the alignment mark on the right bank camshaft timing pulley. Secure the belt with a clip.

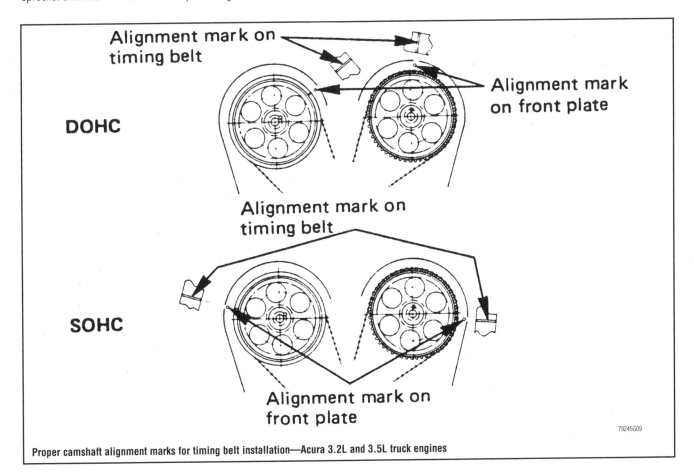

Proper camshaft alignment marks for timing belt installation—Acura 3.2L and 3.5L truck engines

10 TIMING BELTS—ACURA

> ※※ **WARNING**
>
> If any binding is felt when adjusting the timing belt tension by turning the crankshaft, STOP turning the engine, because the pistons may be hitting the valves.

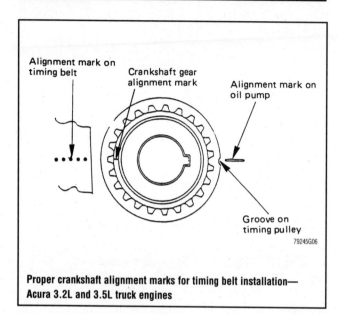

Proper crankshaft alignment marks for timing belt installation—Acura 3.2L and 3.5L truck engines

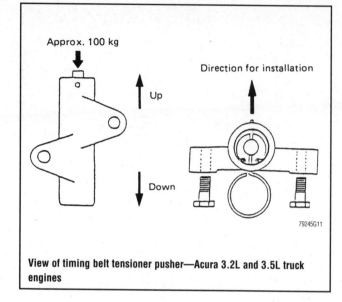

View of timing belt tensioner pusher—Acura 3.2L and 3.5L truck engines

21. Rotate the crankshaft counterclockwise to remove the slack between the crankshaft sprocket and the right camshaft timing belt sprocket.
22. Install the belt around the water pump pulley.
23. Install the belt on the idler pulley.
24. Align the white alignment mark on the timing belt with the alignment mark on the left bank camshaft timing belt sprocket.
25. Install the crankshaft pulley and tighten the center bolt by hand. Rotate the crankshaft pulley clockwise to give slack between the crankshaft timing belt pulley and the right bank camshaft timing belt pulley.
26. Insert a 1.4mm piece of wire through the hole in the pusher to hold the rod in. Install the pusher assembly while pushing the tension pulley toward the belt.
27. Pull the pin out from the pusher to release the rod.
28. Remove the clamps from the sprockets. Rotate the crankshaft pulley clockwise two turns. Measure the rod protrusion to ensure it is between 0.16–0.24 in. (4–6mm).
29. If the tensioner pulley bracket pivot bolt was removed, tighten it to 31 ft. lbs. (42 Nm).
30. Tighten the pusher bolts to 14 ft. lbs. (19 Nm).
31. Remove the crankshaft pulley. Install the lower and upper timing belt covers and tighten their bolts to 12 ft. lbs. (17 Nm).
32. Fit the oil cooler hose onto the timing cover and tighten its mounting bracket bolts to 16 ft. lbs. (22 Nm).
33. Install the crankshaft pulley and tighten the pulley bolt to 123 ft. lbs. (167 Nm).
34. Install fan pulley assembly and tighten the bolts to 16 ft. lbs. (22 Nm).
35. Install and adjust the accessory drive belts.
36. Install the cooling fan assembly and tighten the bolts to 6 ft. lbs. (8 Nm).
37. Install the upper fan shroud.
38. Install the air cleaner assembly and intake air duct.
39. Connect the upper radiator hose to the coolant inlet.
40. Refill and bleed the cooling system.
41. Connect the negative battery cable.

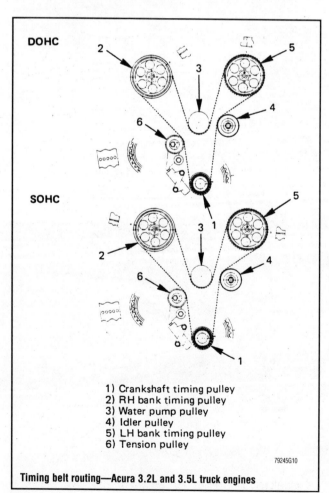

1) Crankshaft timing pulley
2) RH bank timing pulley
3) Water pump pulley
4) Idler pulley
5) LH bank timing pulley
6) Tension pulley

Timing belt routing—Acura 3.2L and 3.5L truck engines

AUDI—TIMING BELTS

Audi

4-CYLINDER ENGINES

1. Disconnect the negative battery cable. Using the large bolt on the crankshaft sprocket, rotate the engine until the No. 1 cylinder is at TDC of the compression stroke. At this point, both valves will be closed and the **0** mark on the flywheel will be aligned with the pointer on the bell housing. If the belt hasn't jumped, the timing mark on the rear face of the camshaft sprocket should be aligned with the upper left edge of the valve cover.
2. Loosen the alternator adjusting bolts, pivot the alternator over and slip the drive belt off.
3. Loosen the air conditioning compressor mounting bolts and remove the drive belt.
4. Remove the valve cover nuts and remove the valve cover and retaining straps.
5. Remove the upper timing belt cover nuts. Note the position of the washers and spacers while removing the cover.
6. Remove the crankshaft pulley retaining bolts. If the sprocket or rear cover is to be serviced, loosen the crankshaft sprocket bolt.

➡ **To remove the crankshaft sprocket bolt, on manual transaxle vehicles, place the vehicle in 5th gear and have an assistant apply the brake. The will stop the engine from rotating while loosening the bolt. On automatic transaxle vehicles, remove the starter and hold the flywheel from turning using a flywheel holding tool VW 10–201 or equivalent.**

7. Remove the crankshaft pulley.
8. Remove the water pump pulley retaining bolts and remove the pulley.
9. Remove the lower cover retaining nuts and remove the cover. Take care not to lose any of the washers or spacers.
10. While holding the large hex nut on the tensioner pulley, loosen the smaller pulley locknut.
11. Turn the tensioner counterclockwise to relieve the tension on the timing belt.
12. Carefully slide the timing belt off the sprockets and remove the belt.

To install:

13. If the engine has moved or jumped timing, use the large bolt on the crankshaft sprocket to rotate the engine until the No. 1 cylinder is at TDC of the compression stroke. At this point, both valves will be closed and the **0** mark on the flywheel will be aligned with the pointer on the bell housing. Rotate the camshaft until the timing mark on the rear face of the camshaft sprocket is aligned with the upper left edge of the valve cover.
14. Install the crankshaft pulley and check that the notch on the pulley is aligned with the mark on the intermediate shaft sprocket. If not, rotate the intermediate shaft until they align.

➡ **If the timing marks are not correctly aligned with the No. 1 piston at TDC of the compression stroke when the belt is installed, valve timing will be incorrect. Poor performance and possible engine damage can result from the improper valve timing.**

15. Remove the crankshaft pulley. Note the pulley location on the crankshaft sprocket so it can be replaced in the same position. Hold the large nut on the tensioner pulley and loosen the smaller locknut.

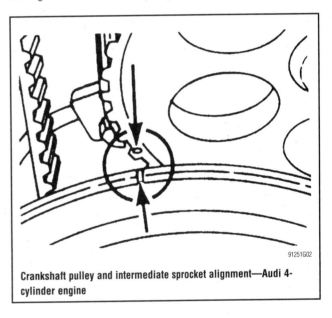

Crankshaft pulley and intermediate sprocket alignment—Audi 4-cylinder engine

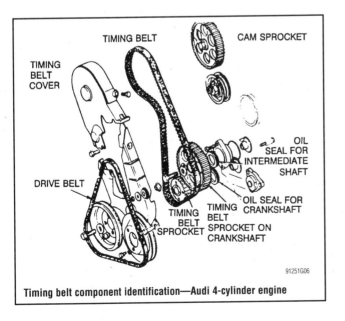

Timing belt component identification—Audi 4-cylinder engine

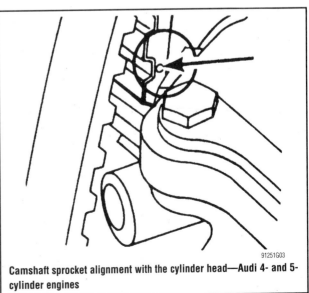

Camshaft sprocket alignment with the cylinder head—Audi 4- and 5-cylinder engines

12 TIMING BELTS—AUDI

The timing belt on all models is correctly tensioned when the belt can be twisted only 90° with your thumb and forefinger

Turn the tensioner counterclockwise to loosen and install the timing belt.

16. Slide the timing belt onto the sprockets and adjust the belt tension. The timing belt tension is correct when the belt can be twisted 90 degrees midway between the camshaft and the intermediate shaft drive sprockets.
17. Install the crankshaft pulley. Tighten the retaining bolts to 15 ft. lbs. (20 Nm).
18. Install the upper and lower timing belt covers.
19. Tighten the crankshaft sprocket bolt to 66 ft. lbs. (89 Nm) plus ½ additional turn.
20. Adjust the drive belt tension when finished.
21. Road test the vehicle for proper operation.

5-CYLINDER ENGINES

1. Disconnect the negative battery cable. Using the large bolt on the crankshaft sprocket, rotate the engine until the No. 1 cylinder is at TDC of the compression stroke. Align the TDC mark **0** with the cast mark on the bell housing. If the belt hasn't jumped teeth, the timing mark on the rear face of the camshaft sprocket should be aligned with the upper left edge of the valve cover.
2. Remove the alternator and air conditioner compressor drive belts.
3. Loosen the alternator adjusting bolts and remove the drive belt.
4. Loosen the power steering pump adjusting bolts and remove the drive belt.
5. Remove the retaining nuts and remove the timing belt cover. Take care not to lose any of the washers or spacers.

6. Loosen the air conditioning compressor mounting bolts and remove the drive belt.
7. Remove the crankshaft balancer center bolt.

➡ **To remove the crankshaft balancer bolt, on manual transaxle vehicles, place the vehicle in 5th gear and have an assistant apply the brake. The will stop the engine from rotating while loosening the bolt. On automatic transaxle vehicles, remove the starter and hold the flywheel from turning, using a flywheel holding tool VW 10–201 or equivalent. This bolt is extremely tight.**

8. Remove the lower timing belt cover bolts and remove the cover.
9. Loosen the water pump bolts only enough to turn the pump clockwise.

➡ **By loosening the water pump bolts, the coolant may drain from the engine at the water pump. If necessary, drain the cooling system, remove the water pump and reinstall it with a new O-ring.**

10. Slide the timing belt off the sprockets.

To install:

11. If necessary, turn the camshaft until the notch on the back of the sprocket is in line with the left side edge of the cylinder head gasket surface.
12. If necessary, align the TDC **0** mark with the with the lug cast on the bell housing.
13. Install the timing belt and turn the water pump counterclockwise to tighten the belt. Tighten the water pump bolts to 15 ft. lbs. (20 Nm).

➡ **The timing belt is correctly tensioned when it can be twisted 90 degrees along the straight run between the camshaft sprocket and water pump. The belt must not be jammed between the oil pump and sprocket when installing the vibration damper.**

14. Install the timing belt covers and tighten the bolts to 89 inch lbs. (10 Nm).
15. Use the same procedure to install the crankshaft center bolt as when removing the bolt. Apply a locking compound on the bolt threads and tighten the bolt to 258 ft. lbs. (350 Nm) in several steps.

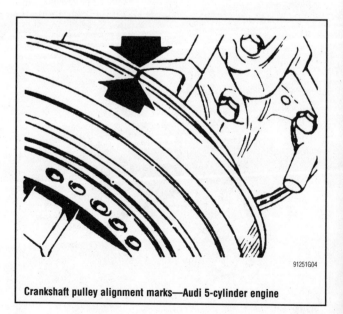

Crankshaft pulley alignment marks—Audi 5-cylinder engine

AUDI—TIMING BELTS

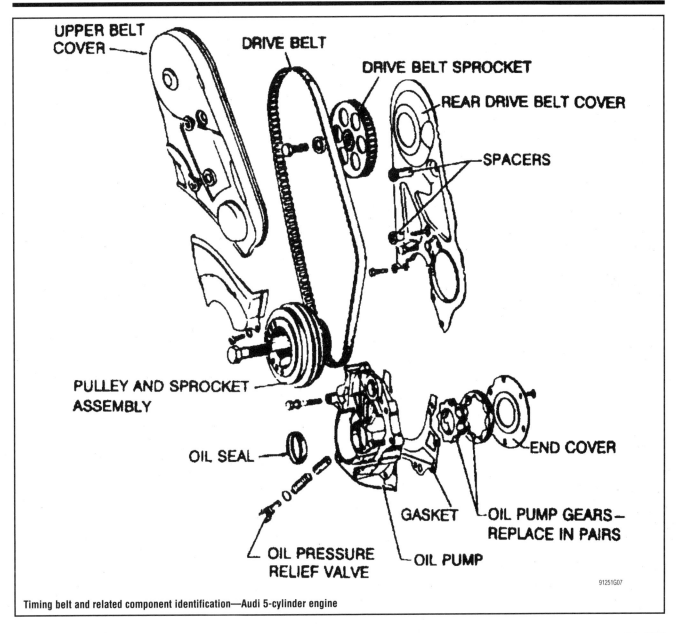

Timing belt and related component identification—Audi 5-cylinder engine

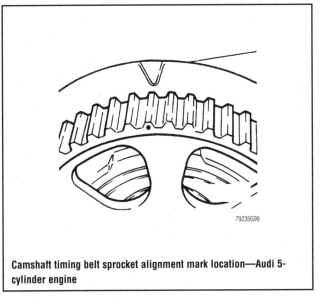

Camshaft timing belt sprocket alignment mark location—Audi 5-cylinder engine

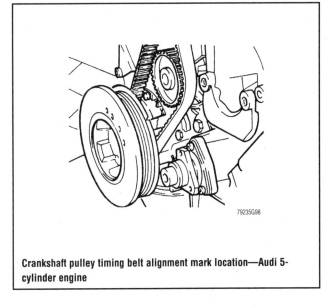

Crankshaft pulley timing belt alignment mark location—Audi 5-cylinder engine

14 TIMING BELTS—AUDI

16. Adjust the drive belt tension when finished.
17. Install the alternator and air conditioning compressor belts. These belts are correctly tensioned when they can be depressed 3/8 in. (9.5mm) along their longest straight run.

8-CYLINDER ENGINES

1. Disconnect the negative battery cable. The battery is under the rear seat.
2. Remove the coolant expansion tank cap. Raise and safely support vehicle. Remove the sound insulator or lower pan. Drain coolant from radiator.
3. Remove the alternator cooling air duct. Loosen the clip and disconnect the coolant temperature sensor harness connector. Remove the coolant hose.
4. Disconnect the outer half of the air duct at the alternator. Remove the screws securing the wiring at the alternator. Note that the alternator wiring must be brought out at the side, not downward. Otherwise, the alternator air duct cannot be mounted.
5. Release tension on the poly-ribbed drive belt and remove the belt in a downward direction. Place a 13mm box wrench on the hexagon guide of the tensioner and pressing the wrench slowly upwards.
6. Remove the alternator mounting bolts and remove the alternator.
7. Working from below, remove the 3 bolts for the toothed belt guard.
8. Remove the bolts securing the bracket for the air intake ducts on the left and right and remove the ducts. Remove the bar-shaped strut brace.
9. Remove the 7 screws securing the upper part of the air cleaner housing, then remove the bolts holding the lower part of the housing. Press towards the rear and lift out.
10. Remove the upper radiator hose and clips.
11. Disconnect the electric fan bolts, lift the fan assembly out and lay to one side, wiring still connected.
12. Remove the bolts securing the supporting clamp for the lower radiator hose and take off the upper part. Disconnect the radiator hose at the thermostat housing. Swing the radiator hose to right at rear.
13. Remove the fan shroud by unscrewing the bolts for the viscous fan at the top. Remove the bolts securing the viscous fan. A spanner wrench or equivalent may be needed to hold the fan hub. Note that this is a left hand thread. Turn to the right to loosen. Lift out the fan and shroud together.
14. Disconnect the engine support at the front. Note any shims. The same thickness shims must be used at installation. Remove the radiator hose. At installation, install the radiator hose first.
15. When loosening the center bolt of the vibration damper, a special tool may be needed to hold the damper from turning. This bolt was installed to over 250 ft. lbs. (340 Nm) tighten and will be difficult to remove. Loosen the center bolt one turn.
16. Remove the bolts securing the left side timing belt cover or guard.
17. Remove the lower part of the supporting clamp for the lower radiator hose. Turn the tensioner for the poly-ribbed accessory drive belt in the loosening direction and insert an appropriate size holding pin in the hole provided.
18. Remove the bolts securing the right side timing belt cover, with the exception of the top bolt. Remove the tensioner holding pin. Remove the belt cover top screw and remove the cover or guard. Carefully lift the guard from the bottom to avoid damaging the radiator.
19. Disconnect the ignition cables at the coils. Remove both distributor caps. Turn the engine to TDC. It may be necessary to temporarily install the damper to align the timing marks. Check that the distributor rotor is pointing to the mark on the housing. If not, turn the crankshaft 1 additional turn. Remove both distributors.
20. Remove the stop plate at the timing belt tensioner. Disconnect the shock-absorber shaped damper at the top bolt.
21. Remove the belt from the tensioning idler pulley (with eccentric). Pulley is on right side of engine. Take the belt off both camshaft sprockets.
22. A special holding tool is available that is installed on the back of the camshafts. It fits the locating pin on the distributor flanges. If necessary, use a special hook wrench tool to turn the camshaft until the pins latch into the special holding tool. Secure the special tool with the distributor mounting bolts.
23. At the camshaft sprocket end, loosen the mounting bolts 2 turns. Using a plastic hammer, tap the edge of the camshaft sprockets loose.

➡**After the sprockets are removed, note the grooves machined in the camshaft ends. Woodruff keys must not be installed in the camshaft sprocket/camshaft connection. Unlike the other engines, this engine does not use cam keys.**

24. Remove the vibration damper which was previously temporarily reinstalled to line up TDC timing marks. A puller can be used. Unscrew 2 opposing bolts of the 4 bolts connecting the vibration damper and toothed belt sprocket. Use a puller in these holes.
25. Remove the timing belt.

To install:
26. Before installing a new belt, the rollers and tensioners must be checked for dirt, rough running and ease of rotation. Clean or replace rollers and tensioners, as necessary.
27. Fit the toothed belt at the crankshaft and install the vibration damper with the belt on the crankshaft.
28. Apply thread locking compound to the center bolt and tighten to 332 ft. lbs. (450 Nm) using hand wrench. A holding tool may be required to keep the crankshaft from turning. It is a must to ensure that the TDC mark on the vibration damper is aligned with the TDC pointer before and after tightening the camshaft sprockets.
29. Guide the toothed belt into position and install the idler pulley with eccentric. Snug the nut enough to hold the eccentric in place but do not fully tighten it yet.
30. Tighten the damper top bolt to 18 ft. lbs. (24 Nm). Engage the damper to the tensioner lever by pressing the lever downward.
31. Perform the basic setting of the toothed belt tensioner by turning the idler pulley eccentric clockwise.
32. A special turning tool may be needed. Turn the eccentric and measure the damper length. Measure the overall length of the barrel not counting the mounting eyes. Turn the eccentric until the damper barrel length is 5.11–5.23 in. (130–133mm).
33. Tighten the idler pulley eccentric to 18 ft. lbs. (24 Nm). Tighten the camshaft sprockets to 33 ft. lbs. (45 Nm).
34. Remove any camshaft and crankshaft locking tool previously installed.
35. Turn engine at least 2 turns. Check the damper length and if necessary, readjust the idler pulley.
36. Reattach the ignition cables at the coils. Install both distributor caps.
37. Install the right cover or guard.

AUDI, BMW—TIMING BELTS

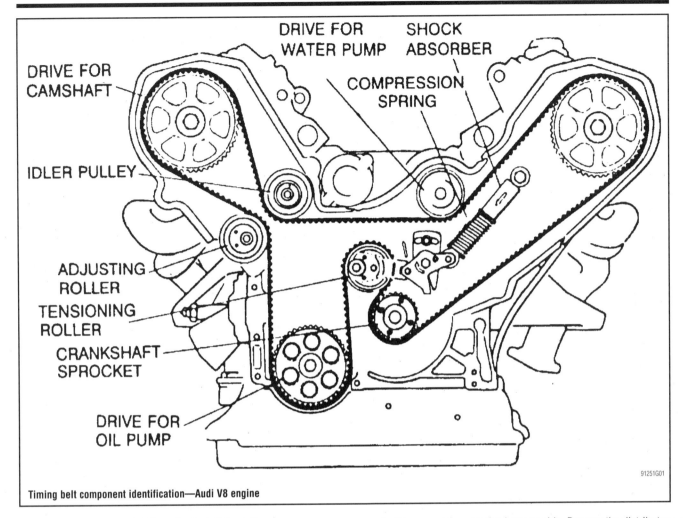

Timing belt component identification—Audi V8 engine

38. Install tensioner holding pin. Install the left cover or guard.
39. Install crankshaft damper. The crankshaft damper center bolt must be torque to 258 ft. lbs. (350 Nm). A holding tool may be needed to keep the crankshaft from turning.
40. Install radiator hoses. Reconnect the engine support.
41. Install the viscous fan hub and shroud assembly. This uses a left hand thread. Turn right to loosen, left to tighten.
42. Install the electric fan and reconnect the wiring.
43. Install the alternator and wiring.
44. Install drive belt.
45. Install the sound insulator or lower pan.
46. Install the coolant expansion tank.
47. Refill and bleed the cooling system.
48. Install the air cleaner housing, air ducts and brace.
49. Check all fluid levels. Operate the engine at normal operating temperatures and check for leaks.
50. Install all remaining components.

BMW

2.5L (M20B25) ENGINE

The timing belt must be replaced every time the belt tensioner is loosened. Do not reinstall a used belt. Reusing a timing belt, even one with low mileage, is not worth the chance of having a belt break and damaging the engine.

1. Disconnect the negative battery cable. Remove the distributor cap and rotor. Remove the inner distributor cover and seal.
2. Remove the 2 distributor guard plate attaching bolts and one nut. Remove the rubber guard and take out the guard plate (upper timing belt cover).

At TDC of No. 1 piston compression stroke, the camshaft sprocket arrow should align directly with the mark on the cylinder head.

3. Rotate the crankshaft to set No. 1 piston at TDC of its compression stroke.
4. On the 3 Series, remove the radiator.
5. Remove the lower splash guard and take off the alternator, power steering and air conditioning belts.
6. Remove the crankshaft pulley and vibration damper.
7. Remove the bolt from the engine end of the alternator bracket. Loosen the alternator adjusting bolt and swing the bracket out of the way.
8. Lift out the TDC transmitter and set it aside.
9. Remove the remaining bolt and lift off the lower timing belt protective cover.
10. Loosen the timing belt tensioner roller bolts and push the roller. Tighten the upper bolt with the roller pushed in. Remove the belt.

To install:

11. Check the alignment of the mark on the camshaft sprocket with the mark on the cylinder head. Check that the crankshaft sprocket mark aligns with the notch in the timing case.

16 TIMING BELTS—BMW

Camshaft and crankshaft sprocket TDC alignment for timing belt service—M20 engine

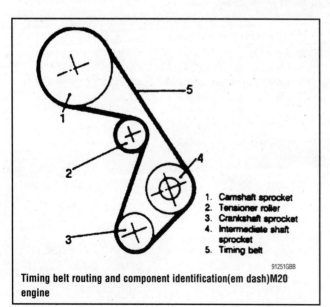

1. Camshaft sprocket
2. Tensioner roller
3. Crankshaft sprocket
4. Intermediate shaft sprocket
5. Timing belt

Timing belt routing and component identification(em dash)M20 engine

12. Install the new timing belt by starting at the crankshaft sprocket and continuing in reverse direction of engine rotation.
13. Loosen the upper timing belt tensioner roller bolt. The spring tension should be enough to move the roller. Turn the engine 1 revolution in the direction of rotation using the crankshaft bolt to tension the belt. Check that the timing marks align exactly at the camshaft sprocket and the crankshaft sprocket. Tighten the roller bolts, top bolt first, then the bottom bolt.
14. Install the lower timing protective cover and tighten the bolt. Install the TDC sender.
15. Replace the alternator bracket.
16. Install the vibration damper and pulley. Tighten the bolts to 17 ft. lbs. (23 Nm).
17. Install the upper cover and nut. Install the rubber guard. Check the condition of the O-ring and install the upper cover. Install the rotor and distributor cap.
18. Install the accessories and belts. Install the splash shield and fill the cooling system with coolant mixture. Bleed the cooling system.
19. Apply a sticker showing the date and mileage of the belt change on the engine.

CHRYSLER CORP.—TIMING BELTS

Chrysler Corporation

DOMESTIC MODELS

1.6L and 2.0L (VIN V) Engines

➡ The 1.6L engine is not equipped with silent shafts. Disregard all instructions pertaining to silent shafts if working on that engine.

1. Disconnect the negative battery cable.
2. Remove the timing belt upper and lower covers.
3. Rotate the crankshaft clockwise and align the timing marks so No. 1 piston will be at TDC of the compression stroke. At this time the timing marks on the camshaft sprocket and the upper surface of the cylinder head should coincide, and the dowel pin of the camshaft sprocket should be at the upper side.

➡ Always rotate the crankshaft in a clockwise direction. Make a mark on the back of the timing belt indicating the direction of rotation so it may be reassembled in the same direction if it is to be reused.

4. Remove the auto tensioner and remove the outermost timing belt.
5. Remove the timing belt tensioner pulley, tensioner arm, idler pulley.
6. Locate the access plug on the side of block. Remove the plug and install a phillips screwdriver. Remove the oil pump sprocket nut, oil pump sprocket, special washer, flange and spacer.
7. Remove the silent shaft (inner) belt tensioner and remove the belt.

To install:

8. Align the timing marks on the crankshaft sprocket and the silent shaft sprocket.
9. Fit the inner timing belt over the crankshaft and silent shaft sprocket. Ensure that there is no slack in the belt.
10. While holding the inner timing belt tensioner with your fingers, adjust the timing belt tension by applying a force towards the center of the belt, until the tension side of the belt is taut. Tighten the tensioner bolt.

➡ When tightening the bolt of the tensioner, ensure that the tensioner pulley shaft does not rotate with the bolt. Allowing it to rotate with the bolt can cause excessive tension on the belt.

11. Check belt for proper tension by depressing the belt on long side with your finger and noting the belt deflection. The desired reading is 0.20–0.28 in. (5–7mm). If tension is not correct, readjust and check belt deflection.
12. Install the flange, crankshaft and washer to the crankshaft. The flange on the crankshaft sprocket must be installed towards the inner timing belt sprocket. Tighten bolt to 80–94 ft. lbs. (110–130 Nm).

➡ There is a possibility to align all timing marks and have the oil pump sprocket out of time, causing an engine vibration during operation. If the following step is not followed exactly, there is a 50 percent chance that the oil pump shaft alignment will be 180 degrees off.

13. Before installing the timing belt, ensure that the oil pump sprocket shaft is in the correct position as follows:
    a. Remove the plug from the rear side of the block and insert a phillips screwdriver with shaft diameter of 0.31 in. (8mm) into the hole.
    b. With the timing marks still aligned, the shaft of the screwdriver must be able to go in at least 2.36 in. (60mm). If the tool can only go in 0.79–0.98 in. (20–25mm), the shaft is not in the correct orientation and will cause a vibration during engine operation. Remove the tool from the hole and turn the oil pump sprocket 1 complete revolution. Realign the timing marks and insert the tool. The shaft of the tool must go in at least 2.36 in. (60mm).
    c. Recheck and realign the timing marks.
    d. Leave the tool in place to hold the oil pump silent shaft while continuing.
14. To install the oil pump sprocket and tighten the nut to 36–43 ft. lbs. (50–60 Nm).
15. Position the auto-tensioner into a vise with soft jaws. The plug at the rear of tensioner protrudes, be sure to use a washer as a spacer to protect the plug from contacting vise jaws.
16. Slowly push the rod into the tensioner until the set hole in rod is aligned with set hole in the auto-tensioner.
17. Insert a 0.055 in. (1.4mm) wire into the aligned set holes. Unclamp the tensioner from vise and install to vehicle. Tighten tensioner to 17 ft. lbs. (24 Nm).
18. When installing the timing belt, turn the 2 camshaft sprockets so their dowel pins are located on top. Align the timing marks facing each other with the top surface of the cylinder head. When you let go of the exhaust camshaft sprocket, it will rotate 1 tooth in the counterclockwise direction. This should be taken into account when installing the timing belts on the sprocket.

➡ The same sprockets are used for the intake and exhaust camshafts and are provided with 2 timing marks. When the sprocket is mounted on the exhaust camshaft, use the timing mark on the right with the dowel pin hole on top. For the intake camshaft sprocket, the timing mark is on the left with the dowel pin hole on top.

19. Align the crankshaft sprocket and oil pump sprocket timing marks.
20. After alignment of the oil pump sprocket timing marks, remove the plug on the cylinder block and insert a Phillips screwdriver with a shaft diameter of 0.31 in. (8mm) through the hole. If the shaft can be inserted 2.4 in. deep, the silent shaft is in the correct position. If the shaft of the tool can only be inserted 0.8–1.0 in. (20–25mm) deep, turn the oil pump sprocket 1 turn and realign the marks. Reinsert the tool making sure it is inserted 2.4 in. deep. Keep the tool inserted in hole for the remainder of this procedure.

➡ The above step assures that the oil pump socket is in correct orientation to the silent shafts. This step must not be skipped or a vibration may develop during engine operation.

21. Install the timing belt as follows:
    a. Install the timing belt around the intake camshaft sprocket and retain it with 2 spring clips or binder clips.
    b. Install the timing belt around the exhaust sprocket, aligning the timing marks with the cylinder head top surface using 2 wrenches. Retain the belt with 2 spring clips.
    c. Install the timing belt around the idler pulley, oil pump sprocket, crankshaft sprocket and the tensioner pulley. Remove the 2 spring clips.
    d. Lift upward on the tensioner pulley in a clockwise direction and tighten the center bolt. Make sure all timing marks are aligned.
    e. Rotate the crankshaft ¼ turn counterclockwise. Then, turn in clockwise until the timing marks are aligned again.

18 TIMING BELTS—CHRYSLER CORP.

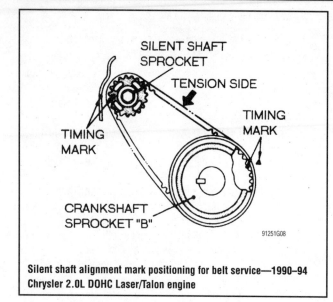

Silent shaft alignment mark positioning for belt service—1990–94 Chrysler 2.0L DOHC Laser/Talon engine

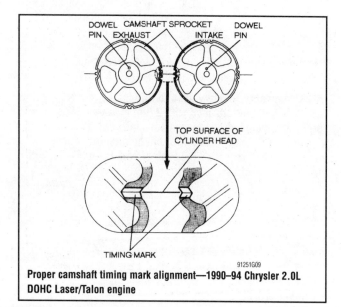

Proper camshaft timing mark alignment—1990–94 Chrysler 2.0L DOHC Laser/Talon engine

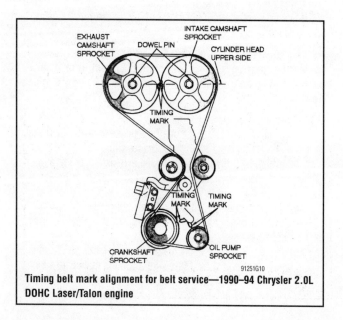

Timing belt mark alignment for belt service—1990–94 Chrysler 2.0L DOHC Laser/Talon engine

22. To adjust the timing (outer) belt, turn the crankshaft ¼ turn counterclockwise, then turn it clockwise to move No. 1 cylinder to TDC.

23. Loosen the center bolt. Using tool MD998738 or equivalent and a torque wrench, apply a torque of 1.88–2.03 ft. lbs. (2.6–2.8 Nm). Tighten the center bolt.

24. Screw the special tool into the engine left support bracket until its end makes contact with the tensioner arm. At this point, screw the special tool in some more and remove the set wire attached to the auto tensioner, if the wire was not previously removed. Then remove the special tool.

25. Rotate the crankshaft 2 complete turns clockwise and let it sit for approximately 15 minutes. Then, measure the auto tensioner protrusion (the distance between the tensioner arm and auto tensioner body) to ensure that it is within 0.15–0.18 in. (3.8–4.5mm). If out of specification, repeat belt adjustment procedure until the specified value is obtained.

26. If the timing belt tension adjustment is being performed with the engine mounted in the vehicle, and clearance between the tensioner arm and the auto tensioner body cannot be measured, the following alternative method can be used:

   a. Screw in special tool MD998738 or equivalent, until its end makes contact with the tensioner arm.

   b. After the special tool makes contact with the arm, screw it in some more to retract the auto tensioner pushrod while counting the number of turns the tool makes until the tensioner arm is brought into contact with the auto tensioner body. Make sure the number of turns the special tool makes conforms with the standard value of 2.5–3 turns.

   c. Install the rubber plug to the timing belt rear cover.

27. Install the timing belt covers and all related items.

28. Connect the negative battery cable.

2.0L (VIN C) SOHC Engine

1. On Cirrus, Stratus, Breeze, and Sebring Convertible models, disconnect the negative battery cable from the left strut tower. The ground cable is equipped with a insulator grommet which should be placed on the stud to prevent the negative battery cable from accidentally grounding.

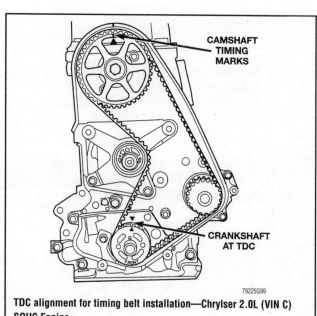

TDC alignment for timing belt installation—Chrysler 2.0L (VIN C) SOHC Engine

CHRYSLER CORP.—TIMING BELTS

2. On Neon models, disconnect the negative battery cable.
3. Remove the drive belts and accessories.
4. Raise and safely support the vehicle.
5. Remove the right inner splash-shield.
6. Remove the crankshaft damper.
7. Remove the right engine mount
8. Place a support under the engine.
9. Remove the engine mount bracket
10. Remove the timing belt cover.
11. Loosen the timing belt tensioner bolts.
12. Remove the timing belt and the tensioner.

➡ **When tensioner is removed from the engine it is necessary to compress the plunger into the tensioner body.**

13. Place the tensioner into a soft-jawed vise to compress the tensioner.
14. After compressing the tensioner place a pin (5/64 in. Allen wrench will work) into the plunger side hole to retain the plunger until installation.

To install:
15. Set the crankshaft sprocket to Top Dead Center (TDC) by aligning the notch on the sprocket with the arrow on the oil pump housing, then back off the sprocket three notches before TDC.
16. Set the camshaft to align the timing marks.
17. Move the crankshaft to ½ notch before TDC.
18. Install the timing belt starting at the crankshaft, around the water pump, then around the camshaft last.
19. Move the crankshaft to TDC to take up the belt slack.
20. Reinstall the tensioner to the block but do not tighten it.
21. Using a torque wrench apply 250 inch lbs. (28 Nm) of torque to the tensioner pulley.
22. With torque being applied to the tensioner pulley, move the tensioner up against the tensioner bracket and tighten the fasteners to 275 inch lbs. (31 Nm).
23. Remove the tensioner plunger pin, the tension is correct when the plunger pin can be removed and replaced easily.
24. Rotate the crankshaft two revolutions and recheck the timing marks.
25. Reinstall the timing belt cover.
26. Reinstall the engine mount bracket.
27. Reinstall the right engine mount.
28. Remove the engine support.
29. Reinstall the crankshaft damper and tighten to 105 ft. lbs. (142 Nm).
30. Reinstall the drive belts and accessories.
31. Reinstall the right inner splash-shield.
32. Perform the crankshaft and camshaft relearn alignment procedure using the DRB scan tool or equivalent.

2.0L (VIN F) Engine

1. Disconnect the negative battery cable.
2. Remove the engine undercover.
3. Remove the engine mount bracket.
4. Remove the drive belts.
5. Remove the belt tensioner pulley.
6. Remove the water pump pulleys.
7. Remove the crankshaft pulley.
8. Remove the stud bolt from the engine support bracket and remove the timing belt covers.
9. Rotate the crankshaft clockwise to line up the camshaft timing marks. Always turn the crankshaft in the forward direction only.
10. Loosen the tension pulley center bolt.

➡ **If the timing belt is to be reused, mark the direction of rotation on the flat side of the belt with an arrow.**

11. Move the tension pulley towards the water pump and remove the timing belt.

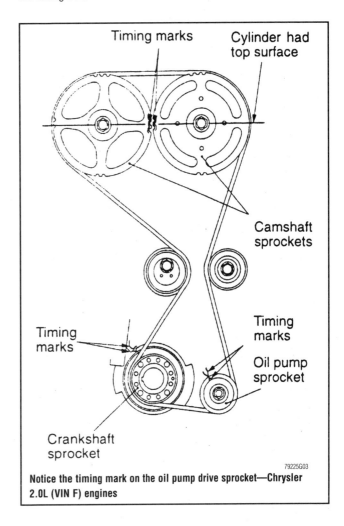

Notice the timing mark on the oil pump drive sprocket—Chrysler 2.0L (VIN F) engines

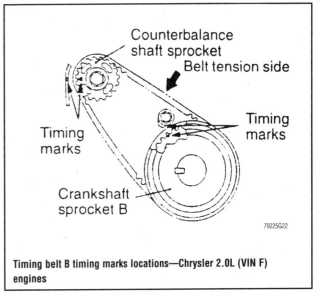

Timing belt B timing marks locations—Chrysler 2.0L (VIN F) engines

20 TIMING BELTS—CHRYSLER CORP.

12. Remove the crankshaft sprocket center bolt using special tool MB990767 to hold the crankshaft sprocket while removing the center bolt. Then, use MB998778 or equivalent puller to remove the sprocket.
13. Mark the direction of rotation on the timing belt B with a arrow.
14. Loosen the center bolt on the tensioner and remove the belt.
15. To remove the camshaft sprocket, remove the cylinder head cover. Use a wrench to hold the hexagonal part of the camshaft and remove the sprocket mounting bolt.

✱✱ WARNING

Do not rotate the camshafts or the crankshaft while the timing belt is removed.

To install:

16. Use a wrench to hold the camshaft ,and install the sprocket and mounting bolt. Tighten the bolt (s) to 65 ft. lbs. (88 Nm).
17. Install the cylinder head cover.
18. Place the crankshaft sprocket on the crankshaft. Use tool MB990767 (or equivalent) to hold the crankshaft sprocket while tightening the center bolt. Tighten the center bolt to 80–94 ft. lbs. (108–127 Nm).
19. Align the timing marks on the crankshaft sprocket B and the balance shaft.
20. Install timing belt B on the sprockets. Position the center of the tensioner pulley to the left and above the center of the mounting bolt.
21. Push the pulley clockwise toward the crankshaft to apply tension to the belt and tighten the mounting bolt to 14 ft. lbs. (19 Nm). Do not let the pulley turn when tightening the bolt because it will cause excessive tension on the belt. The belt should deflect 0.20–0.28 in. (5–7mm) when finger pressure is applied between the pulleys.
22. Install the crankshaft sensing blade and the crankshaft sprocket. Apply engine oil to the mounting bolt and tighten the bolt to 80–94 ft. lbs. (108–127 Nm).
23. Use a press or vise to compress the auto tensioner pushrod. Insert a set pin when the holed are lined up.

✱✱ WARNING

Do not compress the pushrod too quickly, damage to the pushrod can occur.

24. Install the auto tensioner on the engine.
25. Align the timing marks on the camshaft sprocket, crankshaft sprocket and the oil pump sprocket.
26. After aligning the mark on the oil pump sprocket, remove the cylinder block plug and insert a Phillips screwdriver in the hole to check the position of the counter balance shaft. The screwdriver should go in at least 2.36 in. or more. If not, rotate the oil pump sprocket once and realign the timing mark so the screwdriver goes in. Do not remove the screwdriver until the timing belt is installed.
27. Install the timing belt on the intake camshaft and secure it with a clip.
28. Install the timing belt on the exhaust camshaft. Align the timing marks with the cylinder head top surface using two wrenches. Secure the belt with another clip.
29. Install the belt around the idler pulley, oil pump sprocket, crankshaft sprocket and the tensioner pulley.

30. Turn the tension pulley so the pinholes are at the bottom. Press the pulley lightly against the timing belt.
31. Screw the special tool into the left engine support bracket until it contacts the tensioner arm, then screw the tool in a little more and remove the pushrod pin from the auto tensioner. Remove the special tool and Tighten the center bolt to 35 ft. lbs. (48 Nm).
32. Turn the crankshaft ¼ turn counterclockwise, then clockwise until the timing marks are aligned.
33. Loosen the center bolt. Install special tool MD998767 (or equivalent) on the tension pulley. Turn the tension pulley counterclockwise with a torque of 2.6 ft. lbs. (3.5 Nm) and tighten the center bolt to 35 ft. lbs. (48 Nm). Do not let the tension pulley turn with the bolt.
34. Turn the crankshaft two revolutions to the right and align the timing marks. After 15 minutes, measure the protrusion of the pushrod on the auto tensioner. The standard measurement is 0.150–0.177 in (3.8–4.5mm). If the protrusion is out of specification, loosen the tension pulley, apply the proper torque to the belt and retighten the center bolt.
35. Install the crankshaft pulley. Tighten the mounting bolts to 18 ft. lbs. (25 Nm).
36. Install the water pump. Tighten the mounting bolts to 6.5 ft. lbs. (8.8 Nm).
37. Install and adjust the drive belts.
38. Install the engine mount bracket.
39. Install the engine undercover.
40. Connect the negative battery cable.

2.0L (VIN Y) Engine

Valve timing is critical to engine operation. Use care when servicing the timing belt. There are a number of timing marks that must be properly aligned or engine damage will result. If the timing belt has not broken, or jumped teeth, it is recommended that the crankshaft be turned by hand (clockwise) to TDC No.1 cylinder compression stroke (firing position) before beginning work. This should align all the timing marks and serve as a reference for later work. Some technicians will apply a small amount of white paint to all timing marks. This helps make them more visible under the low-light conditions found underhood.

1. Disconnect the negative battery cable.
2. Remove the accessory drive belts.
3. Using C 3281, or equivalent crankshaft holding tool, remove the crankshaft pulley center retaining bolt.
4. Using puller tools 1026 and 6827 or equivalent, remove the crankshaft pulley.
5. Remove the power steering pump from the bracket and position it out of the way. Do not disconnect the hoses.
6. Remove the power steering pump bracket from the engine.
7. Use a floor jack with a piece of wood on it and jack up the engine to take the weight off of the engine mount.
8. Remove the engine mount and bracket.
9. Remove the front timing belt cover.
10. If not done so previously, align the timing marks. Loosen the timing belt tensioner and remove the belt.

✱✱ WARNING

Do not rotate the crankshaft or camshafts after removing the timing belt or valvetrain components may be damaged. Always align the timing marks before removing the timing belt.

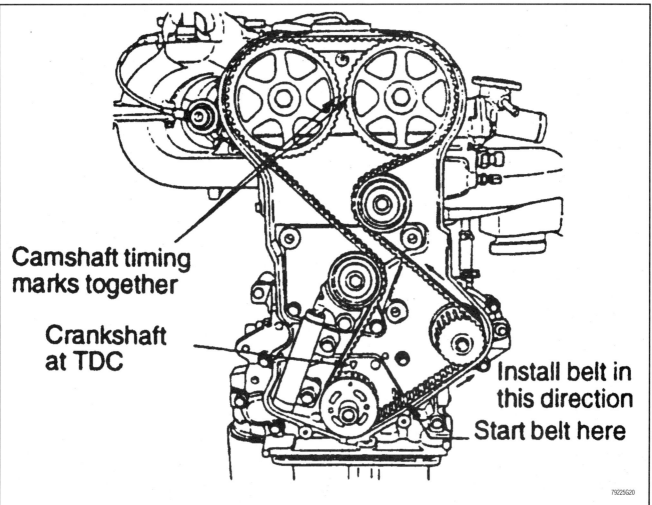

Install the timing belt by starting at the crankshaft sprocket and working around the other pulleys in a counterclockwise direction—Chrysler 2.0L (VIN Y) engine

11. For 1995–97 models:
   a. If the timing belt tensioner is to be replaced, remove the retaining bolts and remove the timing belt tensioner. When the timing belt tensioner is removed from the engine it is necessary to compress the plunger into the tensioner body.
   b. Place the tensioner in a vise and slowly compress the plunger.

➡ **Position the tensioner in the vise the same way it will be installed on the engine. This is to ensure proper pin orientation for when the tensioner is installed on the engine.**

   c. When the plunger is compressed into the tensioner body, install a pin through the body and plunger to hold the plunger in place until the tensioner is installed.
12. For 1998–99 models:
   a. Place an 8mm Allen wrench into the belt tensioner, then using the long end of a 1/8 in. (3mm) Allen wrench, rotate the tensioner counterclockwise until it slides into the locking hole.
13. Remove the timing belt.

To install:
14. Check that all timing marks are still aligned. Bring the crankshaft sprocket to ½ a notch before TDC.
15. Install the timing belt. Starting at the crankshaft, route the belt around the water pump sprocket, idler pulley, camshaft sprockets, then around the tensioner pulley.

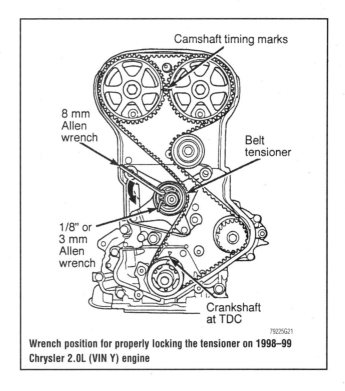

Wrench position for properly locking the tensioner on 1998–99 Chrysler 2.0L (VIN Y) engine

22 TIMING BELTS—CHRYSLER CORP.

16. Move the crankshaft to TDC to take up the slack in the belt. Install the tensioner to the block but do not tighten the retaining bolts.
17. For 1995–97 models:
   a. Using a torque wrench on the tensioner pulley, apply 21 ft. lbs. (28 Nm) of torque to the pulley.
   b. With the torque being applied to the tensioner pulley, move the tensioner up against the tensioner pulley bracket and tighten the retaining bolts to 23 ft. lbs. (31 Nm).
   c. Remove the tensioner plunger pin. Pretension is correct when the pin can be removed and installed.
18. For 1998–99 models, remove the wrenches from the tensioner.
19. Rotate the crankshaft two revolutions and check the timing marks. If the timing marks are not properly aligned, remove the belt and reinstall it as described.
20. Install the front timing belt cover.
21. Lower the engine enough to install the engine mount bracket.
22. Install the bracket and remove the floor jack.
23. Install the power steering pump bracket and pump.
24. Install the crankshaft pulley using C-4685-C or an equivalent pulley installer.
25. Tighten the mounting bolt to 105 ft. lbs. (142 Nm).
26. Install the accessory drive belts.
27. Connect the negative battery cable.
28. Start the engine. Check for leaks and proper engine operation.

2.2L SOHC & 2.5L Engines

1. Disconnect the negative battery cable.
2. Position the engine so the No. 1 piston is at TDC of its compression stroke.
3. Remove the nuts and bolts that attach the upper cover to the valve cover, block or cylinder head.
4. Remove the bolt that attaches the upper cover to the lower cover.
5. Remove the upper cover.
6. Remove the right tire and wheel assembly. Remove the right side inner splash shield.
7. Remove the crankshaft pulley, water pump pulley and the accessory drive belt(s).
8. Remove the lower cover attaching bolts and the cover from the engine.
9. Remove the timing belt tensioner and allow the belt to hang free.
10. Place a floor jack under the engine and separate the right motor mount.

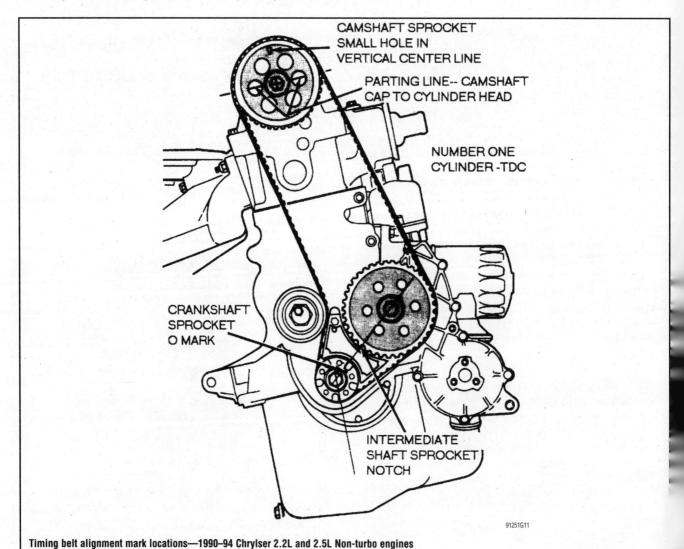

Timing belt alignment mark locations—1990–94 Chrysler 2.2L and 2.5L Non-turbo engines

CHRYSLER CORP.—TIMING BELTS

11. Remove the air conditioning compressor belt idler pulley, if equipped, and remove the mounting stud. Remove the compressor/alternator bracket as follows:

    a. Remove the alternator pivot bolt and remove the alternator from the bracket. Turn the alternator so the wire connections are facing up and disconnect the harness connectors from the rear of the alternator.

    b. Remove the air conditioning compressor belt idler.

    c. Remove the right engine mount yoke screw securing engine mount support strut to the engine.

    d. Remove the 5 side mounting bolts retaining the bracket to the front of the engine.

    e. Remove the front mounting nut. Remove the front bolt and strut and rotate the solid mount bracket away from the engine. Slide the bracket on the stud until free of the mounting studs and remove from the engine.

    f. Remove the timing belt from the vehicle.

To install:

12. Turn the crankshaft sprocket and intermediate shaft sprocket until the marks are in line. Use a straight-edge from bolt to bolt to confirm alignment.

13. Turn the camshaft until the small hole in the sprocket is at the top and the arrows on the hub are in line with the camshaft cap to cylinder head mounting lines. When looking through the hole on top of the camshaft sprocket, the uppermost center nipple of the valve cover end seal should be at the center of the hole. Use a mirror to check the alignment of the arrows so it is viewed straight on and not at an angle from above. Install the belt but let it hang free at this point.

14. Install the air conditioning compressor/alternator bracket, idler pulley and motor mount. Remove the floor jack. Raise the vehicle and support safely. Have the tensioner at an arm's reach because the timing belt will have to be held in position with one hand.

15. To properly install the timing belt, reach up and engage it with the camshaft sprocket. Turn the intermediate shaft counterclockwise slightly, then engage the belt with the intermediate shaft sprocket. Hold the belt against the intermediate shaft sprocket and turn clockwise to take up all tension; if the timing marks are out of alignment, repeat until alignment is correct.

16. Using a wrench, turn the crankshaft sprocket counterclockwise slightly and wrap the belt around it. Turn the sprocket clockwise so there is no slack in the belt between sprockets; if the timing marks are out of alignment, repeat until alignment is correct.

➡ **If the timing marks are in line but slack exists in the belt between either the camshaft and intermediate shaft sprockets or the intermediate and crankshaft sprockets, the timing will be incorrect when the belt is tensioned. All slack must be only between the crankshaft and camshaft sprockets.**

17. Install the tensioner and install the mounting bolt loosely. Place the special tensioning tool C-4703 on the hex of the tensioner so the weight is at about the 9 o'clock position (parallel to the ground, hanging toward the rear of the vehicle) plus or minus 15 degrees.

18. Hold the tool in position and tighten the bolt to 45 ft. lbs. (61 Nm). Do not pull the tool past the 9 o'clock position; this will make the belt too tight and will cause it to howl or possibly break.

19. Lower the vehicle and recheck the camshaft sprocket positioning. If it is correct install the timing belt covers and all related parts.

20. Connect the negative battery cable and road test the vehicle.

2.2L DOHC Engine

1. Position the engine so the No. 1 piston is at TDC of its compression stroke. Disconnect the negative battery cable.
2. Remove the timing belt covers.
3. Remove the timing belt tensioner and allow the belt to hang free.
4. Place a floor jack under the engine and separate the right motor mount.
5. Remove the timing belt from the vehicle.

To install:

6. Turn the crankshaft sprocket and intermediate shaft sprocket until the marks are in line. Use a straight-edge from bolt to bolt to confirm alignment.
7. No. 1 and No. 6 camshaft journals have aligning pin holes to index with the blind holes in the camshaft. Turn the camshafts until the pin holes in the journals align with the aligning holes in the corresponding bearing caps. Install pin punches to secure this timing position. At this position, the sprocket timing holes on the camshaft

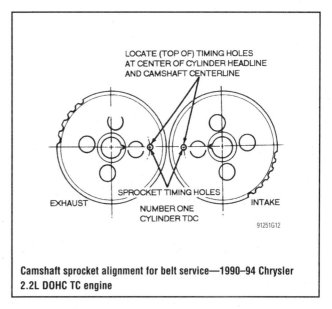

Camshaft sprocket alignment for belt service—1990–94 Chrysler 2.2L DOHC TC engine

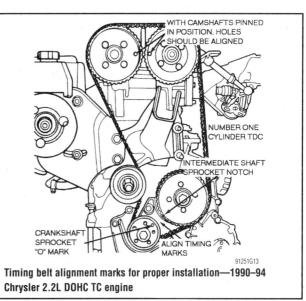

Timing belt alignment marks for proper installation—1990–94 Chrysler 2.2L DOHC TC engine

24 TIMING BELTS—CHRYSLER CORP.

sprockets should both be centered at the cylinder head mounting surface line.

8. Install the motor mount. Remove the floor jack. Raise the vehicle and support safely. Have the tensioner at arm's reach because the timing belt will have to be held in position with one hand.

9. To properly install the timing belt, reach up and engage it with the camshaft sprockets, leaving no tension between sprockets. Turn the intermediate shaft counterclockwise slightly, then engage the belt with the intermediate shaft sprocket. Hold the belt against the intermediate shaft sprocket and turn clockwise to take up all tension; if the timing marks are out of alignment, repeat until alignment is correct.

10. Using a wrench, turn the crankshaft sprocket counterclockwise slightly and wrap the belt around it. Turn the sprocket clockwise so there is no slack in the belt between sprockets; if the timing marks are out of alignment, repeat until alignment is correct.

➡ **If the timing marks are in line but slack exists in the belt anywhere except on the tensioner side, the timing will be incorrect when the belt is tensioned. All slack must be only between the crankshaft and exhaust camshaft sprockets.**

11. Install the tensioner and install the mounting bolt loosely. Remove the pin punches from the camshafts. Place the special tensioning tool C–4703 on the hex of the tensioner so the weight is at about the 9 o'clock position (parallel to the ground, hanging toward the rear of the vehicle) plus or minus 15 degrees.

12. Hold the tool in position and tighten the bolt to 45 ft. lbs. (61 Nm). Do not pull the tool past the 9 o'clock position; this will make the belt too tight and will cause it to howl or possibly break.

13. Rotate the crankshaft 2 full revolutions. With the No. 1 cylinder at TDC, all timing marks must be in line. Repeat the procedure if the timing is not correct.

14. Install the timing belt covers and all related parts.

15. Connect the negative battery cable and road test the vehicle.

2.4L (VIN B) Engine

➡ **You may need DRB scan tool to perform the crankshaft and camshaft relearn alignment procedure.**

1. Disconnect the negative battery cable remote connection, located on the left strut tower.

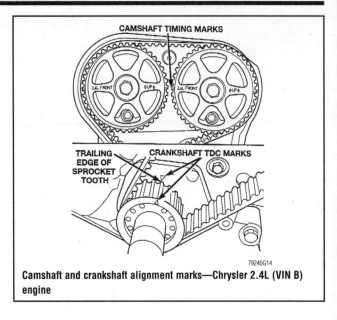
Camshaft and crankshaft alignment marks—Chrysler 2.4L (VIN B) engine

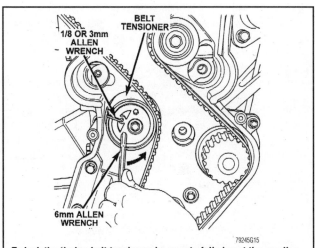
To lock the timing belt tensioner, be sure to fully insert the smaller Allen wrench into the tensioner as shown—Chrysler 2.4L (VIN B) engine

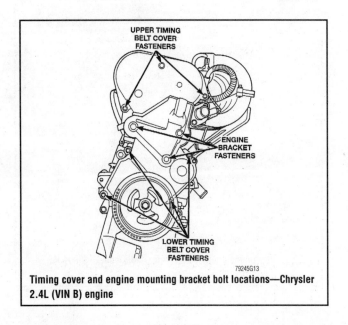
Timing cover and engine mounting bracket bolt locations—Chrysler 2.4L (VIN B) engine

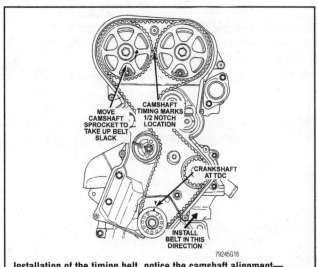
Installation of the timing belt, notice the camshaft alignment—Chrysler 2.4L (VIN B) engine

2. Remove the right inner splash shield.
3. Remove the accessory drive belts.
4. Remove the crankshaft damper.
5. Remove the right engine mount.
6. Place a floor jack under the engine to support it while the engine mount is removed.
7. Remove the engine mount bracket.
8. Remove the timing belt cover.

➡ **This is an interference engine. Do not rotate the crankshaft or the camshafts after the timing belt has been removed. Damage to the valve components may occur. Before removing the timing belt, always align the timing marks.**

9. Align the timing marks of the timing belt sprockets to the timing marks on the rear timing belt cover and oil pump cover. Loosen the timing belt tensioner bolts.
10. Remove the timing belt and the tensioner.
11. If necessary, remove the camshaft timing belt sprockets.
12. If necessary, remove the crankshaft timing belt sprocket using removal tool No. 6793 or equivalent.
13. Place the tensioner into a soft-jawed vise to compress the tensioner.
14. After compressing the tensioner, insert a pin (5/64 in. Allen wrench will also work) into the plunger side hole to retain the plunger until installation.

To install:

15. If necessary, using tool No. 6792 or equivalent, to install the crankshaft timing belt sprocket onto the crankshaft.
16. If necessary, install the camshaft sprockets onto the camshafts. Install and tighten the camshaft sprocket bolts to 75 ft. lbs. (101 Nm).
17. Set the crankshaft sprocket to Top Dead Center (TDC) by aligning the notch on the sprocket with the arrow on the oil pump housing.
18. Set the camshafts to align the timing marks on the sprockets.
19. Move the crankshaft to ½ notch before TDC.
20. Install the timing belt starting at the crankshaft, then around the water pump sprocket, idler pulley, camshaft sprockets and around the tensioner pulley.
21. Move the crankshaft sprocket to TDC to take up the belt slack.
22. Install the tensioner on the engine block but do not tighten.
23. Using a torque wrench on the tensioner pulley, apply 250 inch lbs. (28 Nm) of torque to the tensioner pulley.
24. With torque being applied to the tensioner pulley, move the tensioner up against the tensioner pulley bracket and tighten the fasteners to 23 ft. lbs. (31 Nm).
25. Remove the tensioner plunger pin, the tension is correct when the plunger pin can be removed and reinserted easily.

※※ WARNING

If any binding is felt when adjusting the timing belt tension by turning the crankshaft, STOP turning the engine, because the pistons may be hitting the valves.

26. Rotate the crankshaft two revolutions and recheck the timing marks. Wait several minutes and then recheck that the plunger pin can easily be removed and installed.
27. Install the front timing belt cover.
28. Install the engine mount bracket.
29. Install the right engine mount.
30. Remove the floor jack from under the vehicle.
31. Install the crankshaft damper and tighten it to 105 ft. lbs. (142 Nm).
32. Install the accessory drive belts and adjust to the proper tension.
33. Install the right inner splash shield.
34. Reconnect the negative battery cable.
35. Perform the crankshaft and camshaft relearn alignment procedure using the DRB scan tool or equivalent.

2.4L (VIN X and S) Engines

1. Disconnect the negative battery cable from the left strut tower. The ground cable is equipped with a insulator grommet which should be placed on the stud to prevent the negative battery cable from accidentally grounding.
2. Remove the right inner splash-shield.
3. Remove the accessory drive belts.
4. Remove the crankshaft damper.
5. Remove the right engine mount.

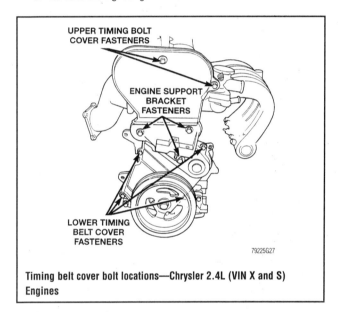

Timing belt cover bolt locations—Chrysler 2.4L (VIN X and S) Engines

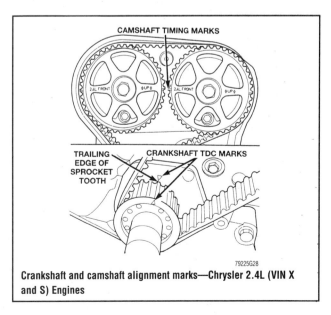

Crankshaft and camshaft alignment marks—Chrysler 2.4L (VIN X and S) Engines

TIMING BELTS—CHRYSLER CORP.

6. Place a suitable floor jack under the vehicle to support the engine.
7. Remove the engine mount bracket
8. Remove the timing belt cover.

➡ Do not rotate the crankshaft or the camshafts after the timing belt has been removed. Damage to the valve components may occur. Before removing the timing belt, always align the timing marks.

9. Align the timing marks of the timing belt sprockets to the timing marks on the rear timing belt cover and oil pump cover. Loosen the timing belt tensioner bolts.
10. Remove the timing belt and the tensioner.
11. Remove the camshaft timing belt sprockets.
12. Remove the crankshaft timing belt sprocket using special removal tool No. 6793 or equivalent.
13. Place the tensioner into a soft-jawed vise to compress the tensioner.
14. After compressing the tensioner, place a pin (5/64 in. Allen wrench will work) into the plunger side hole to retain the plunger until installation.

To install:
15. Using special tool No. 6792, or its equivalent, install the crankshaft timing belt sprocket onto the crankshaft.
16. Install the camshaft sprockets onto the camshafts. Install and tighten the camshaft sprocket bolts to 75 ft. lbs. (101 Nm).
17. Set the crankshaft sprocket to Top Dead Center (TDC) by aligning the notch on the sprocket with the arrow on the oil pump housing.
18. Set the camshafts to align the timing marks on the sprockets.
19. Move the crankshaft to ½ notch before TDC.
20. Install the timing belt starting at the crankshaft, then around the water pump sprocket, idler pulley, camshaft sprockets, then around the tensioner pulley.
21. Move the crankshaft sprocket to TDC to take up the belt slack.
22. Reinstall the tensioner to the block, but do not tighten it at this time.
23. Using a torque wrench on the tensioner pulley, apply 250 inch. lbs. (28 Nm) of torque to the tensioner pulley.
24. With torque being applied to the tensioner pulley, move the tensioner up against the tensioner pulley bracket and tighten the fasteners to 275 inch. lbs. (31 Nm).
25. Remove the tensioner plunger pin, the tension is correct when the plunger pin can be removed and replaced easily.
26. Rotate the crankshaft two revolutions and recheck the timing marks. Wait several minutes, then recheck that the plunger pin can easily be removed and installed.
27. Reinstall the front timing belt cover.
28. Reinstall the engine mount bracket.
29. Reinstall the right engine mount.
30. Remove the floor jack from under the vehicle.
31. Install the crankshaft damper and tighten to 105 ft. lbs. (142 Nm).
32. Install and adjust the accessory drive belts.
33. Install the right inner splash-shield.
34. Reconnect the negative battery cable.
35. Perform the crankshaft and camshaft relearn alignment procedure using the DRB scan tool or equivalent.

2.5L (VIN G) Engine

1. Position the engine so that the No. 1 piston is at TDC.
2. Disconnect the negative battery cable.
3. Remove the timing belt covers.
4. Remove the timing belt tensioner and allow the belt to hang free.
5. Remove the air conditioning compressor belt idler pulley, if equipped, and remove the mounting stud. Unbolt the compressor/alternator bracket and position it to the side.
6. Remove the timing belt from the vehicle.

To install:
7. Turn the crankshaft sprocket and intermediate shaft sprocket until the marks are aligned. Use a straightedge from bolt-to-bolt to confirm alignment.
8. Turn the camshaft until the small hole in the sprocket is at the top and rows on the hub are aligned with the camshaft cap-to-cylinder head mounting lines. Use a mirror to see the alignment so it is viewed straight on and not at an angle from above. Install the belt, but let it hang free at this point.
9. Install the air conditioning compressor/alternator bracket, idler pulley and motor mount. Raise the vehicle and support safely. Have the tensioner within an arm's reach because the timing belt will have to be held in position with one hand.
10. To properly install the timing belt, reach up and engage it with the camshaft sprocket. Turn the intermediate shaft counterclockwise slightly, then engage the belt with the intermediate shaft sprocket. Hold the belt against the intermediate shaft sprocket and turn the sprocket clockwise to take up all tension; if the timing

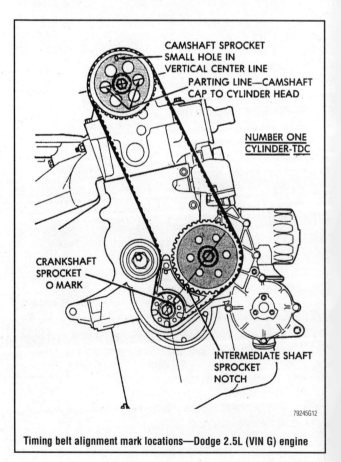

Timing belt alignment mark locations—Dodge 2.5L (VIN G) engine

CHRYSLER CORP.—TIMING BELTS

marks are out of alignment, repeat the installation until alignment is correct.

11. Using a 13mm wrench, turn the crankshaft sprocket counterclockwise slightly and wrap the belt around it. Turn the sprocket clockwise so there is no slack in the belt between the sprockets, if the timing marks are out of alignment, repeat the installation until alignment is correct.

✳✳ WARNING

If any binding is felt when adjusting the timing belt tension by turning the crankshaft, STOP turning the engine, because the pistons may be hitting the valves.

➡ If the timing marks are aligned, but slack exists in the belt between either the camshaft and intermediate shaft sprockets or the intermediate and crankshaft sprockets, the timing will be incorrect when the belt is tensioned. All slack must be between the crankshaft and camshaft sprockets only.

12. Install the tensioner and install the mounting bolt loosely. Place special tensioning tool C-4703, or equivalent, on the hex of the tensioner so the weight is approximately at the 9 o'clock position (parallel to the ground, hanging to the left) plus or minus 15°.

13. Hold the tool in position and tighten the bolt to 45 ft. lbs. (61 Nm). Do not pull the tool past the 9 o'clock position; this will make the belt too tight and will cause it to howl or possibly break during engine use.

14. Lower the vehicle and recheck the camshaft sprocket positioning. If it is correct, install the timing belt covers and all related parts.

15. Connect the negative battery cable and road test the vehicle.

2.5L (VIN H) Engine

1. Disconnect the negative battery cable from the left strut tower. The ground cable is equipped with a insulator grommet which should be placed on the stud to prevent the negative battery cable from accidentally grounding.
2. Raise and safely support the vehicle. Remove the right inner splash-shield.
3. Remove the accessory drive belts.
4. Remove the crankshaft damper.
5. Remove the right engine mount
6. Place a suitable floor jack under the vehicle to support the engine.
7. Remove the right engine mount bracket
8. Remove the timing belt upper left cover, upper right cover and lower cover.
9. Loosen the timing belt tensioner bolts.

➡ Before removing timing belt, be sure to align the sprocket timing marks to the timing marks on the rear timing belt cover.

10. If the present timing belt is going to be reused, mark the running direction of the timing belt for installation. Remove the timing belt and the tensioner.

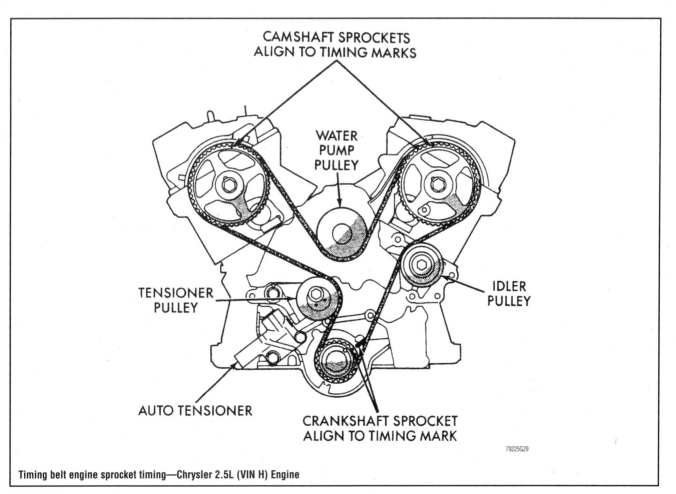

Timing belt engine sprocket timing—Chrysler 2.5L (VIN H) Engine

28   TIMING BELTS—CHRYSLER CORP.

11. Remove the camshaft timing belt sprockets from the camshaft, if necessary.
12. Remove the crankshaft timing belt sprocket and key.
13. Place the tensioner into a soft-jawed vise to compress the tensioner.
14. After compressing the tensioner place a pin into the plunger side hole to retain the plunger until installation.

To install:
15. If removed, reinstall the camshaft sprockets onto the camshaft. Install the camshaft sprocket bolt and tighten to 65 ft. lbs. (88 Nm).
16. If removed, reinstall the crankshaft timing belt sprocket and key onto the crankshaft.
17. Set the crankshaft sprocket to Top Dead Center (TDC) by aligning the notch on the sprocket with the arrow on the oil pump housing, then back off the sprocket three notches before TDC.
18. Set the camshafts to align the timing marks on the sprockets with the marks on the rear timing belt cover.
19. Install the belt on the rear camshaft sprocket first.
20. Install a binder clip on the belt to the sprocket so it won't slip out of position.
21. Keeping the belt taut, install it under the water pump pulley and around the front camshaft sprocket.
22. Install a binder on the front sprocket and belt.
23. Rotate the crankshaft to TDC.
24. Continue routing the belt by the idler pulley and around the crankshaft sprocket to the tensioner pulley.
25. Move the crankshaft sprocket clockwise to TDC to take up the belt slack. Check that all timing marks are in alignment.
26. Reinstall the tensioner to the block but do not tighten it at this time.
27. Using special tool No. MD998767 (or equivalent) and a torque wrench on the tensioner pulley, apply 39 inch lbs. (4.4 Nm) of torque to the tensioner. Tighten the tensioner pulley bolt to 35 ft. lbs. (48 Nm).
28. With torque being applied to the tensioner pulley, move the tensioner up against the tensioner bracket and tighten the fasteners to 17 ft. lbs. (23 Nm).
29. Remove the tensioner plunger pin, the tension is correct when the plunger pin can be removed and replaced easily.
30. Rotate the crankshaft two revolutions clockwise and recheck the timing marks. Check to be sure the tensioner plunger pin can be easily installed and removed. If the pin does not remove and install easily, perform the procedure again.
31. Reinstall the timing belt cover.
32. Reinstall the engine mount bracket.
33. Reinstall the right engine mount.
34. Remove the engine support.
35. Install the crankshaft damper and tighten to 134 ft. lbs. (182 Nm).
36. Reinstall the accessory drive belts and adjust them.
37. Reinstall the right inner splash-shield.
38. Perform the crankshaft and camshaft relearn alignment procedure using the DRB scan tool or equivalent.

2.5L (VIN K) Engine

Working on any engine (especially overhead camshaft engines) requires much care be given to valve timing. It is good practice to set the engine up to TDC No. 1 cylinder firing position. Verify that all timing marks on the crankshaft and camshaft sprockets are properly aligned before removing the timing belt. This serves as a point

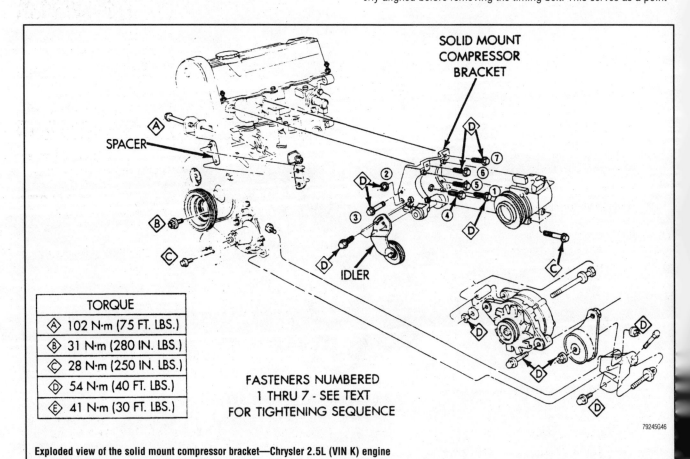

Exploded view of the solid mount compressor bracket—Chrysler 2.5L (VIN K) engine

CHRYSLER CORP.—TIMING BELTS

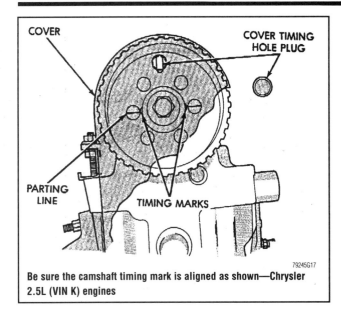

Be sure the camshaft timing mark is aligned as shown—Chrysler 2.5L (VIN K) engines

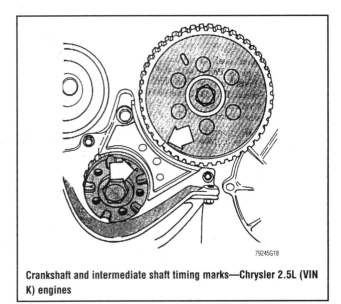

Crankshaft and intermediate shaft timing marks—Chrysler 2.5L (VIN K) engines

of reference for all work that follows. Valve timing is most important and engine damage will result if the work is incorrect.

➡ This is an interference engine. Do not rotate the crankshaft or the camshafts after the timing belt has been removed. Damage to the valve components may occur. Before removing the timing belt, always align the timing marks.

1. Disconnect the negative battery cable.
2. Remove the accessory drive belts.
3. Remove the right engine mount yoke screw.
4. Remove the air conditioning compressor and set it aside. Remove the solid mount compressor bracket mounting bolts.
5. Turn the solid mount bracket away from the engine and slide it on the No. 2 stud until it is free. The front bolt and spacer will be removed with the bracket.
6. Remove the alternator and the drive belt idler.
7. Raise and safely support the vehicle. Remove the right inner fender splash shield.
8. Loosen and remove the three water pump pulley mounting bolts and remove the pulley.

9. Remove the four crankshaft pulley retaining bolts and the crankshaft pulley.
10. Remove the nuts at the upper portion of the timing cover and the bolts from the lower portion, then remove both halves of the cover.
11. Remove the timing belt covers.
12. Position a jack under the engine.
13. Separate the right engine mount and raise the engine slightly.
14. Loosen the timing belt tensioner bolt, rotate the hex nut, and remove timing belt.
15. Remove the timing belt tensioner, if necessary.
16. Remove the crankshaft sprocket with a suitable puller tool and a bolt approximately 6 in. (15 cm) long.
17. Remove the camshaft sprocket and intermediate shaft sprocket, if necessary.

To install:

18. Clean all parts well. A small amount of white paint on the sprocket timing marks may make alignment easier.
19. Install the crankshaft sprocket. Tighten the crankshaft sprocket bolt to 85 ft. lbs. (115 Nm).
20. If necessary, turn the crankshaft and intermediate shaft until markings on both sprockets are aligned.
21. Rotate the camshaft so the arrows on the hub are in line with the No. 1 camshaft cap-to-cylinder head line. The small hole in the cam sprocket should be centered in the vertical center line.
22. If removed, install the timing belt tensioner.
23. Install the timing belt over the drive sprockets and adjust.
24. Tighten the tensioner by turning the tensioner hex to the right. Tension should be correct when the belt can be twisted 90 degrees with the thumb and forefinger, midway between the camshaft and intermediate sprocket.

✶✶ WARNING

If any binding is felt when adjusting the timing belt tension by turning the crankshaft, STOP turning the engine, because the pistons may be hitting the valves.

25. Turn the engine clockwise from TDC, 2 complete revolutions with the crankshaft bolt. Check the timing marks for correct alignment.

✶✶ WARNING

Do not use the camshaft or intermediate shaft to rotate the engine. Do not allow oil or solvent to contact the timing belt as they will deteriorate the belt and cause slipping.

26. Tighten the locknut on the tensioner, while holding the weighted wrench (tool C-4503 or equivalent) in position, to 45 ft. lbs. (61 Nm).
27. Lower the engine onto the right engine mount and install the fasteners. Remove the support from the engine.
28. Some engines use a foam stuffer block inside the timing belt housing. Inspect the foam block's condition and position. The stuffer block should be intact and secure within the engine bracket tunnel.
29. Install the timing belt cover. Secure the upper section to the cylinder head with nuts and the lower section to the cylinder block with screws. Tighten all of the timing belt cover fasteners to 40 inch lbs. (4 Nm).
30. Check valve timing again. With the timing belt cover installed, and with No. 1 cylinder at TDC, the small hole in the sprocket must be centered in the timing belt cover hole. If the hole is not aligned correctly, perform the timing belt installation procedure again.

TIMING BELTS—CHRYSLER CORP.

31. Install the water pump pulley and the crankshaft pulley. Tighten the water pump pulley bolts to 250 inch lbs. (28 Nm). Tighten the crankshaft pulley bolts to 280 inch lbs. (31 Nm).
32. Install the inner fender splash shield. Lower the vehicle.
33. Install the solid mount compressor bracket. The bracket mounting fasteners must be tightened to 40 ft. lbs. (54 Nm).
34. Install the alternator and drive belt idler. Tighten mounting bolts to 40 ft. lbs. (54 Nm).
35. Install the right engine mount yoke bolt and tighten to 100 ft. lbs. (133 Nm).
36. Install the accessory drive belts and adjust them to the proper tension.

➡ With the timing belt cover installed and the piston in the No. 1 cylinder at TDC, the small hole in the cam sprocket should be centered in timing belt cover hole.

37. Reconnect the negative battery cable. Road test the vehicle.

2.5L (VIN N) Engine

1. Disconnect the negative battery cable.
2. Remove the accessory drive belts.
3. Using Crankshaft Holding Tools MB990767 and MB998754, or their equivalents, remove the crankshaft bolt and remove the pulley.
4. Remove the heated oxygen sensor connection.
5. Remove the power steering pump with the hose attached and position it aside.
6. Remove the power steering pump bracket.
7. Place a floor jack under the engine oil pan, with a block of wood in between, and jack up the engine so that the weight of the engine is no longer being applied to the engine support bracket.
8. Remove the upper engine mount. Spraying lubricant, slowly remove the reamer (alignment) bolt and remaining bolts and remove the engine support bracket.

➡ The reamer bolt is sometimes heat-seized on the engine support bracket

9. Remove the front timing belt covers.
10. If the timing belt is to be reused, draw an arrow indicating the direction of rotation on the back of the belt for reinstallation.

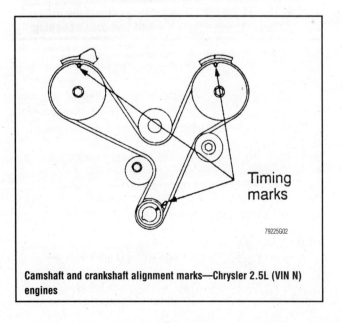

Camshaft and crankshaft alignment marks—Chrysler 2.5L (VIN N) engines

11. Align the timing marks by turning the crankshaft with MD998769 Crankshaft Turning tool or its equivalent. Loosen the center bolt on the timing belt tensioner pulley and remove the belt.

✱✱ WARNING

Do not rotate the crankshaft or camshaft after removing the timing belt, or valvetrain components may be damaged. Always align the timing marks before removing the timing belt.

12. Check the belt tensioner for leaks and check the pushrod for cracks.
13. If the timing belt tensioner is to be replaced, remove the retaining bolts and remove the timing belt tensioner. When the timing belt tensioner is removed from the engine it is necessary to compress the plunger into the tensioner body.
14. Place the tensioner in a vise and slowly compress the plunger. Take care not to damage the pushrod.

➡ Position the tensioner in the vise the same way it will be installed on the engine. This is to ensure proper pin orientation for when the tensioner is installed on the engine.

15. When the plunger is compressed into the tensioner body, install a pin through the body and plunger to hold the plunger in place until the tensioner is installed.

To install:
16. Install the timing belt tensioner and tighten the retaining bolts to 17 ft. lbs. (24 Nm), but do not remove the pin at this time.
17. Check that all timing marks are still aligned.
18. Use bulldog clips (large paper binder clips) or other suitable tool to secure the timing belt and to prevent it from slacking. Install the timing belt. Starting at the crankshaft, go around the idler pulley, then the front camshaft sprocket, the water pump pulley, the rear camshaft sprocket, then around the tensioner pulley.
19. Be sure the belt is tight between the crankshaft and front camshaft sprocket, between the camshaft sprockets and the water pump. Gently raise the tensioner pulley, so that the belt does not sag, and temporarily tighten the center bolt.
20. Move the crankshaft ¼ turn counterclockwise,, then turn it clockwise to the position where the timing marks are aligned.
21. Loosen the center bolt of the tensioner pulley. Using MD998767, or equivalent tensioner tool, and a torque wrench apply 3.3 ft. lbs. (4.4 Nm) tensional torque to the timing belt and tighten the center bolt to 35 ft. lbs. (48 Nm). When tightening the bolt, be sure that the tensioner pulley shaft does not rotate with the bolt.
22. Remove the tensioner plunger pin. Pretension is correct when the pin can be removed and installed easily. If the pin cannot be easily removed and installed it is still satisfactory as long as it is within its standard value.
23. Check that the tensioner pushrod is within the standard value. When the tensioner is engaged the pushrod should measure 0.149–0.177 in. (3.8–4.5mm).
24. Rotate the crankshaft two revolutions and check the timing marks. If the timing marks are not properly aligned remove the belt and repeat Steps 17 through 23.
25. Install the timing belt covers.
26. Install the engine mounting bracket.
27. Lower the engine enough to install the engine mount onto bracket and remove the floor jack.
28. Install the power steering pump bracket and pump.
29. Install the crankshaft pulley and tighten the retaining bolt to 13 ft. lbs. (18 Nm).

CHRYSLER CORP.—TIMING BELTS

30. Install the accessory drive belts.
31. Properly fill the cooling system.
32. Connect the negative battery cable.
33. Check for leaks and proper engine and cooling system operation.

3.0L (VIN 3) Engine

1990—94 MODELS

1. Position the engine so the No. 1 cylinder is at TDC of its compression stroke. Disconnect the negative battery cable.
2. Remove the engine undercover.
3. Remove the cruise control actuator.
4. Remove the accessory drive belts.
5. Remove the air conditioner compressor tension pulley assembly.
6. Remove the tension pulley bracket.
7. Using the proper equipment, slightly raise the engine to take the weight off the side engine mount. Remove the engine mounting bracket.
8. Disconnect the power steering pump pressure switch connector. Remove the power steering pump with hoses attached and wire aside.
9. Remove the engine support bracket.
10. Remove the crankshaft pulley.
11. Remove the timing belt cover cap.
12. Remove the timing belt upper and lower covers.
13. If the same timing belt will be reused, mark the direction of the timing belt's rotation for installation in the same direction. Make sure the engine is positioned so the No. 1 cylinder is at the TDC of its compression stroke and the timing marks are aligned with the engine's timing mark indicators.
14. Loosen the timing belt tensioner bolt and remove the belt. If the tensioner is not being removed, position it as far away from the center of the engine as possible and tighten the bolt.
15. If the tensioner is being removed, paint the outside of the spring to ensure that it is not installed backwards. Unbolt the tensioner and remove it along with the spring.

To install:

16. Install the tensioner, if removed, and hook the upper end of the spring to the water pump pin and the lower end to the tensioner in exactly the same position as originally installed. If not already done, position both camshafts so the marks align with those on the rear. Rotate the crankshaft so the timing mark aligns with the mark on the front cover.
17. Install the timing belt on the crankshaft sprocket and while keeping the belt tight on the tension side, install the belt on the front camshaft sprocket.
18. Install the belt on the water pump pulley, then the rear camshaft sprocket and the tensioner.
19. Rotate the front camshaft counterclockwise to tension the belt between the front camshaft and the crankshaft. If the timing marks became misaligned, repeat the procedure.
20. Install the crankshaft sprocket flange.
21. Loosen the tensioner bolt and allow the spring to apply tension to the belt.
22. Turn the crankshaft 2 full turns in the clockwise direction until the timing marks align again. Now that the belt is properly tensioned, torque the tensioner lock bolt to 21 ft. lbs. (29 Nm). Measure the belt tension between the rear camshaft sprocket and the crankshaft with belt tension gauge. The specification is 46–68 lbs. (210–310 N).
23. Install the timing covers. Make sure all pieces of packing are positioned in the inner grooves of the covers when installing.
24. Install the crankshaft pulley. Tighten the bolt to 108–116 ft. lbs. (150–160 Nm).
25. Install the engine support bracket.
26. Install the power steering pump and reconnect wire harness at the power steering pump pressure switch.
27. Install the engine mounting bracket and remove the engine support fixture.
28. Install the tension pulleys and drive belts.
29. Install the cruise control actuator.
30. Install the engine undercover.
31. Connect the negative battery cable and road test the vehicle.

1995–99 MODELS

The timing belt can be inspected by removing the upper front outer timing belt cover.

Working on any engine (especially overhead camshaft engines) requires much care be given to valve timing. It is good practice to set the engine up at TDC No. 1 cylinder firing position before beginning work. Verify that all timing marks on the crankshaft and camshaft sprockets are properly aligned before removing the timing belt and starting camshaft service. This serves as a point of reference for all work that follows. Valve timing is very important and engine damage will result if the work is incorrect.

1. Disconnect the negative battery cable.
2. Remove the accessory drive belts. Remove the engine mount insulator from the engine support bracket.
3. Remove the engine support bracket. Remove the crankshaft pulleys and torsional damper. Remove the timing belt covers.
4. Rotate the crankshaft until the sprocket timing marks are aligned. The crankshaft sprocket timing mark should align with the oil pump timing mark. The rear camshaft sprocket timing mark should align with the generator bracket timing mark and the front camshaft sprocket timing mark should align with the inner timing belt cover timing mark.
5. If the belt is to be reused, mark the direction of rotation on the belt for installation reference.
6. Loosen the timing belt tensioner bolt and remove the timing belt.
7. If necessary, remove the timing belt tensioner.
8. Remove the crankshaft sprocket flange shield and crankshaft sprocket.
9. Hold the camshaft sprocket using spanner tool MB990775 or equivalent, and remove the camshaft sprocket bolt and washer. Remove the camshaft sprocket.

To install:

10. Install the camshaft sprocket on the camshaft with the retaining bolt and washer. Hold the camshaft sprocket using spanner tool MB990775 or equivalent, and tighten the bolt to 70 ft. lbs. (95 Nm).
11. Install the crankshaft sprocket.
12. If removed, install the timing belt tensioner and tensioner spring. Hook the spring upper end to the water pump pin and the lower end to the tensioner bracket with the hook out.
13. Turn the timing belt tensioner counterclockwise full travel in the adjustment slot and tighten the bolt to temporarily hold it in this position.
14. Rotate the crankshaft sprocket until its timing mark is aligned with the oil pump timing mark.
15. Rotate the rear camshaft sprocket until its timing mark is aligned with the timing mark on the generator bracket.

32 TIMING BELTS—CHRYSLER CORP.

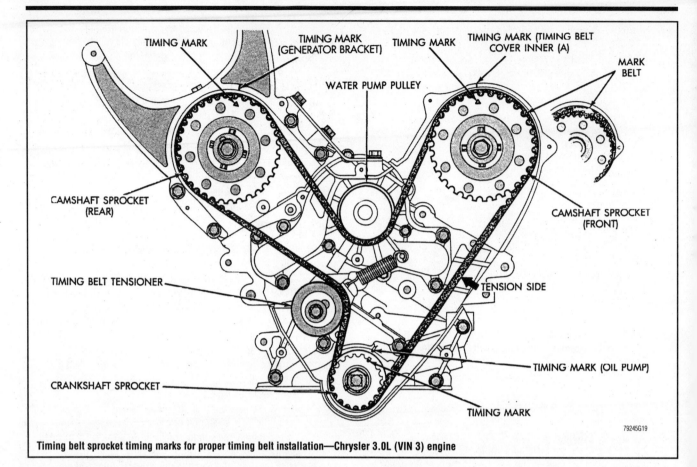

Timing belt sprocket timing marks for proper timing belt installation—Chrysler 3.0L (VIN 3) engine

16. Rotate the front (radiator side) camshaft sprocket until its mark is aligned with the timing mark on the inner timing belt cover.

17. Install the timing belt on the crankshaft sprocket while keeping the belt tight on the tension side.

➡️ **If the original belt is being reused, be sure to install it in the same rotational direction.**

18. Position the timing belt over the front camshaft sprocket (radiator side). Next, position the belt under the water pump pulley, then over the rear camshaft sprocket and finally over the tensioner.

❋❋ WARNING

If any binding is felt when adjusting the timing belt tension by turning the crankshaft, STOP turning the engine, because the pistons may be hitting the valves.

19. Apply rotating force in the opposite direction to the front camshaft sprocket (radiator side) to create tension on the timing belt tension side. Check that all timing marks are aligned.

20. Install the crankshaft sprocket flange.

21. Loosen the tensioner bolt and allow the tensioner spring to tension the belt.

22. Rotate the crankshaft two full turns in a clockwise direction. Turn the crankshaft smoothly and in a clockwise direction only.

23. Again line up the timing marks. If all marks are aligned, tighten the tensioner bolt to 250 inch lbs. (28 Nm). Otherwise repeat the installation procedure.

24. Install the timing belt covers. Install the engine support bracket. Tighten the support bracket mounting bolts to 35 ft. lbs. (47 Nm).

25. Install the engine mount insulator, torsional damper and crankshaft pulleys. Tighten the crankshaft pulley bolt to 112 ft. lbs. (151 Nm).

26. Install the accessory drive belts and adjust them to the proper tension.

27. Reconnect the negative battery cable.

28. Run the engine and check for proper operation. Road test the vehicle.

3.0L (VIN J) Engine

1. Position the engine so the No. 1 cylinder is at TDC of its compression stroke. Disconnect the negative battery cable.

2. Remove the engine undercover.

3. Remove the cruise control actuator.

4. Remove the alternator. Remove the air hose and pipe.

5. Remove the belt tensioner assembly and the power steering belt.

6. Remove the crankshaft pulley.

7. Disconnect the brake fluid level sensor.

8. Remove the timing belt upper cover.

9. Using the proper equipment, slightly raise the engine to take the weight off the side engine mount. Remove the engine mount bracket.

10. Remove the alternator/air conditioner idler pulley.

11. Remove the engine support bracket. The mounting bolts are different lengths; mark them for proper installation.

12. Remove the timing belt lower cover. Timing bolt cover mounting bolts are different in length, note their position during removal.

13. If the same timing belt will be reused, mark the direction of the timing belt's rotation for installation in the same direction. Make

CHRYSLER CORP.—TIMING BELTS

sure the engine is positioned so the No. 1 cylinder is at the TDC of its compression stroke and the timing marks are aligned with the engine's timing mark indicators on the valve covers or head.

14. Loosen the center bolt of tensioner pulley and unbolt auto-tensioner assembly. The auto-tensioner assembly must be reset to correctly adjust belt tension. Remove the timing belt.

15. Remove and position the auto-tensioner into a vise with soft jaws. The plug at the rear of tensioner protrudes, be sure to use a washer as a spacer to protect the plug from contacting vise jaws.

16. Slowly push the rod into the tensioner until the set hole in rod is aligned with set hole in the auto-tensioner.

17. Insert a 0.055 in. (1.4mm) wire into the aligned set holes. Unclamp the tensioner from vise and install to vehicle. Tighten tensioner to 17 ft. lbs. (24 Nm).

18. On 1991 DOHC 3.0L engines, clean and inspect both auto tensioner mounting bolts. Coat the threads of the old bolts with Mopar thread sealer 4318034 or equivalent. If new bolts are installed, inspect the heads of the new bolts. If there is white paint on the bolt head, no sealer is required. If there is no paint on the head of the bolt, apply a coat of thread sealer to the bolt. Install both bolts and tighten to 17 ft. lbs. (24 Nm).

To install:

> ※※ **WARNING**
>
> **Turning the camshaft sprocket when the timing belt is removed could cause the valves to interfere with the pistons.**

19. Align the mark on the crankshaft sprocket with the mark on the front case. Then move the sprocket 3 teeth clockwise to lower the piston so the valves do not touch the piston if the camshafts are being moved.

20. Turn each camshaft sprocket 1 at a time to align the timing marks with the mark on the valve cover or head. If the intake and exhaust valves of the same cylinder are opened simultaneously, they could interfere with each other. Therefore, if any resistance is felt, turn the other camshaft to move the valve.

21. Using paper clips to secure the timing belt to sprockets, install the timing belt in the following order. Be sure camshafts to cylinder heads and crankshaft to front cover timing marks are aligned.

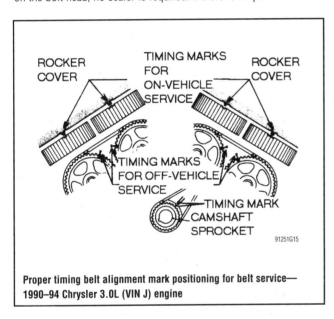

Proper timing belt alignment mark positioning for belt service—1990–94 Chrysler 3.0L (VIN J) engine

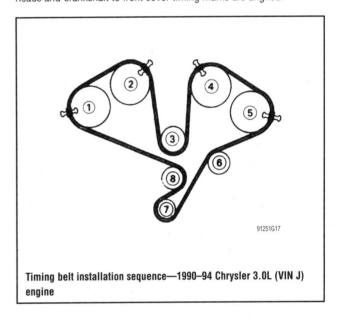

Timing belt installation sequence—1990–94 Chrysler 3.0L (VIN J) engine

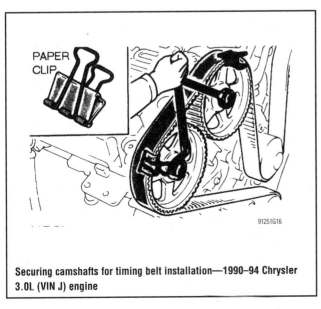

Securing camshafts for timing belt installation—1990–94 Chrysler 3.0L (VIN J) engine

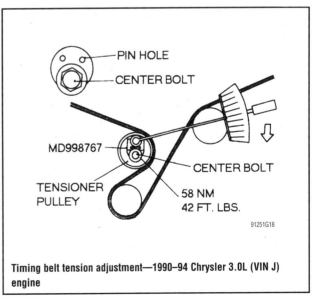

Timing belt tension adjustment—1990–94 Chrysler 3.0L (VIN J) engine

34 TIMING BELTS—CHRYSLER CORP.

    a. Exhaust camshaft sprocket (front bank).
    b. Intake camshaft sprocket (front bank).
    c. Water pump pulley.
    d. Intake camshaft sprocket (rear bank).
    e. Exhaust camshaft sprocket (rear bank).
    f. Idler pulley.
    g. Crankshaft pulley.
    h. Tensioner pulley.

➡ **Since the camshaft sprockets turn easily, secure them with box wrenches to install timing belt.**

22. Align all timing marks and raise tensioner pulley against belt to remove slack, snug tensioner bolt.
23. Loosen the center bolt on the tensioner pulley. Using tool MD998767 or equivalent and a torque wrench, apply a torque of 7.2 ft. lbs. (10 Nm). Tighten the tensioner bolt to 42 ft. lbs. (58 Nm) and make sure the tensioner does not rotate with the bolt.
24. Remove the set wire attached to the auto tensioner.
25. Rotate the crankshaft 2 complete turns clockwise and let it sit for approximately 5 minutes. Then, check that the set pin can easily be inserted and removed from the hole in the auto tensioner.

➡ **Even if the set pin cannot be easily inserted, the auto tensioner is normal if its rod protrusion is within specification.**

26. Measure the auto tensioner protrusion (the distance between the tensioner arm and auto tensioner body) to ensure that it is within 0.15–0.18 in. (3.8–4.5mm). If out of specification, repeat adjustment procedure until the specified value is obtained.
27. Check again that the timing marks on all sprockets are in proper alignment.
28. Make sure all pieces of packing are positioned in the inner grooves of the lower cover, position cover on engine and install mounting bolts in their original location.
29. Install the engine support bracket and secure using mounting bolts in their original location. Lubricate the reaming area of the reamer bolt and tighten slowly.
30. Install the idler pulley.
31. Install the engine mount bracket. Remove the engine support fixture.
32. Make sure all pieces of packing are positioned in the inner grooves of the upper cover and install.
33. Connect the brake fluid level sensor.
34. Install the crankshaft pulley. Tighten the bolt to 130–137 ft. lbs. (180–190 Nm).
35. Install the belt tensioner assembly and the power steering belt.
36. Install the air hose and pipe.
37. Install the alternator.
38. Install the cruise control actuator.
39. Install the engine undercover.
40. Connect the negative battery cable.

3.2L (VIN J) and 3.5L (VIN F) Engines

Use care when servicing a timing belt. Valve timing is absolutely critical to engine performance. If the valve timing marks on all drive sprockets are not properly aligned, engine damage will result. If only the belt and tensioner are being serviced, do not loosen the camshaft drive sprockets unless they are to be replaced. The sprockets have oversized openings and can be rotated several degrees in each direction on their shafts. This means the sprockets must be retimed, requiring some special tools.

✱✱ CAUTION

Fuel injection systems remain under pressure, even after the engine has been turned off. The fuel system pressure must be relieved before disconnecting any fuel lines. Failure to do so may result in fire and/or personal injury.

1. Disconnect the negative battery cable.
2. Rotate the engine to Top Dead Center (TDC) on the compression stroke for cylinder No. 1.
3. Release the fuel system pressure using the recommended procedure.
4. Place a pan under the radiator and drain the coolant.
5. Remove the radiator and cooling fan assemblies.
6. Remove the accessory drive belts.
7. Remove the upper radiator hose.
8. Remove the crankshaft damper with a quality puller tool gripping the inside of the pulley.
9. Remove the stamped steel cover. Do not remove the sealer on the cover; it may be reusable.
10. Remove the left side cast cover. If necessary, remove the lower belt cover, located behind the crankshaft damper.
11. If the timing belt is to be reused, mark the timing belt with the running direction for installation.
12. Align the camshaft sprockets with the marks on the rear covers.
13. Remove the timing belt and tensioner.
14. If it is necessary to service the camshaft sprockets, use the following procedure:
    a. Hold the camshaft sprocket with a 36mm box end wrench, loosen and remove the sprocket retaining bolt and washer.

➡ **To remove the camshaft sprocket retainer bolt while the engine is in the vehicle, it may be necessary to raise that side of the engine due to the length of the retainer bolt. The right bolt is 8 ⅜-in. (213mm) long, while the left bolt is 10in. (254mm) long. These bolts are not interchangeable and their original location during removal should be noted.**

    b. Remove the camshaft sprocket from the camshaft. The camshaft sprockets are not interchangeable from side-to-side.
    c. Remove the crankshaft sprocket using Puller L-4407A, or equivalent.

To install:
15. If it was necessary to remove the camshaft sprockets, use the following procedure:

➡ **This procedure can only be used when the camshaft sprockets have been loosened or removed from the camshafts. Each sprocket has a D-shaped hole that allows it to be rotated several degrees in each direction on its shaft. This design must be timed with the engine to ensure proper performance.**

    a. Install the crankshaft sprocket, using tool C-4685-C1, thrust bearing, washer and 12mm bolt.
    b. When the camshaft sprockets are loosened or removed, the camshafts must be timed to the engine. Install the Camshaft Alignment tools 6642-A, or their exact equivalents, to the rear of the cylinder heads. These tools lock the camshafts in the proper position.
16. Preload the belt tensioner as follows:
    a. Place the tensioner in a vise the same way it is mounted on the engine.
    b. Slowly compress the plunger into the tensioner body.

CHRYSLER CORP.—TIMING BELTS

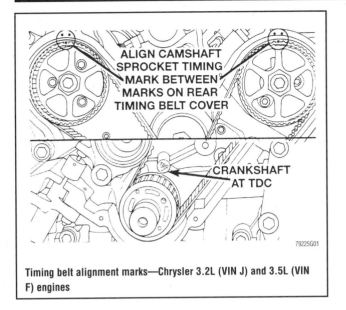

Timing belt alignment marks—Chrysler 3.2L (VIN J) and 3.5L (VIN F) engines

  c. When the plunger is compressed into the tensioner body, install a pin through the body and plunger to retain the plunger in place until the tensioner is installed.

  17. Install both camshaft sprockets to the appropriate shafts. The left camshaft sprocket has the DIS pick-up as part of the sprocket.

➥ **The right bolt is 8 3/8 in. (213mm) long, while the left bolt is 10in. (254mm) long. These bolts are not interchangeable.**

  18. Apply Loctite®271 or equivalent, to the threads of the camshaft sprocket retainer bolts and install to the appropriate shafts. Do not tighten the bolts at this time.
  19. Align the camshaft sprockets between the marks on the rear belt covers.
  20. Align the crankshaft sprocket with the TDC mark on the oil pump cover.
  21. Install the timing belt, starting at the crankshaft sprocket and going in a counterclockwise direction. After the belt is installed on the right sprocket, keep tension on the belt until it is past the tensioner pulley.
  22. Holding the tensioner pulley against the belt, install the timing belt tensioner into the housing and tighten to 21 ft. lbs. (28 Nm).
  23. When the tensioner is in place, pull the retainer pin out to allow tensioner to extend to the pulley bracket.

➥ **Be sure that the timing marks on the cam sprockets are still between the marks on the rear cover.**

  24. Remove the spark plug in the No.1 cylinder and install a dial indicator to check for Top Dead Center (TDC) of the piston. Rotate the crankshaft until the piston is exactly at TDC.
  25. Hold the camshaft sprocket hex with a 36mm wrench and tighten the right camshaft sprocket bolt to 75 ft. lbs. (102 Nm) plus an additional 90 degree turn. Tighten the left camshaft sprocket bolt to 85 ft. lbs. (115 Nm) plus an additional 90 degree turn.
  26. Remove the dial indicator. Install the spark plug and tighten to 20 ft. lbs. (28 Nm).
  27. Remove the camshaft alignment tools from the back of the cylinder heads and install the cam covers with new O-rings.
  28. Tighten the fasteners to 20 ft. lbs. (27 Nm). Repeat this procedure on the other camshaft.

  29. Rotate the crankshaft sprocket two revolutions and check for proper alignment of the timing marks on the camshaft and the crankshaft. If the timing marks do not line up, repeat the procedure.
  30. Before installing, inspect the sealer on the stamped steel cover. If some sealer is missing, use MOPAR Silicone Rubber Adhesive sealant or equivalent to replace the missing sealer.
  31. Install the lower belt cover behind the crankshaft damper, if necessary.
  32. Install the stamped steel cover and the left side cast cover. Tighten the 6mm bolts to 105 in.. lbs. (12 Nm), the 8mm bolts to 250 inch lbs. (28 Nm) and the 10mm bolts to 40 ft. lbs. (54 Nm).
  33. Install the crankshaft damper using special tool L-4524, a 5.9 in. long bolt, thrust bearing and washer or equivalent damper installation tools. Tighten the center bolt to 85 ft. lbs. (115 Nm).
  34. Install the upper radiator hose.
  35. Install the accessory drive belts and adjust them to the proper tension.
  36. Install the radiator and cooling fan assemblies.
  37. Refill and bleed the cooling system.
  38. Connect the negative battery cable.
  39. With the radiator cap off so coolant can be added, run the engine. Watch for leaks and listen for unusual engine noises.

IMPORT MODELS

1.5L and 1.8L Engines

1990–94 MODELS

  1. Disconnect the negative battery cable. Remove the engine under cover.
  2. Rotate crankshaft clockwise and position engine at TDC compression stroke.
  3. Raise and safely support the weight of the engine using the appropriate equipment. Remove the front engine mount bracket and accessory drive belts.
  4. On Summit Wagon, remove the coolant reservoir tank.
  5. Remove the drive belts, tension pulley brackets, water pump pulley and crankshaft pulley.
  6. Remove all attaching screws and remove the upper and lower timing belt covers.
  7. Make a mark on the back of the timing belt indicating the direction of rotation so it may be reassembled in the same direction if it is to be reused. Loosen the timing belt tensioner and remove the timing belt.

➥ **If coolant or engine oil comes in contact with the timing belt, they will drastically shorten its life. Also, do not allow engine oil or coolant to contact the timing belt sprockets or tensioner assembly.**

  8. Remove the tensioner spacer, tensioner spring and tensioner assembly.
  9. Inspect the timing belt for cracks on back surface, sides, bottom and check for separated canvas. Check the tensioner pulley for smooth rotation.

To install:
  10. Position the tensioner, tensioner spring and tensioner spacer on engine block.
  11. Align the timing marks on the camshaft sprocket and crankshaft sprocket. This will position No. 1 piston on TDC on the compression stroke.

TIMING BELTS—CHRYSLER CORP.

12. Position the timing belt on the crankshaft sprocket and keeping the tension side of the belt tight, set it on the camshaft sprocket.

13. Apply counterclockwise force to the camshaft sprocket to give tension to the belt and make sure all timing marks are aligned.

14. Loosen the pivot side tensioner bolt and the slot side bolt. Allow the spring to remove the slack.

15. Tighten the slot side tensioner bolt and then the pivot side bolt. If the pivot side bolt is tightened first, the tensioner could turn with bolt, causing over tension.

16. Turn the crankshaft clockwise. Loosen the pivot side tensioner bolt and then the slot side bolt to allow the spring to take up any remaining slack. On 1.8L engine, tighten the adjuster bolt to 18 ft. lbs. (24 Nm). On 1.5L engine, tighten the slot bolt and then the pivot side bolt to 14–20 ft. lbs. (20–27 Nm).

17. Install the timing belt covers and all related items.

18. Connect the negative battery cable.

1995–99 MODELS

1. Rotate the crankshaft clockwise and position the engine at Top Dead Center (TDC) on the compression stroke for the No. 1 cylinder.

2. Remove the drive belts, tension pulley brackets, water pump pulley and crankshaft pulley.

3. Remove all attaching screws and remove the upper and lower timing belt covers.

4. Make a mark on the back of the timing belt indicating the direction of rotation so it may be reassembled in the same direction if it is to be reused. Loosen the timing belt tensioner and remove the timing belt.

5. For the 1.5L engine, loosen the timing belt tensioner and move the tensioner to provide slack to the timing belt. Tighten the tensioner in this position.

6. For the 1.8L engine, loosen the timing belt tensioner, insert a thin prytool into the tensioner and release tension by prying against the spring tension. Temporarily tighten the tensioner bolt to provide slack.

7. Remove the timing belt.

❈❈ WARNING

Coolant and engine oil will damage the rubber in the timing belt, drastically reducing its life. Do not allow engine oil or coolant to contact the timing belt, the sprockets or tensioner assembly.

8. If defective, remove the tensioner spacer, tensioner spring and tensioner assembly.

➡ **It is recommended that the timing belt be replaced at least every 60,000 miles (96,000 km).**

9. Inspect the timing belt for cracks or wear. Check the tensioner pulley for smooth rotation.

To install:

10. If removed, position the tensioner, tensioner spring and tensioner spacer on the engine block.

11. Align the timing marks on the camshaft sprocket and crankshaft sprocket. This will position the No. 1 piston at TDC on its compression stroke.

12. For the 1.5L engine, position the timing belt on the crankshaft sprocket, keeping the tension side of the belt tight, set it on the camshaft sprocket, then the tensioner sprocket.

13. For the 1.8L engine, position the timing belt on the crankshaft sprocket, water pump sprocket, camshaft sprocket and the tensioner sprocket, keeping the tension side of the belt tight.

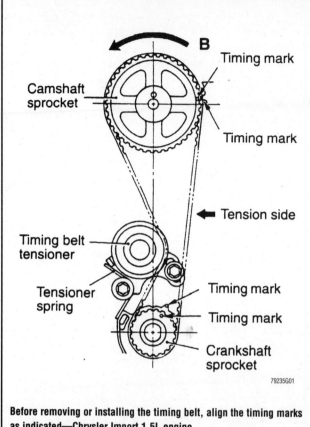

Before removing or installing the timing belt, align the timing marks as indicated—Chrysler Import 1.5L engine

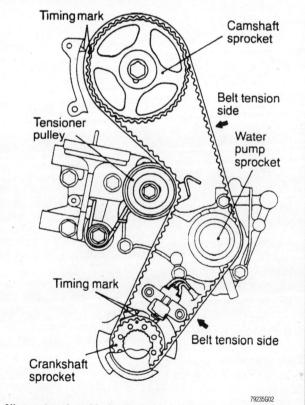

Alignment mark positioning for timing belt service—Chrysler Import 1.8L engine

14. Apply a slight counterclockwise force to the camshaft sprocket to give tension to the belt, and be sure all timing marks are aligned.
15. Loosen the pivot side tensioner bolt and the slot side bolt. Allow the spring to remove any slack in the timing belt.
16. For the 1.5L engine, tighten the slot side tensioner bolt, then the pivot side bolt. If the pivot side bolt is tightened first, the tensioner could turn with the bolt, causing over tension.
17. For the 1.8L engine, turn the crankshaft clockwise two rotations, then tighten the adjuster bolt to 18 ft. lbs. (24 Nm) and the pivot (spring) bolt to 35 ft. lbs. (45 Nm).
18. Turn the crankshaft clockwise. Loosen the pivot side tensioner bolt, then the slot side bolt to allow the spring to take up any remaining slack. Tighten the slot bolt, then the pivot side bolt to 17 ft. lbs. (24 Nm).
19. Install the timing belt covers, and tighten the cover bolts to 96 inch lbs. (11 Nm). Install all applicable components.

1.6L DOHC Engine

→Special tools MD998752 tension pulley torque adapter and MD998738 tension pulley locker or equivalents, are required.

1. Bring the engine to No. 1 piston at TDC (top dead center) timing marks aligned. Disconnect the negative battery cable.
2. Raise the vehicle and support it safely. Remove the under engine splash shield.
3. Place a piece of wood on a suitable floor jack and support the engine. Remove the engine mount bracket.
4. Remove the alternator and power steering drive belts. Remove the air conditioner drive belt and tensioner assembly.
5. Remove the water pump pulley and the crankshaft pulley.
6. Remove the upper and lower timing belt covers.
7. Remove the engine center cover. Remove the breather hose from the rear of the rocker cover. Remove the PCV hose. Disconnect the spark plug cables from the plugs.
8. Remove the rocker cover and rear half-moon seal.
9. Confirm the engine is still at No. 1 TDC. The timing marks on the camshaft sprocket and the upper surface of the cylinder head should coincide. The dowel pin on the front of the camshafts should be in the 12 o'clock position. Remove the automatic belt tensioner. Loosen the tensioner pulley center bolt.
10. If the timing belt is to be reused, mark an arrow, on the belt, in the direction of rotation, for installation reference. Remove the timing belt.

To install:
11. Install the automatic tensioner, after reset.

→To reset the tensioner: Keep the adjuster level and clamp it in a soft jawed vise. Clamp with the extended adjuster on one side and the end mounting a plug on the other side. If the plug extends out of the adjuster body, place a suitable hole sized washer over the plug so the vise jaw pushes on the washer, not the plug. Close the vise slowly, forcing the adjuster back into the body. When the hole in the adjuster boss aligns with the adjuster rod, insert a snug fitting pin or wire into the holes to keep the rod in the compressed position. With the locking pin or wire in place, install the tensioner.

12. Align the timing marks on the camshaft sprockets. Align the crankshaft timing marks. Align the oil pump timing marks. Place the timing belt around the intake camshaft and secure it to the sprocket

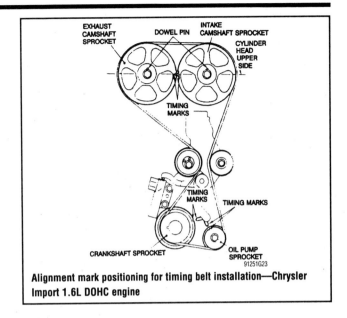

Alignment mark positioning for timing belt installation—Chrysler Import 1.6L DOHC engine

with a stationary binder spring clip. Install the timing belt around the exhaust camshaft sprocket, check sprocket marks for alignment and secure the belt with a second binder clip on the exhaust sprocket.
13. Install the timing belt around the idler pulley, oil pump sprocket, crankshaft sprocket and the tensioner pulley.
14. Lift up the tensioner pulley against the belt and tighten the center bolt to hold it in position.
15. Check to see that all of the timing marks are aligned. Remove the binder clips. Rotate the crankshaft a quarter turn counter clockwise. Then turn the crankshaft clockwise until the timing marks are aligned.
16. Place special tool MD998752 or equivalent, on a torque wrench. Insert the tool into the place provided on the tension pulley. Loosen the center pulley bolt and apply 2.2 ft. lbs. (3.0 Nm) of pressure against the timing belt with the tension pulley. While holding the required torque, tighten the center bolt. Screw in special tool MD998738 or equivalent, through the left engine support bracket until it contacts the tensioner arm bracket. Turn the tool a little more to secure the tensioner and remove the locking wire place into the automatic adjuster when it was reset.
17. Remove the special tool. Rotate the crankshaft 2 complete turns clockwise and allow it to sit for about 15 minutes. Then measure the protrusion of the automatic adjuster. It should be 0.0150.018 in. (0.3810.457mm). If the proper amount of protrusion is not present, repeat the tensioning process.

2.4L Engine

1990–92 MODELS

1. Position the engine so the No. 1 piston is at TDC of compression stroke.
2. Disconnect the negative battery cable. Remove the coolant reservoir and the power steering and air conditioner hose clamp bolt.
3. Remove the drive belts, tension pulley brackets, water pump pulley and crankshaft pulley.
4. Remove all attaching screws and remove the upper and lower timing belt covers.
5. Remove the timing belt tensioner pulley, tensioner arm, idler pulley.

TIMING BELTS—CHRYSLER CORP.

6. Locate the access plug on the side of block. Remove the plug and install a phillips screwdriver. Remove the oil pump sprocket nut, oil pump sprocket, special washer, flange and spacer.

7. Remove the outer crankshaft sprocket and flange.

8. Remove the silent shaft (inner) belt tensioner and remove the belt.

To install:

9. Align the timing marks of the silent shaft sprockets and the crankshaft sprocket with the timing marks on the front case. Wrap the timing belt around the sprockets so there is no slack in the upper span of the belt and the timing marks are still aligned.

10. Install the tensioner pulley and move the pulley by hand so the long side of the belt deflects 0.20–0.28 in. (5–7mm).

11. Hold the pulley tightly so the pulley cannot rotate when the bolt is tightened. Tighten the bolt to 15 ft. lbs. (20 Nm) and recheck the deflection amount.

12. Install the timing belt tensioner fully toward the water pump and tighten the bolts. Place the upper end of the spring against the water pump body.

13. Align the timing marks of the camshaft, crankshaft and oil pump sprockets with their corresponding marks on the front case or rear cover.

➡ **There is a possibility to align all timing marks and have the oil pump sprocket out of time, causing an engine vibration during operation. If the following step is not followed exactly, there is a 50 percent chance that the oil pump shaft alignment will be 180 degrees off.**

14. Before installing the timing belt, ensure that the oil pump sprocket is in the correct position as follows:

　a. Remove the plug from the rear side of the block and insert a phillips screwdriver with shaft diameter of 0.31 in. (8mm) into the hole.

　b. With the timing marks still aligned, the shaft of the tool must be able to go in at least 2.36 in. (60mm). If the tool can only go in 0.79–0.98 in. (20–25mm), the shaft is not in the correct orientation and will cause a vibration during engine operation. Remove the tool from the hole and turn the oil pump sprocket 1 complete revolution. Realign the timing marks and insert the tool. The shaft of the tool must go in at least 2.36 in. (60mm).

　c. Recheck and realign the timing marks.

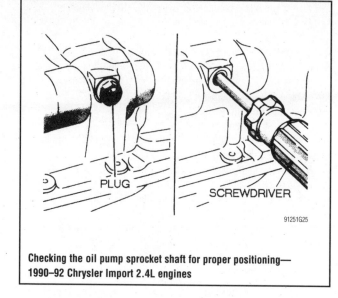

Checking the oil pump sprocket shaft for proper positioning—1990–92 Chrysler Import 2.4L engines

　d. Leave the tool in place to hold the oil pump shaft while continuing.

15. Install the belt to the crankshaft sprocket, oil pump sprocket, then camshaft sprocket. While doing so, make sure there is no slack between the sprocket except where the tensioner is installed.

16. Tighten oil pump sprocket bolt to 26–29 ft. lbs. (34–40 Nm) and tighten crankshaft bolt to 80–94 ft. lbs. (110–130 Nm).

17. Recheck the timing mark alignment. If all are aligned, loosen the tensioner mounting bolt and allow the tensioner to apply tension to the belt.

18. Remove the tool that is holding the silent shaft and rotate the crankshaft a distance equal to 2 teeth on the camshaft sprocket. This will allow the tensioner to automatically apply the proper tension on the belt. Do not manually overtighten the belt or it will howl.

19. Tighten the lower mounting bolt first, then the upper spacer bolt.

20. To verify correct belt tension, check that the deflection at the longest span of the belt has 0.40 in. (12mm) clearance from the belt cover.

21. Install the timing belt covers and all related items.

22. Connect the negative battery cable.

1993–94 MODELS

1. Disconnect the negative battery cable.
2. Remove the timing belt upper and lower covers.
3. Rotate the crankshaft clockwise and align the timing marks so No. 1 piston will be at TDC of the compression stroke. At this time the timing marks on the camshaft sprocket and the upper surface of the cylinder head should coincide, and the dowel pin of the camshaft sprocket should be at the upper side.

➡ **Always rotate the crankshaft in a clockwise direction. Make a mark on the back of the timing belt indicating the direction of rotation so it may be reassembled in the same direction if it is to be reused.**

4. Remove the auto tensioner and remove the outermost timing belt.

5. Remove the timing belt tensioner pulley, tensioner arm, idler pulley.

6. Locate the access plug on the side of block. Remove the plug and install a phillips screwdriver. Remove the oil pump sprocket nut, oil pump sprocket, special washer, flange and spacer.

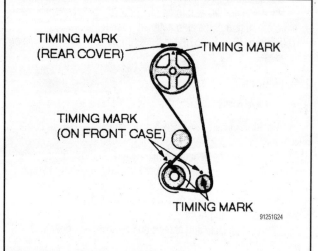

Proper timing belt mark alignment—1990–92 Chrysler Import 2.4L engines

CHRYSLER CORP.—TIMING BELTS

7. Remove the silent shaft (inner) belt tensioner and remove the belt.

To install:

8. Align the timing marks on the crankshaft sprocket and the silent shaft sprocket. Fit the inner timing belt over the crankshaft and silent shaft sprocket. Ensure that there is no slack in the belt.

9. While holding the inner timing belt tensioner with your fingers, adjust the timing belt tension by applying a force towards the center of the belt, until the tension side of the belt is taut. Tighten the tensioner bolt.

➥ **When tightening the bolt of the tensioner, ensure that the tensioner pulley shaft does not rotate with the bolt. Allowing it to rotate with the bolt can cause excessive tension on the belt.**

10. Check belt for proper tension by depressing the belt on long side with your finger and noting the belt deflection. The desired reading is 0.20–0.28 in. (5–7mm). If tension is not correct, readjust and check belt deflection.

11. Install the flange, crankshaft and washer to the crankshaft. The flange on the crankshaft sprocket must be installed towards the inner timing belt sprocket. Tighten bolt to 80–94 ft. lbs. (110–130 Nm).

➥ **There is a possibility to align all timing marks and have the oil pump sprocket out of time, causing an engine vibration during operation. If the following step is not followed exactly, there is a 50 percent chance that the oil pump shaft alignment will be 180 degrees off.**

12. Before installing the timing belt, ensure that the oil pump sprocket is in the correct position as follows:

   a. Remove the plug from the rear side of the block and insert a phillips screwdriver with shaft diameter of 0.31 in. (8mm) into the hole.

   b. With the timing marks still aligned, the shaft of the tool must be able to go in at least 2.36 in. (60mm). If the tool can only go in 0.79–0.98 in. (20–25mm), the shaft is not in the correct orientation and will cause a vibration during engine operation. Remove the tool from the hole and turn the oil pump sprocket 1 complete revolution. Realign the timing marks and insert the tool. The shaft of the tool must go in at least 2.36 in. (60mm).

   c. Recheck and realign the timing marks.

   d. Leave the tool in place to hold the silent shaft while continuing.

13. To install the oil pump sprocket and tighten the nut to 36–43 ft. lbs. (50–60 Nm).

14. Position the auto-tensioner into a vise with soft jaws. The plug at the rear of tensioner protrudes, be sure to use a washer as a spacer to protect the plug from contacting vise jaws.

15. Slowly push the rod into the tensioner until the set hole in rod is aligned with set hole in the auto-tensioner.

16. Insert a 0.055 in. (1.4mm) wire into the aligned set holes. Unclamp the tensioner from vise and install to vehicle. Tighten tensioner to 17 ft. lbs. (24 Nm).

17. When installing timing belt, the camshaft sprocket dowel pin should be located on top. Align all timing marks.

18. Align the crankshaft sprocket, camshaft sprocket and oil pump sprocket timing marks.

19. Install the timing belt as follows:

   a. Install the timing belt around the idler pulley, oil pump sprocket, crankshaft sprocket, camshaft and the tensioner pulley.

   b. Lift upward on the tensioner pulley in a clockwise direction and tighten the center bolt. Make sure all timing marks are aligned.

   c. Rotate the crankshaft ¼ turn counterclockwise. Then, turn in clockwise until the timing marks are aligned again.

20. Loosen the center bolt. Using tool No. MD998752 or equivalent and a torque wrench, apply a torque of 1.88–2.03 ft. lbs. (2.6–2.8 Nm). Tighten the center bolt.

21. Screw the tool No. MD998738 into the engine left support bracket until its end makes contact with the tensioner arm and tighten tensioner pulley to 35 ft. lbs. (48 Nm). At this point, screw the special tool in some more and remove the set wire attached to the auto tensioner. Then remove the special tool.

22. Rotate the crankshaft 2 complete turns clockwise and let it sit for approximately 15 minutes. Then, measure the auto tensioner protrusion (the distance between the tensioner arm and auto tensioner body) to ensure that it is within 0.15–0.18 in. (3.8–4.5mm). If out of specification, repeat belt adjustment procedure until the specified value is obtained.

23. If the timing belt tension adjustment is being performed with the engine mounted in the vehicle, and clearance between the ten-

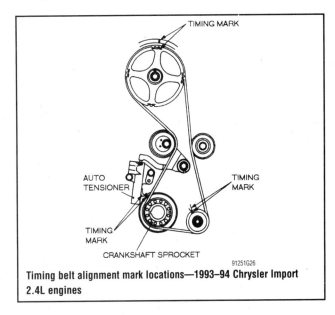

Timing belt alignment mark locations—1993–94 Chrysler Import 2.4L engines

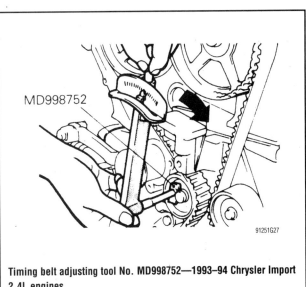

Timing belt adjusting tool No. MD998752—1993–94 Chrysler Import 2.4L engines

40 TIMING BELTS—CHRYSLER CORP.

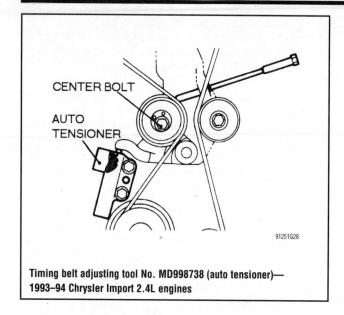

Timing belt adjusting tool No. MD998738 (auto tensioner)—1993–94 Chrysler Import 2.4L engines

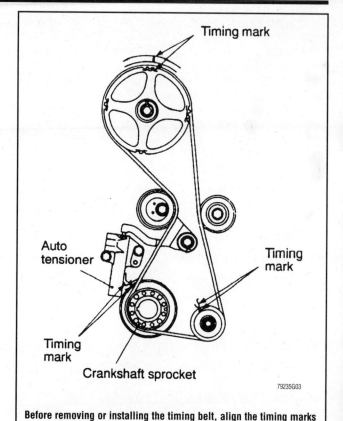

Before removing or installing the timing belt, align the timing marks as indicated—1995–99 Chrysler Import 2.4L engine

sioner arm and the auto tensioner body cannot be measured, the following alternative method can be used:

    a. Screw in tool No. MD998738 or equivalent, until its end makes contact with the tensioner arm.

    b. After the tool makes contact with the arm, screw it in some more to retract the auto tensioner pushrod while counting the number of turns the tool makes until the tensioner arm is brought into contact with the auto tensioner body. Make sure the number of turns the tool makes conforms with the standard value of 2.5–3 turns.

    c. Install the rubber plug to the timing belt rear cover.

24. Install the timing belt covers and all related items.
25. Connect the negative battery cable.

1995–99 MODELS

1. Rotate the crankshaft so that the No. 1 piston is at Top Dead Center (TDC) on its compression stroke.
2. Remove the timing belt covers.
3. To loosen the timing (outer) belt tensioner, install Special Tool MD998738 (or equivalent), to the slot, then screw inward to move the tensioner toward the water pump. Once the tension has been relieved, remove the outer timing belt.

➡ **If the timing belts are going to be reused, mark the direction of their rotation on the belts. This will ensure the belt is reinstalled in same direction, extending belt life.**

4. Remove the outer crankshaft sprocket and flange.
5. Loosen the silent shaft (inner) belt tensioner and remove the belt.

To install:

6. Turn both tensioner pulleys and check for any signs of bearing wear.
7. Align the timing marks of the silent shaft sprockets and the crankshaft sprocket with the timing marks on the front case. Route the timing belt around the sprockets so there is no slack in the upper span of the belt and the timing marks are still aligned.
8. Install the tensioner pulley and move the pulley by hand so the long side of the belt deflects approximately ¼ in. (6mm).
9. Hold the pulley tightly so the pulley cannot rotate when the bolt is tightened. Tighten the bolt to 14 ft. lbs. (19 Nm) and recheck the deflection amount.

10. Align the timing marks of the camshaft, crankshaft and oil pump sprockets with their corresponding marks on the front case or rear cover.

➡ **There is a possibility to align all timing marks and have the oil pump sprocket and silent shaft out of time, causing an engine vibration during operation. If the following step is not followed exactly, there is a 50 percent chance that the silent shaft alignment will be 180 degrees (½ turn) off.**

11. Before installing the timing belt, ensure that the left side (rear) silent shaft (oil pump sprocket) is in the correct position as follows:

    a. Remove the plug from the rear side of the block and insert a tool with an outer shaft diameter of 0.31 in. (8mm) into the hole.

    b. With the timing marks still aligned, the shaft of the tool must be able to go in at least 2 ½ (63.5mm). If the tool can only go in approximately 1 in. (25mm), the silent shaft is not in the correct orientation and will cause a vibration during engine operation. Remove the tool from the hole and turn the oil pump sprocket 1 complete revolution. Realign the timing marks and insert the tool. The shaft of the tool should now go in at least 2 ½ (63.5mm)

    c. Recheck and realign the timing marks.

    d. Leave the tool in place to hold the silent shaft while continuing.

12. Install the belt on the crankshaft sprocket, the oil pump sprocket, then the camshaft sprocket—in that order. While doing so, be sure there is no slack between the sprockets, except where the tensioner is installed.

13. To adjust the timing (outer) belt perform the following steps:

    a. Turn the crankshaft ¼ turn counterclockwise, then turn it clockwise to move the No. 1 cylinder to TDC.

CHRYSLER CORP., FORD MOTOR COMPANY—TIMING BELTS 41

b. Loosen the center bolt. Using tool MD998752 (or equivalent) and a torque wrench, apply 2.6 ft. lbs. (3.6 Nm) to the tensioner, then tighten the center bolt.

c. Thread the special tool into the engine left support bracket until its end makes contact with the tensioner arm. At this point, thread the special tool in some more and remove the set wire attached to the auto-tensioner, if the wire was not previously removed. Remove the tool.

d. Rotate the crankshaft two complete turns clockwise and let it sit for approximately 15 minutes. Then, measure the auto-tensioner protrusion (the distance between the tensioner arm and auto-tensioner body) to ensure that it is within 0.15–0.18 in. (3.8–4.5mm). If out of specification, repeat Substeps a through d until the specified value is obtained.

➡ Do not manually overtighten the belt or it will howl.

14. Install the timing belt covers and all related items.

Ford Motor Company

1.3L (VIN H) ENGINE

1990–94 Models

1. Disconnect the negative battery cable. Remove the drive belts.
2. Remove the 3 water pump pulley attaching bolts and remove the water pump pulley.
3. Raise and safely support the vehicle.
4. Remove the right front wheel and tire assembly and the right inner fender panel.
5. Remove the 4 attaching bolts and the screws from the crankshaft pulley. Remove the spacer and outer pulley, if equipped. Remove the inner spacer, inner pulley and the baffle or guide plates, as required.
6. Remove the attaching bolts and the upper and lower covers.
7. Mark the direction of rotation of the timing belt, if the belt is to be reused.
8. Remove the timing belt tensioner spring and retaining bolt. Remove the timing belt.

To install:

9. Align the camshaft and crankshaft timing marks with the marks located on the cylinder head and oil pump housing.
10. If reusing the original timing belt, install the timing belt with the mark made indicating the direction of rotation.
11. Install the timing belt tensioner spring and cover on the pulley. Position the tensioner and spring assembly on the engine and install the attaching bolt. Do not tighten the bolt at this time.
12. Reconnect the free end of the spring to the spring anchor. Torque the tensioner bolt to 14–19 ft. lbs. (19–26 Nm).
13. Install the upper and lower covers. Install the attaching bolts and tighten to 69–95 inch lbs. (8–11 Nm).
14. Install the crankshaft pulley baffle with the curved lip facing outward or install the large guide plate and then the small guide plate, as required.
15. Install the inner pulley with the deep recess facing outward. Install the spacer and then the outer pulley, spacer and screws. Install the pulley bolts and tighten to 109–152 inch lbs. (12–17 Nm).

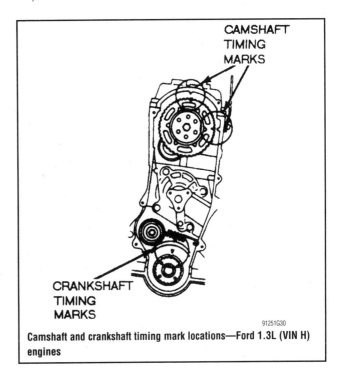

Camshaft and crankshaft timing mark locations—Ford 1.3L (VIN H) engines

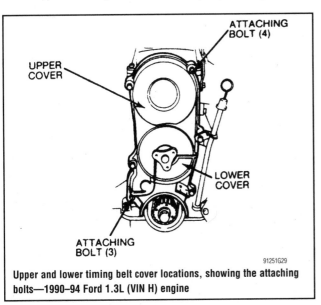

Upper and lower timing belt cover locations, showing the attaching bolts—1990–94 Ford 1.3L (VIN H) engine

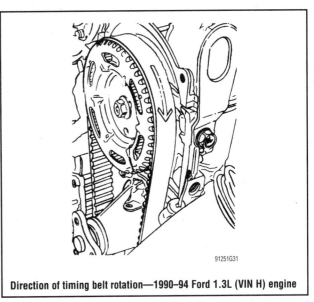

Direction of timing belt rotation—1990–94 Ford 1.3L (VIN H) engine

42 TIMING BELTS—FORD MOTOR COMPANY

16. Install the inner fender panel and the wheel and tire assembly. Lower the vehicle.
17. Install the water pump pulley and tighten the bolts to 36–45 ft. lbs. (49–61 Nm).
18. Install the drive belts. Connect the negative battery cable.

1995–99 Models

1. Disconnect the negative battery cable.
2. Remove the accessory drive belts.
3. Remove the three water pump pulley attaching bolts and remove the water pump pulley.
4. Raise and safely support the vehicle on jack stands.
5. Remove the right front wheel and tire assembly and the right inner fender panel.
6. Remove the four attaching bolts and the screws from the crankshaft pulley. Remove the spacer and outer pulley, if equipped. Remove the inner spacer, inner pulley and the baffle or guide plates, as required.
7. Remove the attaching bolts and the upper and lower covers.
8. Rotate the crankshaft until the sprocket timing marks are aligned.
9. Remove the timing belt tensioner spring, spring cover and timing belt tensioner bolt. Remove the timing belt.

➡ **If the timing belt is to be reused, mark the direction of rotation on the belt, using a crayon, so the belt can be reinstalled in the same direction.**

10. If the camshaft sprocket requires removal, proceed as follows:
    a. Hold the camshaft stationary with an open end wrench and remove the camshaft sprocket retaining bolt.
    b. Pull the camshaft sprocket with the dowel pin off of the camshaft. Use care not to drop the dowel pin.
11. If the crankshaft sprocket requires removal, proceed as follows:
    a. Remove the crankshaft pulley retaining bolt.
    b. Pull the crankshaft pulley hub, sprocket and key from the crankshaft. Be sure not to drop the crankshaft key.

To install:

12. If removed, install the crankshaft sprocket as follows:
    a. Install the sprocket with the key onto the crankshaft.
    b. Install the crankshaft pulley hub.
    c. Clean the threads of the crankshaft pulley bolt and coat with a non-hardening sealer.
    d. Install the bolt and tighten to 80–85 ft. lbs. (108–118 Nm).
13. If removed, install the camshaft sprocket as follows:
    a. Position the sprocket and dowel pin to the camshaft and install the retaining bolt.
    b. Hold the camshaft stationary with an open end wrench and tighten the retaining bolt to 36–45 ft. lbs. (49–61 Nm).
14. Align the camshaft and crankshaft timing marks with the marks located on the cylinder head and oil pump housing.
15. If reusing the original timing belt, install the timing belt with the mark made indicating the direction of rotation.
16. Install the timing belt tensioner spring and cover on the tensioner. Position the tensioner and spring assembly on the engine and install the attaching bolt. Do not tighten the bolt at this time.
17. Rotate the crankshaft two turns in the direction of normal rotation and align the timing marks. Ensure all marks are still correctly aligned.
18. Reconnect the free end of the spring to the spring anchor. Tighten the tensioner bolt to 14–19 ft. lbs. (19–26 Nm).
19. Install the upper and lower covers. Install the attaching bolts and tighten to 71–97 inch lbs. (8–11 Nm).
20. Install the crankshaft pulley baffle with the curved lip facing outward, or install the large guide plate, then the small guide plate, as required.
21. Install the inner pulley with the deep recess facing outward. Install the spacer, then the outer pulley, spacer and screws. Install the pulley bolts and tighten to 109–152 inch lbs. (12–17 Nm).
22. Install the inner fender panel.
23. Install the wheel and tire assembly. Tighten the lug bolts to 65–87 ft. lbs. (88–118 Nm).
24. Lower the vehicle.
25. Install the water pump pulley and tighten the bolts to 36–45 ft. lbs. (49–61 Nm).
26. Install the accessory drive belts.
27. Connect the negative battery cable.
28. Run the engine and check for proper operation.

1.6L (VIN Z & 6) ENGINES

1. Disconnect the negative battery cable. Raise and safely support the vehicle.
2. Remove the right front wheel and tire assembly and remove the right splash guard. Lower the vehicle.
3. Remove the spark plugs. Set the engine position to TDC on the No. 1 cylinder.
4. Remove the alternator and power steering belts. Remove the oil dipstick and the water pump pulley.
5. Remove the crankshaft pulley, damper and baffle plate.
6. Remove the upper, center and lower timing belt covers.
7. Remove the timing belt tension spring and loosen the timing belt tension pulley.
8. Support the engine with a floor jack and remove the right engine mount.
9. Mark the timing belt rotation direction and remove the timing belt.
10. Inspect the timing belt and timing sprockets for wear and/or damage and replace as necessary.
11. Check the free length of the timing belt tension spring. It should be 2.315 in. (58.8mm). Replace if out of specification.

To install:

Make sure the timing marks are properly positioned on the camshafts and crankshaft. The intake camshaft should have the letter **I** aligned with the arrow on the belt cover. The exhaust camshaft should have the letter **E** aligned with the arrow on the belt covers.

12. The crankshaft key should align with the arrow.
13. Tighten the tension pulley with the tension spring fully extended.
14. Install the timing belt. Keep tension on the opposite side of the tensioner as tight as possible. Make sure the rotation mark on the belt is correct.
15. Turn the crankshaft 2 full turns. Check the alignment marks. If any mark is not aligned, remove the timing belt and reset the timing.
16. Loosen the tension pulley retaining bolt to allow the tension spring to tighten the belt.
17. Tighten the tension pulley retaining bolt to 27–38 ft. lbs. (37–52 Nm). Rotate the engine 2 full turns. Make sure the timing marks are aligned.

Ford Motor Company—Timing Belts

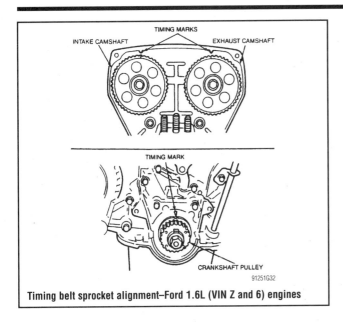

Timing belt sprocket alignment—Ford 1.6L (VIN Z and 6) engines

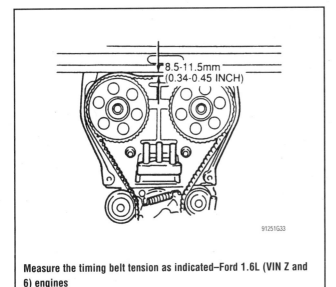

Measure the timing belt tension as indicated—Ford 1.6L (VIN Z and 6) engines

18. Measure the timing belt tension between the camshaft pulleys. Belt deflection should be 0.33–0.45 in. (8.5–11.5mm). If incorrect, loosen the tension pulley and repeat the procedure. If proper tension cannot be achieved, replace the tension spring.
19. Install the lower, center and upper timing belt covers. Tighten the retaining bolts to 71–97 inch lbs. (8–11 Nm).
20. Install the right engine mount and lower the engine. Tighten the retaining nuts to 44–63 ft. lbs. (60–85 Nm).
21. Install the crankshaft pulley, damper and baffle. Tighten the baffle and damper retaining screws to 109–152 inch lbs. (12–17 Nm). Tighten the pulley retaining bolts to 109–152 inch lbs. (12–17 Nm).
22. Install the water pump pulley. Tighten the retaining bolts to 69–95 inch lbs. (8–11 Nm).
23. Install the alternator and power steering belts. Install the dipstick. Raise and safely support the vehicle.
24. Install the splash guard and the right front wheel and tire assembly.
25. Lower the vehicle and install the spark plugs. Start the engine and check for proper operation.

1.8L (VIN 8) ENGINE

1. Disconnect the negative battery cable.
2. Remove the timing belt upper cover and gasket.
3. Remove the accessory drive belts.
4. Remove the water pump pulley bolts and remove the pulley.
5. Raise and safely support the vehicle on jack stands.
6. Remove the right front wheel and tire assembly.
7. Remove the right upper and lower splash-shields.
8. Remove the timing belt middle and lower covers along with the gaskets.
9. Remove the crankshaft pulley hub bolt and hub.
10. Rotate the crankshaft and align the timing marks located on the camshaft sprockets and seal plate.
11. Check that the crankshaft sprocket and the oil pump are aligned.

➡ **If the timing belt is to be reused, mark an arrow on the belt to indicate it's rotational direction for installation reference.**

12. Loosen the timing belt tensioner bolt.
13. Turn the timing belt tensioner counterclockwise and hand-tighten the tensioner bolt to relieve the tension on the timing belt.
14. Remove the timing belt.
15. If the camshaft sprockets are to be removed, continue as follows:
    a. Disconnect and tag the ignition wires and vacuum lines blocking the removal of the cylinder head cover.
    b. Remove the cylinder head cover retaining bolts and remove the cover and gasket.
    c. While holding the camshaft with a wrench, remove the camshaft sprocket retaining bolt.
    d. Remove the camshaft sprocket.
    e. If removing both camshaft sprockets, tag the sprockets for identification at reassembly.
16. If removing the crankshaft sprocket, remove the crankshaft pulley bolt and hub, if not already done. Slide the crankshaft sprocket off the crankshaft.
17. Inspect the timing belt tensioner and spring, replace if necessary.

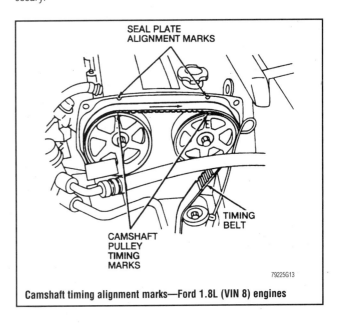

Camshaft timing alignment marks—Ford 1.8L (VIN 8) engines

44 TIMING BELTS—FORD MOTOR COMPANY

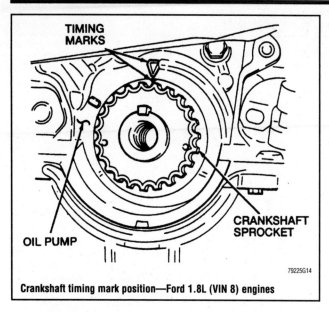

Crankshaft timing mark position—Ford 1.8L (VIN 8) engines

To install:

18. If the crankshaft sprocket was removed, install the crankshaft key with the tapered end facing the oil pump. Install the crankshaft sprocket onto the crankshaft while making sure to match the alignment grooves.
19. If removed, install the camshaft sprockets as follows
    a. Turn the camshaft until the dowel pins face straight up.
    b. Install the camshaft sprocket with the **I** mark straight up for the intake camshaft or with the **E** mark straight up for the exhaust camshaft.
    c. Align the camshaft sprockets with the timing marks on the seal plate.
    d. While holding each camshaft with a wrench, install the camshaft sprocket retaining bolts. Tighten the bolts to 36–45 ft. lbs. (49–61 Nm).
    e. Install a new cylinder head cover gasket onto the cylinder head.
    f. Place the cylinder head cover into its mounting position and install the retaining bolts. Tighten the cylinder head cover bolts to 43–78 inch lbs. (4.9–8.8 Nm).
    g. Install the ignition wires to the spark plugs and connect the vacuum hoses to the cylinder head cover.
20. Temporarily secure the timing belt tensioner in the far left position.
21. Verify that the timing marks on the crankshaft sprocket and the oil pump are aligned.
22. Verify that the timing marks on the camshaft sprockets and the seal plate are aligned.
23. Install the timing belt in a counterclockwise motion. Be sure there is no looseness on the idler side of the timing belt or between the camshaft sprockets.

➡ **If using the old timing belt, be sure to install the belt in the same direction of travel as it was removed.**

24. Loosen the timing belt tensioner bolt. Allow the tensioner spring to apply tension to the timing belt.
25. Rotate the crankshaft 1 5/6 turns clockwise and align the timing belt pulley mark with the tension set mark which is located at approximately the 10 o'clock position.
26. Turn the crankshaft two turns clockwise and align the crankshaft sprocket with the tension set mark on the oil pump.
27. Verify that all timing marks are aligned. If not, remove the timing belt and repeat the installation procedures.
28. Apply tension to the timing belt tensioner and tighten the tensioner lockbolt to 27–38 ft. lbs. (37–52 Nm).
29. Rotate the crankshaft 2 1/6 (780 degrees) turns clockwise and verify that the camshaft and crankshaft timing marks are aligned.
30. Measure the timing belt deflection by applying 22 lbs. (98 N) of pressure on the timing belt between the camshaft sprockets. The timing belt deflection should be 0.35–0.45 in. (9–11.5mm). If necessary to adjust the timing belt deflection, rotate the crankshaft two turns clockwise and ensure that the timing marks are still aligned. If the timing marks are not aligned, repeat the installation procedure.
31. Install the crankshaft pulley hub and tighten the retaining bolt to 80–87 ft. lbs. (108–118 Nm).
32. Install the crankshaft pulley and washer. Install the retaining bolts and tighten to 109–152 inch lbs. (12–17 Nm).
33. Install the timing belt middle and lower covers with the gaskets. Tighten the middle and lower timing belt cover retaining bolts to 65–95 inch lbs. (7.8–11 Nm,).
34. Install the power steering drive belt.
35. Install the water pump pulley and retaining bolts. Tighten the bolts to 69–95 inch lbs. (7.8–11.0 Nm).
36. Install the alternator/water pump drive belt.
37. Install the splash-shields. Tighten the bolts to 69–95 inch lbs. (7.8–11.0 Nm).
38. Install the right wheel and tire assembly. Tighten the lug nuts to 65–87 ft. lbs. (88–118 Nm).
39. Lower the vehicle.
40. Install the timing belt upper cover and gasket. Tighten the bolts to 69–95 inch lbs. (7.8–11.0 Nm).
41. Connect the negative battery cable.
42. Run the engine and check for leaks.
43. Road test the vehicle and check for proper engine operation.

1.9L (VIN J) ENGINE

1990 Models

1. Disconnect the negative battery cable.
2. Remove the timing belt cover.
3. Align the timing mark on the camshaft sprocket with the timing mark on the cylinder head.
4. Install the timing belt cover and confirm that the timing mark on the crankshaft damper aligns with the **TDC** on the front cover.
5. Remove the timing belt cover.
6. Loosen both timing belt tensioner attaching bolts.
7. Pry the belt tensioner away from the belt as far as possible and tighten 1 of the tensioner attaching bolts.
8. Remove crankshaft damper as follows:
    a. Properly support the engine and remove the right side engine mount bolt.
    b. Lower the engine at the right side until the crankshaft damper bolt clears the frame rail and remove the damper bolt.
    c. Raise the engine and remove the damper.
9. Remove the timing belt. If the belt is to be reused, mark the direction of rotation on the belt so it can be reinstalled in the same direction.

➡ **With the timing belt removed and the No. 1 piston at TDC, Do not rotate the camshaft. If the camshaft must be rotated, align the crankshaft dampener 90 degrees BTDC.**

FORD MOTOR COMPANY—TIMING BELTS

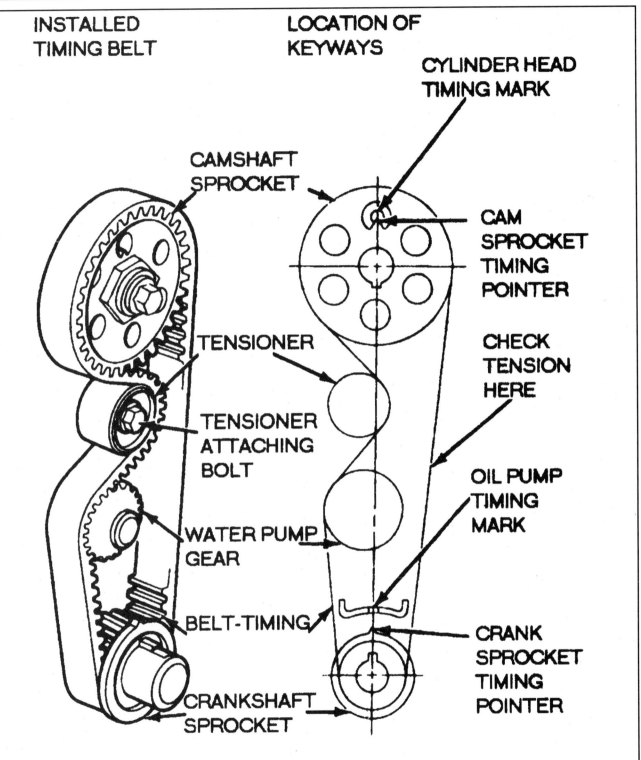

Timing belt sprocket alignment for belt service—Ford 1.9L (VIN J) Engines

46 TIMING BELTS—FORD MOTOR COMPANY

To install:

10. Install the timing belt over the sprockets in a counterclockwise direction starting at the crankshaft. Keep the belt span from the crankshaft to the camshaft tight as the belt is installed over the remaining sprocket.

11. Loosen belt tensioner attaching bolts and allow the tensioner to snap against the belt.

12. Tighten 1 of the tensioner attaching bolts.

13. Install the crankshaft damper, driveplate and damper attaching bolt. Hold the crankshaft damper stationary and tighten the attaching bolt to 81–96 ft. lbs. (110–130 Nm).

14. To seat the belt on the sprocket teeth, proceed as follows:
   a. Connect the negative battery terminal.
   b. Crank engine several revolutions.
   c. Disconnect the negative battery terminal.
   d. Turn camshaft, as necessary, to align the timing pointer on the cam sprocket with the timing mark on the cylinder head.

➡**Do not turn the engine counterclockwise to align the timing marks.**

   e. Position the timing belt cover on the engine and check to see that the timing mark on the crankshaft aligns with the TDC pointer on the cover. If the timing marks do not align, remove the belt, align the timing marks and return to Step 10.

15. Loosen the belt tensioner attaching bolt tightened in Step 12. The tensioner spring will apply the proper load on the belt. Tighten the belt tensioner bolt.

➡**The engine must be at room temperature. Do not set belt tension on a hot engine.**

16. Install timing belt cover.
17. Install accessory drive belts.
18. Connect negative battery cable.

1991–96 Models

1. Disconnect the negative battery cable.
2. Remove the accessory drive belt automatic tensioner and the accessory drive belt.
3. Remove the timing belt cover.
4. Align the timing mark on the camshaft sprocket with the timing mark on the cylinder head.
5. Confirm that the timing mark on the crankshaft sprocket is aligned with the timing mark on the oil pump housing.
6. Loosen the belt tensioner attaching bolt, pry the tensioner away from the timing belt and retighten the bolt.
7. Remove the spark plugs. Remove the right engine mount.
8. Raise and safely support the vehicle on jack stands.
9. Remove the right side splash-shield.

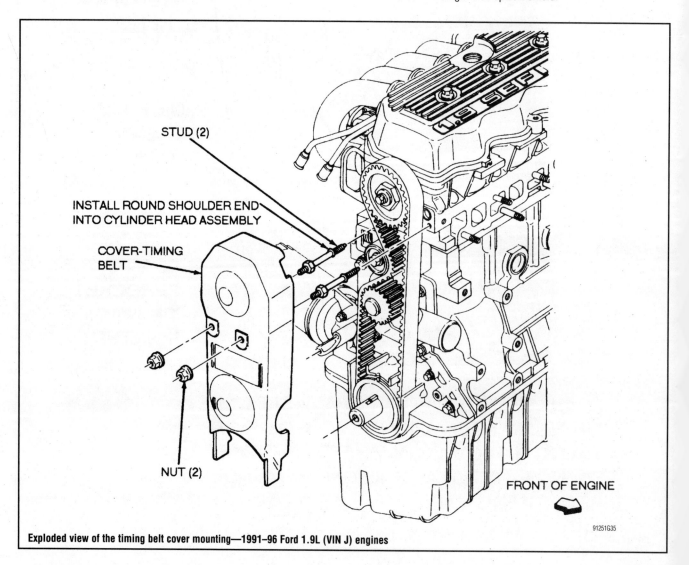

Exploded view of the timing belt cover mounting—1991–96 Ford 1.9L (VIN J) engines

FORD MOTOR COMPANY—TIMING BELTS

10. Remove the flywheel inspection shield.
11. Use a suitable tool to hold the flywheel in place.
12. Remove the crankshaft damper bolt and washer and remove the bolt.
13. Remove the timing belt.

➡ **With the timing belt removed and the No. 1 piston at TDC, do not rotate the camshaft. If the camshaft must be rotated, align the crankshaft damper 90 degrees BTDC.**

To install:

14. Install the timing belt over the sprockets in a counterclockwise direction starting at the crankshaft. Keep the belt span from the crankshaft to the camshaft tight while installing over the remaining sprocket.
15. Loosen the belt tensioner attaching bolt, allowing the tensioner to snap against the belt.
16. Rotate the crankshaft clockwise two complete revolutions, stopping at TDC. This will allow the tensioner spring to load the timing belt.

➡ **Do not turn the engine counterclockwise to align the timing marks. Do not rotate the crankshaft with the spark plugs installed.**

17. Recheck the camshaft and crankshaft timing marks for alignment, to be sure the timing belt has not skipped a tooth during rotation. Repeat the procedure if the timing marks are not aligned.
18. Tighten the tensioner attaching bolt to 17–22 ft. lbs. (23–30 Nm).
19. Install the crankshaft dampener and the bolt and washer. Tighten the bolt to 81–96 ft. lbs. (110–130 Nm).
20. Install the flywheel inspection shield.
21. Install the splash-shield and lower the vehicle.
22. Install the right engine mount. Install the spark plugs.
23. Install the timing belt cover.
24. Install the accessory drive belt automatic tensioner and the accessory drive belt.
25. Connect the negative battery cable.

2.0L (VIN 3) ENGINE

When installing a timing belt, tensioner spring (6L277) and retaining bolt (W700001-S309) must be purchased and properly installed on the engine. First check to see if these parts are already installed. The tensioner spring will adjust the timing belts tension and should not require further adjustments.

1. Disconnect the negative battery cable.
2. Remove the engine air intake resonators.
3. Label and remove the ignition wires from the spark plugs. Move the ignition wires aside.
4. Remove the spark plugs.
5. Manually rotate the crankshaft to Top Dead Center (TDC) for the No. 1 piston on its compression stroke. Be sure to align the timing marks.
6. Disconnect the retaining bracket for the power steering pressure hose from the engine lifting eye.
7. Install the Three Bar Engine Support D88L-6000-A or equivalent, onto the engine lifting eyes and slightly raise the engine.
8. Remove the upper camshaft timing belt cover retaining bolts and the cover from the engine.

➡ **Mark the location of the upper front engine support bracket before removing it from the engine support bracket.**

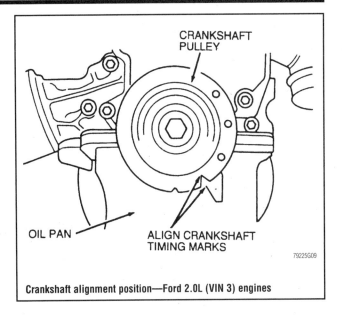

Crankshaft alignment position—Ford 2.0L (VIN 3) engines

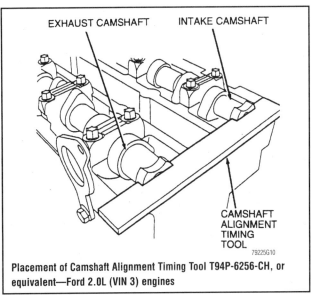

Placement of Camshaft Alignment Timing Tool T94P-6256-CH, or equivalent—Ford 2.0L (VIN 3) engines

9. Remove the upper front engine support bracket retainer nuts, the bracket and the upper front engine support insulator.
10. If equipped, remove the wiring harness connector from the low coolant level sensor at the radiator coolant recovery reservoir.
11. Remove the radiator coolant recovery reservoir retainers and move the reservoir aside.
12. Remove the upper front engine support insulator.
13. Set the coolant recovery reservoir back into position temporarily.
14. Loosen the water pump pulley retaining bolts. Do not remove the bolts completely.
15. Remove the accessory drive belt.
16. Remove the drive belt idler pulley retaining bolt and pulley from the alternator mounting bracket.
17. Finish removing the water pump retaining bolts and remove the water pump pulley.
18. Remove the center camshaft timing belt cover retaining bolts and the cover from the engine.
19. Raise and safely support the vehicle on jack stands.

48 TIMING BELTS—FORD MOTOR COMPANY

20. Remove the crankshaft pulley.
21. Remove the lower camshaft timing belt cover bolts and the cover from the engine.
22. Remove the valve cover as follows:
    a. Disconnect the crankcase ventilation tube from the valve cover.
    b. Remove the retaining bolt and nut for the power steering pressure hose retaining bracket and move the hose aside.
    c. Remove the valve cover retaining bolts in a standard removal sequence starting from the outside of the valve cover and working toward the inside of the valve cover.
    d. Remove the valve cover and gasket from the engine.
23. Place Camshaft Alignment Timing Tool T94P-6256-CH or equivalent, into the slots of both camshafts at the rear of the cylinder head to lock the camshafts into position.
24. Loosen the camshaft timing belt tensioner pulley retaining bolt and move the tensioner pulley to relieve the tension on the timing belt.
25. Temporarily tighten the tensioner in this position.

➡ **If the timing belt is to be reused, mark the belt for the direction of rotation before removing to prevent premature wear or failure.**

26. Remove the timing belt.
27. If required, remove the sprockets as follows:
    a. Hold the camshaft with the Camshaft Sprocket Holding Tool T74P-6256-B or equivalent.
    b. Loosen and remove the camshaft sprocket retaining bolt.
    c. Remove the sprocket from the camshaft.
    d. Repeat the procedure for the 2nd camshaft sprocket.
    e. Remove the crankshaft sprocket.

To install:
28. Slide the crankshaft sprocket onto the crankshaft aligning the keyway.
29. Align the camshafts using the Camshaft Alignment Timing Tool T94P-6256-CH.
30. Reinstall the sprockets onto the camshafts and loosely install the camshaft retaining bolts.
31. Tighten the camshaft sprocket retaining bolts to 47–53 ft. lbs. (64–72 Nm).
32. Loosely install the crankshaft pulley to verify that the engine is at TDC. Realign the marks if they have moved.
33. Verify that the camshafts are aligned.

➡ **It is recommended to purchase a tensioner spring and retaining bolt through the dealer parts to apply the proper tension for used or new belt installations. The spring is bolted to the tensioner assembly and becomes a part of the engine. Ignore this notice if the tensioner spring is already installed.**

34. Reinstall the retaining bolt (W700001-S309) into the hole provided in the cylinder block and place the tensioner spring (6L277) between the bolt and the camshaft timing belt tensioner pulley.
35. Tighten the retainer bolt to 71–97 inch lbs. (8–11 Nm).
36. Remove the crankshaft pulley and install the timing belt onto the crankshaft sprocket, then onto the camshaft sprockets working in a counterclockwise direction.
37. Tighten the camshaft sprocket retaining bolts to 47–53 ft. lbs. (64–72 Nm).
38. Be sure that the span of the camshaft timing belt between the crankshaft sprocket and the exhaust camshaft sprocket is not loose.
39. Be sure that the camshaft timing belt is securely aligned on all sprockets.
40. Reinstall the lower timing belt cover and tighten the retaining bolts to 53–71 inch lbs. (6–8 Nm).
41. Apply silicone sealer to the keyway of the crankshaft pulley and install. Tighten the retaining bolt to 81–89 ft. lbs. (110–120 Nm).
42. Inspect the timing mark on the crankshaft pulley to verify that the engine is still at TDC.
43. Loosen the camshaft timing belt tensioner pulley retaining bolt and allow the tensioner spring attached to the pulley to draw the tensioner pulley against the camshaft timing belt.
44. Remove the camshaft alignment timing tool from the camshafts at the rear of the engine.
45. Turn the crankshaft two revolutions in a clockwise direction.
46. Tighten the camshaft timing belt tensioner pulley retaining bolt to 26–30 ft. lbs. (35–40 Nm).
47. Recheck that the crankshaft timing mark is at TDC for the No. 1 piston, and that both camshafts are in alignment using the camshaft alignment timing tool.

➡ **A slight adjustment of the camshafts to allow the insertion of the camshaft alignment timing tool is permissible as long as the crankshaft stays at the TDC location.**

48. Camshaft Sprocket Holding Tool T74P-6256-b can be used to move the camshaft sprocket (s) if a slight adjustment is required.
49. If a camshaft is not properly aligned, perform the following procedure:
    a. Loosen the retaining bolt securing the sprocket to the camshaft while holding the camshaft sprocket from turning with the sprocket holding tool.
    b. Turn the camshaft until the camshaft alignment timing tool can be installed.
    c. Verify that the crankshaft timing mark is at TDC for the No. 1 cylinder.
    d. While holding the camshaft sprocket with the camshaft sprocket holding tool, tighten the retaining bolt to 47–53 ft. lbs. (64–72 Nm).
    e. Remove the tool and rotate the crankshaft two revolutions (clockwise).
    f. Verify that the camshafts are aligned and that the crankshaft is at TDC for the No. 1 cylinder.
50. Reinstall the valve cover as follows:
    a. Clean the gasket sealing surfaces.
    b. Inspect the valve cover gasket and O-rings; replace as required.
    c. Reinstall the valve cover retaining bolts and tighten them in a standard sequence starting from the center and working towards the outside of the valve cover to 53–71 inch lbs. (6–8 Nm).
    d. Reinstall the power steering hose retaining bracket and the power steering hose.
    e. Reinstall the crankcase ventilation tube to the valve cover.
51. Position the center camshaft timing belt cover.
52. Reinstall the center camshaft timing belt cover retaining bolts and tighten them to 53–71 inch lbs. (6–8 Nm).
53. Reinstall the water pump pulley and the retaining bolts. Reinstall the bolts, finger-tight.
54. Reinstall the drive belt idler pulley.
55. Reinstall the drive belt idler pulley retaining bolt and tighten it to 35 ft. lbs. (48 Nm).
56. Reinstall the accessory drive belt.

FORD MOTOR COMPANY—TIMING BELTS

57. Tighten the water pump pulley retaining bolts to 89–124 inch lbs. (10–14 Nm).
58. Move the radiator coolant recovery reservoir aside.
59. Reinstall the upper front engine support insulator.
60. Position the radiator coolant recovery reservoir and install the retainers.
61. If equipped, install the wiring harness to the low coolant level sensor on the coolant recovery reservoir.
62. Reinstall the upper front engine support bracket to the engine and the upper front engine support insulator using the mark made during the removal procedure for reference.
63. Install the upper camshaft timing belt cover.
64. Reinstall the upper camshaft timing belt cover retaining bolts and tighten to 27–44 inch lbs. (3–5 Nm).
65. Remove the engine support.
66. Reinstall the retaining bracket for the power steering pressure hose to the engine lifting eye.
67. Reinstall the spark plugs and the ignition wires.
68. Reinstall the engine air intake resonators.
69. Replace the engine oil.
70. Reconnect the negative battery cable.
71. Run the engine and check for leaks and proper operation.

2.0L (VIN A) ENGINE

1. Disconnect the negative battery cable.
2. Label and disconnect the spark plug wires and clips from the cylinder head cover. Remove the ignition distributor with wiring and set it aside.
3. Remove the power steering hose brackets from the cylinder head cover. If necessary disconnect the crankshaft position sensor.
4. Disconnect the breather tube and PCV valve from the cylinder head cover.
5. Loosen the cylinder head cover bolts in 2–3 steps. Remove the cylinder head cover.
6. Remove the power steering belt shield. Loosen the power steering adjusting bolt, lockbolt and through-bolt and remove the power steering belt.
7. Loosen the alternator adjusting bolt and upper mounting bolt. Remove the alternator belt.
8. Support the engine with Engine Support Tool 014–00750 or equivalent. Raise the engine slightly with a jack and remove the right side engine support insulator (mount).
9. Remove the oil level indicator bolt and four upper timing belt cover bolts and remove the upper timing belt cover.

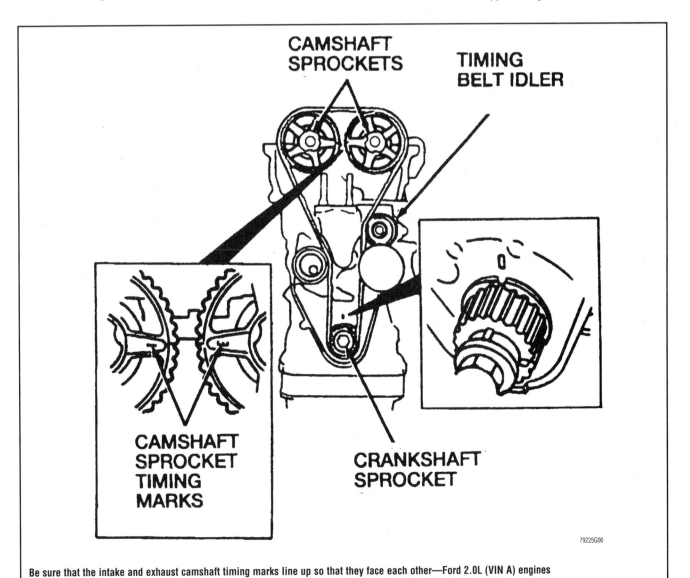

Be sure that the intake and exhaust camshaft timing marks line up so that they face each other—Ford 2.0L (VIN A) engines

50 TIMING BELTS—FORD MOTOR COMPANY

10. Raise and safely support the vehicle.
11. Remove the splash-shields. Using Holder Tool T92C-6316-AH or equivalent, hold the crankshaft pulley and remove the pulley bolt. Use a suitable puller to remove the pulley, then remove the guide plate.
12. Remove the four lower timing belt cover bolts and remove the lower timing belt cover.
13. Temporarily install the crankshaft pulley bolt.
14. Turn the crankshaft until the timing mark on the crankshaft sprocket aligns with the timing mark on the oil pump and the camshaft sprocket timing marks, E and I, line up on the camshaft sprockets.
15. Lower the vehicle.
16. Insert camshaft sprocket holding tool T92C-6256-AH or equivalent, between the camshaft sprockets.
17. Turn the timing belt tensioner with an Allen wrench and remove the tensioner spring from the tensioner spring pin.
18. If the timing belt is to be reused, mark the direction of rotation on the timing belt. Remove the timing belt.
19. If it is necessary to remove the sprockets, remove the camshaft sprocket holding tool. Hold the camshaft by placing a suitable wrench on the hexagon which is cast into the camshaft. Place another wrench onto the camshaft sprocket retaining bolt and loosen the bolt.

➡ **Before removing the camshaft sprocket (s) , be sure that the camshafts are still in alignment and tag each sprocket to the camshaft from which it was removed.**

20. Remove the camshaft sprocket bolt and the camshaft sprocket from the camshaft.
21. Repeat the camshaft sprocket removal procedure for the opposite camshaft if required.
22. Raise and safely support the vehicle.
23. Slide off the crankshaft sprocket and remove the crankshaft key.

To install:
24. Install the crankshaft key and slide the crankshaft sprocket into position.
25. Lower the vehicle.
26. Install the camshaft sprocket onto the proper camshaft, making sure to align the dowel pin.
27. Be sure that the I and E are in alignment.
28. Install the camshaft sprocket bolt. Hold the hexagon on the camshaft with a suitable wrench and tighten the sprocket bolt to 35–48 ft. lbs. (47–65 Nm). Be sure the camshaft sprockets are still properly aligned and reinstall the sprocket holding tool.
29. Be sure the timing marks on the camshaft and crankshaft sprockets are still aligned.
30. Install the timing belt. If reusing the original timing belt, be sure it is installed in the same direction of rotation.
31. Turn the tensioner clockwise with an Allen wrench and install the tensioner spring. Remove the holding tool from between the camshaft sprockets.
32. Rotate the crankshaft clockwise two turns and align the timing marks. Be sure all marks are still correctly aligned.

➡ **The timing chain tensioner automatically adjusts the tension on the timing belt.**

33. Raise and safely support vehicle on jack stands.
34. Install the timing belt lower cover and tighten the four bolts to 71–88 inch lbs. (8–10 Nm).
35. Install the guide plate, crankshaft pulley and pulley bolt. Secure the pulley with the holder tool and tighten the bolt to 116–123 ft. lbs. (157–167 Nm).
36. Install the splash-shields and lower the vehicle.
37. Raise the engine slightly with the jack and install the right side engine mount. Tighten the mount through-bolt to 63–86 ft. lbs. (86–116 Nm) and the mount attaching nuts to 54–75 ft. lbs. (74–103 Nm). Remove the engine support tool.
38. Install the upper timing belt cover and tighten the bolts to 71–88 inch lbs. (8–10 Nm).
39. Clean the cylinder head and valve cover mating surfaces thoroughly.
40. Apply silicone sealant to the cylinder head surface in the area adjacent to the front camshaft bearing caps. Apply sealant to a new gasket and install it on the cylinder head cover.
41. Install the cylinder head cover and tighten the bolts.
42. Install the power steering hose brackets and tighten the bolts to 71–88 inch lbs. (8–10 Nm). Connect the spark plug wires and wire clips. Connect the breather tube and PCV valve. If necessary, connect the crankshaft position sensor.
43. Install the alternator belt and adjust the tension. Tighten the upper mounting bolt to 14–18 ft. lbs. (19–25 Nm) and the lower through-bolt to 27–38 ft. lbs. (37–52 Nm).
44. Install the power steering belt and adjust the tension. Tighten the through-bolt to 32–45 ft. lbs. (43–61 Nm) and the lockbolt to 23–34 ft. lbs. (31–46 Nm). Install the power steering belt shield and tighten the bolts to 61–86 inch lbs. (7–9 Nm).
45. Connect the negative battery cable.
46. Run the engine and check for leaks and proper engine operation.

2.2L (VIN C) ENGINE

1. Bring the No. 1 cylinder piston to Top Dead Center (TDC) on the compression stroke. The notch on the crankshaft damper should align with the TDC mark on the front cover.
2. Disconnect the negative battery cable.
3. Loosen the air conditioning compressor and alternator adjusting and pivot bolts, rotate the compressor and alternator toward the engine and remove the drive belts.
4. Raise and safely support the vehicle.
5. Remove the right front wheel and tire assembly and the right inner fender panel. Remove the 6 bolts, the crankshaft pulley and baffle plate.
6. Lower the vehicle.
7. Support the engine with a floor jack. Remove the 2 nuts and dowels from the right engine mount and remove the mount.
8. Remove the 7 bolts that retain the timing belt covers and remove the covers.
9. Remove the timing belt tensioner spring and retaining bolt. Remove the idler pulley retaining bolt.
10. If the timing belt is to be reused, mark the direction of rotation so it can be reinstalled in the same direction.
11. Remove the timing belt.

To install:
12. Align the camshaft and crankshaft sprockets with the marks on the cylinder head front housing and the oil pump housing.
13. Install the timing belt. If reusing the old belt, observe the direction of rotation mark made during the removal procedure.
14. Place the timing belt tensioner and spring in position. Temporarily secure the tensioner with the spring fully extended. Make

FORD MOTOR COMPANY—TIMING BELTS

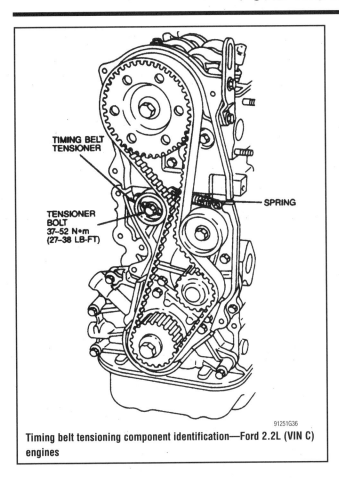

Timing belt tensioning component identification—Ford 2.2L (VIN C) engines

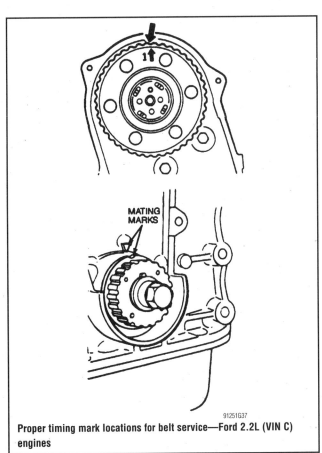

Proper timing mark locations for belt service—Ford 2.2L (VIN C) engines

sure the timing belt is installed so there is no looseness at the water pump pulley at the idler side.

15. Loosen the idler bolt. Turn the crankshaft twice in the direction of rotation; align the timing marks.

➡**Always turn the crankshaft in the correct direction of rotation only. If the crankshaft is turned in the opposite direction, the timing belt may lose tension and correct belt timing may be lost.**

16. Check to see that the timing marks are correctly aligned. If they are not aligned, remove the timing belt and align the timing marks, then repeat Steps 8–11.

17. Tighten the tensioner bolt to 27–38 ft. lbs. (37–52 Nm).

18. Measure the belt deflection between the crankshaft and camshaft pulleys. The correct deflection should be 0.30–0.33 (7.5–8.5mm) at 22 ft. lbs. (98 Nm) of pressure. If the deflection is not correct, loosen the tensioner bolt and repeat Steps 10 and 11.

19. Install the lower cover gasket and the lower cover. Tighten the bolts to 61–87 inch lbs. (7–10 Nm).

20. Install the upper cover gasket and the upper cover. Tighten the bolts to 61–87 inch lbs. (7–10 Nm).

21. Position the engine mount on the engine and install the 2 nuts and dowels. Remove the floor jack.

22. Install the crankshaft sprocket baffle with the curved outer lip facing outward. Install the crankshaft pulley with the deep recess facing out and install the 6 bolts. Tighten the bolts to 109–152 inch lbs. (12–17 Nm).

23. Install the drive belts. Adjust the belt tension and tighten the adjusting and pivot bolts.

24. Install the right inner fender panel and wheel and tire assembly. Connect the negative battery cable.

2.3L & 2.5L (VIN C) ENGINES

1990–94 Models

1. Disconnect the negative battery cable and drain the cooling system. Remove the 4 water pump pulley bolts.

2. Remove the automatic belt tensioner and accessory drive belt. Remove the upper radiator hose.

3. Remove the crankshaft pulley bolt and pulley. Remove the thermostat housing and gasket.

4. Remove the timing belt outer cover retaining bolt(s). Release the cover interlocking tabs, if equipped, and remove the cover.

5. Loosen the belt tensioner adjustment screw, position belt tensioner tool T74P–6254–A or equivalent, on the tension spring roll pin and release the belt tensioner. Tighten the adjustment screw to hold the tensioner in the released position.

6. On 1991–93 vehicles, remove the bolts holding the timing sensor in place and pull the sensor assembly free of the dowel pin.

7. Remove the crankshaft pulley, hub and belt guide. Remove the timing belt. If the belt is to be reused, mark the direction of rotation so it may be reinstalled in the same direction.

To install:

8. Position the crankshaft sprocket to align with the TDC mark and the camshaft sprocket to align with the camshaft timing pointer. On 1990 vehicles, remove the distributor cap and set the rotor to the No. 1 firing position by turning the auxiliary shaft.

9. Install the timing belt over the crankshaft sprocket and then counterclockwise over the auxiliary and camshaft sprockets. Align the belt fore-and-aft on the sprockets.

52 TIMING BELTS—Ford Motor Company

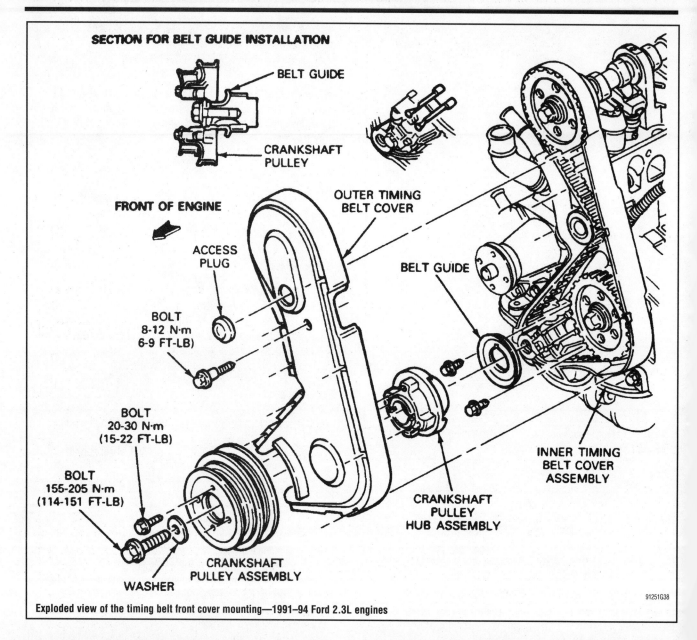

Exploded view of the timing belt front cover mounting—1991–94 Ford 2.3L engines

10. Loosen the tensioner adjustment bolt to allow the tensioner to move against the belt. If the spring does not have enough tension to move the roller against the belt, it may be necessary to manually push the roller against the belt and tighten the bolt.

11. To make sure the belt does not jump time during rotation in Step 10, remove a spark plug from each cylinder.

12. Rotate the crankshaft 2 complete turns in the direction of normal rotation to remove the slack from the belt. Tighten the tensioner adjustment to 29–40 ft. lbs. (40–55 Nm) and pivot bolts to 14–22 ft. lbs. (20–30 Nm). Check the alignment of the timing marks.

13. Install the crankshaft belt guide.

14. On 1990 vehicles, install the crankshaft pulley and tighten the retaining bolt to 103–133 ft. lbs. (140–180 Nm). On 1991–93 vehicles, proceed as follows:

   a. Install the timing sensor onto the dowel pin and tighten the 2 longer bolts to 14–22 ft. lbs. (20–30 Nm).

   b. Rotate the crankshaft 45 degrees counterclockwise and install the crankshaft pulley and hub assembly. Tighten the bolt to 114–151 ft. lbs. (155–205 Nm).

   c. Rotate the crankshaft 90 degrees clockwise so the vane of the crankshaft pulley engages with timing sensor positioner tool T89P–6316–A or equivalent. Tighten the 2 shorter sensor bolts to 14–22 ft. lbs. (20–30 Nm).

   d. Rotate the crankshaft 90 degrees counterclockwise and remove the sensor positioner tool.

   e. Rotate the crankshaft 90 degrees clockwise and measure the outer vane to sensor air gap. The air gap must be 0.018–0.039 in. (0.458–0.996mm).

15. Position the timing belt front cover. Snap the interlocking tabs into place, if necessary. Install the timing belt outer cover retaining bolt(s) and tighten to 71–106 inch lbs. (8–12 Nm).

16. Install the thermostat housing and a new gasket. Install the upper radiator hose.

17. Install the crankshaft pulley and retaining bolt. Tighten to 103–133 ft. lbs. (140–180 Nm) on 1990 vehicles or 114–151 ft. lbs. (155–205 Nm) on 1991–93 vehicles.

18. Install the water pump pulley and the automatic belt tensioner. Install the accessory drive belt.

FORD MOTOR COMPANY—TIMING BELTS

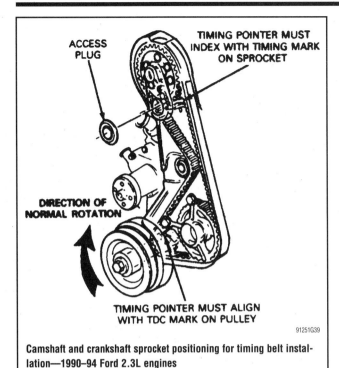

Camshaft and crankshaft sprocket positioning for timing belt installation—1990–94 Ford 2.3L engines

19. Install the spark plugs and remaining components.
20. Connect the negative battery cable, start the engine and check the ignition timing.

1995–99 Models

1. Rotate the engine so that No. 1 cylinder is at TDC on the compression stroke. Check that the timing marks are aligned on the camshaft and crankshaft pulleys. An access plug is provided in the cam belt cover so that the camshaft timing can be checked without removal of the cover or any other parts. Set the crankshaft to TDC by aligning the timing mark on the crank pulley with the TDC mark on the belt cover. Look through the access hole in the belt cover to be sure that the timing mark on the cam drive sprocket is lined up with the pointer on the inner belt cover.

➥ **Always turn the engine in the normal direction of rotation. Backward rotation may cause the timing belt to jump time, due to the arrangement of the belt tensioner.**

2. Drain cooling system. Remove the upper radiator hose as necessary. Remove the fan blade and water pump pulley bolts.

✱✱ CAUTION

When draining the coolant, keep in mind that cats and dogs are attracted by ethylene glycol antifreeze, and are quite likely to drink any that is left in an uncovered container or in puddles on the ground. This will prove fatal in sufficient quantity. Always drain the coolant into a sealable container. Coolant should be reused unless it is contaminated or several years old.

3. Loosen the alternator retaining bolts and remove the drive belt from the pulleys. Remove the water pump pulley.
4. Remove the power steering pump and set it aside.
5. Remove the four timing belt outer cover retaining bolts and remove the cover. Remove the crankshaft pulley and belt guide.
6. Loosen the belt tensioner pulley assembly, then position a camshaft belt adjuster tool (T74P-6254-A or equivalent) on the tension spring rollpin and retract the belt tensioner away from the timing belt. Tighten the adjustment bolt to lock the tensioner in the retracted position.
7. If the belt is to be reused, mark the direction of rotation on the belt for installation reference.
8. Remove the timing belt.

To install:

9. Install the new belt over the crankshaft sprocket and then counterclockwise over the auxiliary and camshaft sprockets, making sure the lugs on the belt properly engage the sprocket teeth on the pulleys. Be careful not to rotate the pulleys when installing the belt.
10. Release the timing belt tensioner pulley, allowing the tensioner to take up the belt slack. If the spring does not have enough tension to move the roller against the belt (belt hangs loose), it might be necessary to manually push the roller against the belt and tighten the bolt.

➥ **The spring cannot be used to set belt tension; a wrench must be used on the tensioner assembly.**

✱✱ WARNING

If any binding is felt when adjusting the timing belt tension by turning the crankshaft, STOP turning the engine, because the pistons may be hitting the valves.

11. Rotate the crankshaft two complete turns by hand (in the normal direction of rotation) to remove slack from the belt, then tighten the tensioner adjustment to 26–33 ft. lbs. (35–45 Nm) and pivot bolts to 30–40 ft. lbs. (40–55 Nm). Be sure the belt is seated

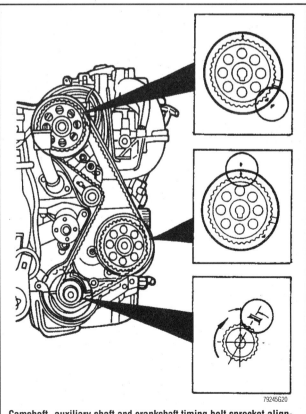

Camshaft, auxiliary shaft and crankshaft timing belt sprocket alignment mark locations—1995–99 Ford 2.3L and 2.5L (VIN C) engines

54 TIMING BELTS—FORD MOTOR COMPANY

properly on the pulleys and that the timing marks are still in alignment when No. 1 cylinder is again at TDC/compression.

12. Install the crankshaft pulley and belt guide.
13. Install the timing belt cover.
14. Install the water pump pulley and fan blades. Install the upper radiator hose if necessary. Refill the cooling system.
15. Install the accessory drive belts.
16. Start the engine and check the ignition timing. Adjust the timing, if necessary.

2.5L (VIN B) ENGINE

1. Disconnect the negative battery cable.
2. Label and disengage the electrical connectors from the coolant elbow. Label and remove the electrical connectors from the knock sensor (if required) and crankshaft position sensor.
3. Loosen the drive belt tensioner locknuts and adjusting bolts. Remove the accessory drive belts.
4. Raise and safely support the vehicle on jack stands.
5. Remove the lower bolt from the A/C and alternator tensioner bracket.
6. Remove the right wheel and splash-shields.
7. Hold the crankshaft pulley (damper) with Holder Tool T92C-6316-AH or equivalent, and remove the crankshaft pulley bolt. Remove the crankshaft pulley, using a puller if needed.
8. Remove the 5 front timing belt cover bolts.
9. Hold the water pump pulley with Holder Tool T92C-6312-AH or equivalent, remove the four bolts and the water pump pulley.
10. Hold the power steering pump pulley with Strap Wrench D85L-6000-A or equivalent and remove the power steering pump pulley nut and pulley.
11. Lower the vehicle.
12. Remove the upper bolt from the belt idler bracket and remove the bracket.
13. Remove the engine oil dipstick tube retaining bolt and the tube.
14. Remove the 8 rear timing belt cover retaining bolts and remove the timing belt covers.
15. Temporarily reinstall the crankshaft pulley bolt.
16. Remove the three nuts and through-bolt from the right-hand engine support insulator and remove the support insulator. Remove the support insulator bracket.
17. Raise and safely support the vehicle on jack stand.
18. Turn the crankshaft to TDC No. 1 cylinder in the direction of normal rotation. Be sure that the timing mark on the crankshaft sprocket aligns with the timing mark on the oil pump.
19. Remove the two bolts from the timing belt tensioner arm, removing the lower bolt first.
20. Remove the timing belt tensioner arm.
21. If the timing belt is to be reused, mark the direction of rotation on the timing belt.
22. Loosen the Allen bolt on the timing belt tensioner.
23. Remove the timing belt.
24. If the timing belt sprockets are to be removed, proceed as follows:
   a. Remove the intake manifold.
   b. Label and disconnect the necessary hoses from the cylinder head covers.
   c. Label and disconnect the spark plug wires from the spark plugs.
   d. Remove the cylinder head cover retaining bolts and remove the cylinder head covers.
   e. Hold the camshaft using a suitable wrench on the hexagon cast into the camshaft. Remove the camshaft sprocket bolts and the camshaft sprockets.

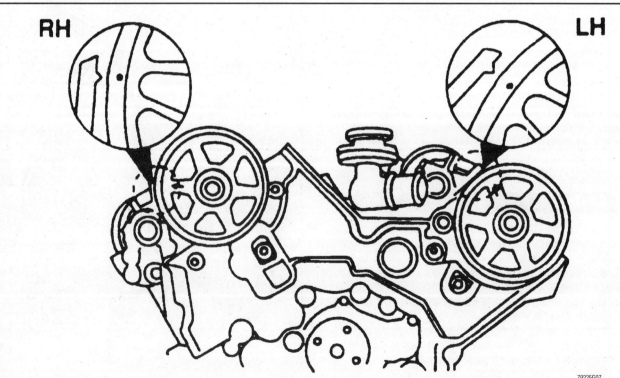

Left and right camshaft timing position—Ford 2.5L (VIN B) engines

FORD MOTOR COMPANY—TIMING BELTS

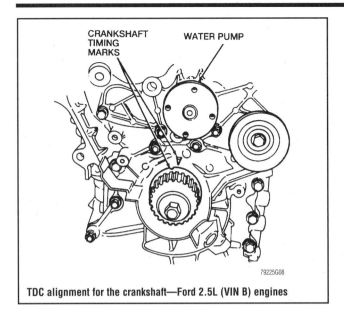

TDC alignment for the crankshaft—Ford 2.5L (VIN B) engines

f. Use Crankshaft Damper Puller T74P-6316-A or equivalent, to remove the crankshaft sprocket. Remove the crankshaft sprocket key.

To install:

25. If the timing belt sprockets were removed, proceed as follows:

   a. Install the crankshaft sprocket key and crankshaft sprocket.

   b. Install the camshaft sprockets on the camshafts with the retaining bolts.

   c. Hold the camshaft using a suitable wrench on the hexagon cast into the camshaft. Tighten the camshaft sprocket bolts to 90–103 ft. lbs. (123–140 Nm).

   d. Be sure the cylinder head cover and cylinder head contact surfaces are clean and free of dirt, oil and old sealant and gasket material.

   e. Apply silicone sealant to the cylinder heads in the area adjacent to the front and rear camshaft caps. Install new gaskets on the cylinder heads.

   f. Install the cylinder head covers and tighten the retaining bolts.

   g. Connect the spark plug wires to the spark plugs and connect the hoses to the cylinder head covers.

   h. Install the intake manifold.

26. Position the timing belt tensioner arm in a suitable press.
27. Compress the tensioner until the hole in the piston is aligned with the 2nd hole in the tensioner case. Insert a 0.060 in. (1.6mm) diameter wire or pin through the 2nd hole to keep the piston compressed.
28. Align the camshaft sprockets to TDC.
29. Turn the crankshaft counterclockwise until the crankshaft sprocket is offset from TDC by one tooth.
30. Install the timing belt.
31. If the original belt is being reused, be sure it is installed in the same direction of rotation.
32. Turn the crankshaft in the direction of normal engine rotation until the crankshaft sprocket timing mark is at TDC. This should place all of the belt slack in the timing belt tensioner portion of the timing belt.
33. Install the timing belt tensioner arm and two bolts. Tighten the bolts to 14–18 ft. lbs. (19–25 Nm).
34. Remove the wire or pin from the tensioner.

➡ **When properly timed, the crankshaft timing marks will line up and the crankshaft sprocket timing mark will no longer be one tooth off.**

35. Rotate the crankshaft two complete turns in the direction of normal rotation and align the timing marks. Be sure all marks are still correctly aligned. This will also set the timing belt tension.

➡ **The timing belt tensioner will automatically adjust the timing belt tension.**

36. Tighten the timing belt tensioner Allen bolt to 28–32 ft. lbs. (35–51 Nm).
37. Install the right-hand engine support insulator. Tighten the three nuts to 54–76 ft. lbs. (74–103 Nm) and the through-bolt to 50–68 ft. lbs. (67–93 Nm).
38. Remove the crankshaft damper bolt.
39. Install the timing belt covers with the rear 8 bolts. Tighten to 71–88 inch lbs. (8–10 Nm).
40. Install the engine oil dipstick tube and retaining nut.
41. Install the belt idler bracket and the upper retaining bolt.
42. Raise and safely support the vehicle on jack stands.
43. Install the power steering pump pulley and nut. Tighten the nut to 36–43 ft. lbs. (49–59 Nm) while holding the pulley with a strap wrench.
44. Install the water pump pulley and four bolts. Secure the pulley with the holder tool and tighten the bolts to 71–88 inch lbs. (8–10 Nm).
45. Install the 5 front timing belt cover bolts and tighten to 71–88 inch lbs. (8–10 Nm).
46. Install the crankshaft pulley (damper) with the bolt. Hold the crankshaft pulley with the holding tool and tighten to 116–122 ft. lbs. (157–166 Nm).
47. Install the splash-shields.
48. Install the wheel and tighten the lug nuts to 65–87 ft. lbs. (88–118 Nm).
49. Install the lower bolt into the A/C and alternator tensioner bracket.
50. Lower the vehicle.
51. Install the accessory drive belts and adjust the tension.
52. Engage the electrical connectors to the sensors at the coolant elbow and the knock sensor and crankshaft position sensor.
53. Connect the negative battery cable.
54. Run the engine and check for leaks and proper engine operation.

3.0L (VIN W) ENGINE

On this vehicle, right side refers to the "rear" components (near the firewall) and left side refers to the "front" components (near the radiator).

1. If the timing belt is to be removed, it is good practice to turn the crankshaft until the engine is at Top Dead Center (TDC) of the number one cylinder, compression stroke (firing position), before beginning work. This should align all timing marks and serve as a reference for all work that follows. After verifying that the engine is at TDC for the number one cylinder, do not crank the engine or allow the crankshaft or camshaft sprockets to be turned otherwise engine timing will be lost.
2. Drain the cooling system.
3. Disconnect the negative battery cable.
4. Remove the alternator drive belt, water pump and power steering pump belt and the A/C compressor belt (if equipped), using the recommended drive belt removal procedure.

56 TIMING BELTS—FORD MOTOR COMPANY

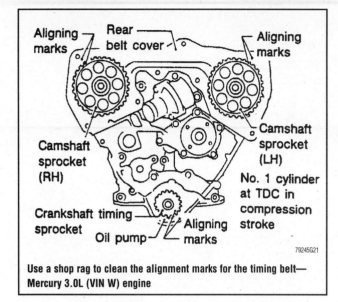

Use a shop rag to clean the alignment marks for the timing belt—Mercury 3.0L (VIN W) engine

5. If equipped with A/C, remove the three A/C compressor drive belt idler pulley bolts and remove the idler pulley.

6. Remove the upper radiator hose bracket bolt. Remove the upper hose with the bracket from the vehicle.

7. Remove the water bypass hose from between the thermostat housing and the lower water hose connection.

8. Remove the main wiring harness from the upper engine front cover.

9. Remove the eight upper engine front cover bolts and remove the upper cover.

10. Raise and safely support the vehicle.

11. Remove the right side front wheel and tire assembly.

12. Remove the four right side engine and transmission splash shield bolts and two screws, and remove the right side outer engine and transaxle splash shield.

13. Use a strap wrench to hold the water pump pulley. Remove the four pulley bolts, and the water pump pulley.

14. Use a strap wrench to hold the crankshaft pulley. Remove the center pulley bolt, and the crankshaft pulley using a harmonic balancer (damper) puller to draw the pulley from the front of the crankshaft.

15. Remove the five lower engine front cover bolts, then remove the lower engine front cover.

16. Be sure that the timing marks between the crankshaft sprocket and the oil pump housing line up.

17. If the timing belt is to be reused, mark an arrow on the belt indicating the direction of rotation. The directional arrow is necessary to ensure that the timing belt, if it to be reused, can reinstalled in the same direction.

18. Loosen the timing belt tensioner nut and slip the timing belt off of the sprockets.

19. If necessary, the camshaft sprockets can be removed. A special spanner tool is designed to hold the sprocket to keep it from turning while the center bolt is being loosened. Use care if using substitutes.

➡ The sprockets are not interchangeable.

20. If necessary, the crankshaft sprocket can be removed. The outer timing belt guide (looks like a large washer) and the crankshaft sprocket simply pull off the front of the crankshaft.

➡ Be careful, there are two crankshaft keys. Use care not to loose them.

To install:

21. Clean all parts well. If removed, inspect the crankshaft sprocket for warping or abnormal wear. Check the sprocket teeth for wear, deformation, chipping or other damage. Replace as necessary. Clean the sprocket mounting surface to ease installation. Install the key. Slip the sprocket onto the crankshaft. Tap it in place with a suitably-sized socket.

22. If removed, inspect the camshaft sprockets for damage and wear. Replace as required. The sprockets should be marked **L3** to designate the front, or left side camshaft and **R3** to designate the rear, or right side camshaft. Use care to install the sprockets properly. A special spanner tool is designed to hold the sprocket to keep it from turning while the center bolt is being tightened. Use care if using a substitute. Tighten the camshaft sprocket center bolts to 58–65 ft. lbs. (78–88 Nm). Verify that the timing marks on the camshaft sprockets and the timing marks on the rear cover (called the seal plate) are aligned.

23. Use an Allen wrench to turn the timing belt tensioner clockwise until the belt tensioner spring is fully extended. Temporarily tighten the tensioner nut to 32–43 ft. lbs. (43–58 Nm).

24. If a new timing belt is to be installed, look for a printed arrow on the belt. Be sure the arrow is pointing away from the engine. If the original timing belt is to be reused, be sure that the directional arrow that was marked at disassembly is facing the correct direction.

25. A new Original Equipment Manufacture (OEM) timing belt should have three white timing marks on it that indicate the correct timing positions of the camshafts and the crankshaft. These marks are to help ensure that the engine is properly timed. When the engine is properly timed, each white timing mark on the timing belt will be aligned with the corresponding camshaft and crankshaft timing mark on the sprocket. Because the white timing marks are not evenly spaced, the technician needs to use care in installing the belt. There should be 40 timing belt teeth between the timing marks on the front and rear camshaft sprockets and 43 teeth between the timing mark on the front camshaft sprocket and the timing mark on the crankshaft sprocket.

26. Verify that the camshaft timing marks are aligned with the timing marks on the rear cover (seal plate) and that the crankshaft sprocket timing mark is aligned with the timing mark on the oil pump housing.

27. Install the timing belt starting at the crankshaft sprocket and moving around the camshaft sprockets following a counter-clockwise path. Do not allow any slack in the timing belt between the sprockets. After all of the timing marks are matched up with the timing belt installed, slip the timing belt onto the belt tensioner.

28. While holding the timing belt tensioner with an Allen wrench, loosen the tensioner nut. Allow the tensioner to put pressure on the timing belt. Use an Allen wrench to turn the timing belt tensioner 70–80 degrees clockwise and tighten the timing belt tensioner nut to 32–43 ft. lbs. (43–58 Nm).

✱✱ WARNING

If any binding is felt when adjusting the timing belt tension by turning the crankshaft, STOP turning the engine, because the pistons may be hitting the valves.

29. Rotate the crankshaft clockwise twice and align the number one piston to TDC on the compression stroke (firing position).

30. Apply 22 lbs. (10kg) of force on the timing belt between the rear camshaft sprocket and the timing belt tensioner. An assistant

FORD MOTOR COMPANY—TIMING BELTS

may be needed. While holding the timing belt tensioner steady with an Allen wrench, loosen the timing belt tensioner nut. Remove the Allen wrench and adjust the timing belt tensioner using the following procedure:

    a. Install a 0.0138 in. (0.35mm) thick and 0.500 in. (12.7mm) wide feeler gauge where the timing belt just starts to go around the tensioner (approximately the four o'clock position, looking at the tensioner).

    b. Turn the crankshaft sprocket clockwise, which should force the feeler gauge between the timing belt and the tensioner, up to a position on the tensioner of about 1 o'clock.

    c. Tighten the timing belt tensioner nut to 32–43 ft. lbs. (43–58 Nm).

    d. Turn the crankshaft clockwise to rotate the feeler gauge out from between the timing belt tensioner and the timing belt.

31. Rotate the crankshaft clockwise twice, and once again align the number one piston to TDC on the compression stroke (firing position).

32. Apply 22 lbs. (10kg) of force on the timing belt between the front and rear camshaft sprockets. Measure the amount of belt deflection. Belt deflection should be between 0.51–0.59 in. (13–15mm). If belt deflection is out of specification, repeat Steps 29 through 33. If the timing belt deflection cannot be adjusted into specification, the timing belt will have to be replaced.

33. Position the lower engine front cover and install the five lower cover bolts. Do not overtighten. Tighten to 27–44 inch lbs. (3–5 Nm).

34. Install the outer timing belt guide next to the crankshaft sprocket with the dished side facing away from the cylinder block. Install the crankshaft pulley. Use a strap wrench to keep the crankshaft pulley from turning and tighten the center bolt to 90–98 ft. lbs. (123–132 Nm).

35. Position the water pump pulley on the pump. Install the four bolts. Use a strap wrench to keep the water pump pulley from turning and tighten the four water pump pulley bolts to 12–15 ft. lbs. (16–21 Nm).

36. Position the right side outer engine and transaxle splash shield, and secure with the four bolts and two screws.

37. Install the right side front wheel and tire assembly. Tighten the lug nuts to 72–87 ft. lbs. (98–118 Nm).

38. Lower the vehicle.

39. Position the upper engine timing belt front cover, and tighten the eight bolts to 27–44 inch lbs. (3–5 Nm).

40. Install the main wiring harness on the upper engine front cover.

41. Position the water bypass hose between the thermostat housing and water connection. Install the upper radiator hose between the radiator and the water hose connection. Secure the hoses with clamps. Install the upper radiator hose bracket. Tighten the bracket bolt to 34–58 ft. lbs. (46–65 Nm).

42. If equipped, position the A/C compressor drive belt idler pulley and install the three bolts. Tighten to 15 ft. lbs. (21 Nm).

43. Install and adjust the alternator drive belt, the water pump and power steering pump drive belt and the A/C compressor drive belt (of equipped).

44. Connect the battery cable.

45. Fill the cooling system.

46. Start the engine and allow it to warm to operating temperature. Check and adjust the ignition timing. Road test to verify correct engine operation.

3.0L (VIN Y) SHO ENGINE

1990–94 Models

1. Disconnect the battery cables.
2. Remove the battery.
3. Remove the right-hand engine roll damper.
4. Disconnect the wiring to the ignition module.
5. Remove the intake manifold crossover tube bolts. Loosen the intake manifold tube hose clamps. Remove the intake manifold crossover tube.
6. Loosen the alternator/air conditioning belt tensioner pulley and relieve the tension on the belt by backing out the adjustment screw. Remove the alternator/air conditioning belt.
7. Loosen the water pump/power steering belt tensioner pulley and relieve the tension on the belt by backing out the adjustment screw. Remove the water pump/power steering belt.
8. Remove the alternator/air conditioning belt tensioner pulley and bracket assembly.
9. Remove the water pump/power steering belt tensioner pulley only.
10. Remove the upper timing belt cover.
11. Disconnect the crankshaft sensor connectors.
12. Place the gear selector in **N**.
13. Rotate the crankshaft until the No. 1 cylinder piston is at TDC on the compression stroke. Make sure the white mark on the crankshaft damper aligns with the **0** degree index mark on the lower timing belt cover and the marks on the intake camshaft sprockets align with the index marks on the metal timing belt cover.
14. Raise the vehicle and support safely.
15. Remove the right front wheel and tire assembly.
16. Loosen the fender splash shield and place it aside.
17. Using a suitable puller, remove the crankshaft damper.
18. Remove the lower timing belt cover.
19. Remove the center timing belt cover and disconnect the crankshaft sensor wire and grommet from the slot in the cover and the stud on the water pump.

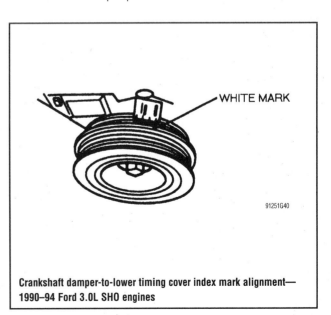

Crankshaft damper-to-lower timing cover index mark alignment—1990–94 Ford 3.0L SHO engines

58 TIMING BELTS—FORD MOTOR COMPANY

20. Loosen the timing belt tensioner, rotate the pulley 180 degrees clockwise and tighten the tensioner nut to hold the pulley in the unload position.
21. Lower the vehicle and remove the timing belt.

To install:

➡ Before installing the timing belt, inspect it for cracks, wear or other damage and replace, if necessary. Do not allow the timing belt to come into contact with gasoline, oil, water, coolant or steam. Do not twist or turn the belt inside out.

22. Make sure the engine is at TDC on the No. 1 cylinder. Check that the camshaft sprocket marks line up with the index marks on the upper steel belt cover and that the crankshaft sprocket aligns with the index mark on the oil pump housing.

➡ The timing belt has 3 yellow lines. Each line aligns with the index marks.

23. Install the timing belt over the crankshaft and camshaft sprockets. The lettering on the belt **KOA** should be readable from the rear of the engine; top of the lettering to the front of the engine. Make sure the yellow lines are aligned with the index marks on the sprockets.
24. Release the tensioner locknut and leave the nut loose.
25. Raise the vehicle and support safely.
26. Install the center timing belt cover. Make sure the crankshaft sensor wiring and grommet are installed and routed properly. Tighten the mounting bolts to 60–90 inch lbs. (7–11 Nm).
27. Install the lower timing belt cover. Tighten the bolts to 60–90 inch lbs. (7–11 Nm).
28. Using a suitable tool, install the crankshaft damper. Tighten the damper attaching bolt to 113–126 ft. lbs. (152–172 Nm).
29. Rotate the crankshaft 2 revolutions in the clockwise direction until the yellow mark on the damper aligns with the **0** degree mark on the lower timing belt cover.
30. Remove the plastic door in the lower timing belt cover. Tighten the tensioner locknut to 25–37 ft. lbs. (33–51 Nm) and install the plastic door.
31. Rotate the crankshaft 60 degrees more in the clockwise direction until the white mark on the damper aligns with the **0** degree mark on the lower timing belt cover.
32. Lower the vehicle.
33. Make sure the index marks on the camshaft sprockets align with the marks on the rear metal timing belt cover.
34. Route the crankshaft sensor wiring and connect with the engine wiring harness.
35. Install the upper timing belt cover. Tighten the bolts to 60–90 inch lbs. (7–11 Nm).
36. Install the water pump/power steering tensioner pulley. Tighten the nut to 11–17 ft. lbs. (15–23 Nm).
37. Install the alternator/air conditioning tensioner pulley and bracket assembly. Tighten the bolts to 11–17 ft. lbs. (15–23 Nm).
38. Install the water pump/power steering and alternator/air conditioning belts and set the tension. Tighten the idler pulley nut to 25–36 ft. lbs. (34–50 Nm).
39. Install the intake manifold crossover tube. Tighten the bolts to 11–17 ft. lbs. (15–23 Nm).
40. Install the engine roll damper and the battery.
41. Connect the wiring to the ignition module.
42. Connect the battery cables.
43. Raise the vehicle and support safely.
44. Install the splash shield and the right front wheel and tire assembly.
45. Lower the vehicle.

1995–99 Models

1. Disconnect both battery cables, negative cable first. Remove the battery.
2. Remove the right engine roll damper. Disconnect the wiring to the ignition module.
3. Remove the intake manifold crossover tube bolts. Loosen the intake manifold tube hose clamps. Remove the intake manifold crossover tube.
4. Loosen the alternator/air conditioning belt tensioner pulley and remove the drive belt.
5. Loosen the water pump/power steering belt tensioner pulley and remove the drive belt.
6. Remove the alternator/air conditioning belt tensioner pulley and bracket assembly.
7. Remove the water pump/power steering belt tensioner pulley only. Remove the upper timing belt cover.
8. Unplug the Crankshaft Position (CKP) sensor electrical connector. Place the transaxle gear selector in **N** (Neutral).
9. Rotate the crankshaft until the piston for the No. 1 cylinder is at TDC on its compression stroke. Be sure the white mark on the crankshaft damper aligns with the **0** degree index mark on the lower timing belt cover and the marks on the intake camshaft sprockets align with the index marks on the metal timing belt cover.
10. Raise and safely support the vehicle on jack stands. Remove the right front wheel and tire assembly. Loosen the fender splash-shield and place it aside.
11. Remove the crankshaft damper and pulley retaining bolt. Using Puller T67L-3600-A or equivalent, remove the crankshaft damper and pulley.

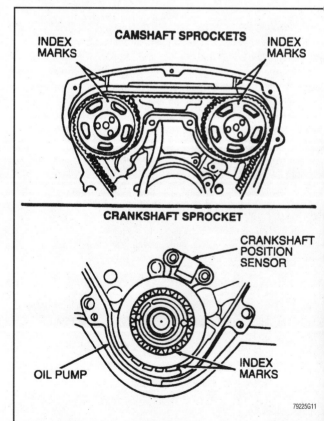

Timing marks for the camshaft and crankshaft sprockets—3.0L (VIN Y) engines

Ford Motor Company—Timing Belts

12. Remove the lower timing belt cover.
13. Remove the center timing belt cover and disconnect the CKP sensor wire and grommet from the slot in the cover and the stud on the water pump.
14. Loosen the timing belt tensioner idler pulley. Rotate the idler pulley 180 degrees clockwise and tighten the tensioner nut to hold the pulley in an unloaded position.
15. Lower the vehicle.
16. Remove the timing belt. If the belt is to be reused, use crayon to mark an arrow on the belt to indicate the direction of rotation, for installation reference.
17. If removing one or more camshaft sprockets, remove two retaining bolts securing each camshaft sprocket and remove the sprocket, noting the location of the dowel pin.
18. If removing the crankshaft sprocket, install Puller T67L-3600-A, or equivalent and pull the crankshaft sprocket off the crankshaft using care not to damage the pulse wheel.

To install:

19. If the crankshaft sprocket was removed, install the crankshaft sprocket by aligning the keyway and pushing the sprocket on by hand.
20. If one or more camshaft sprockets was removed, install each camshaft sprocket by aligning the timing marks on the sprockets with the camshaft using the dowel pin as a guide. Install two retaining bolts and tighten to 10–13 ft. lbs. (14–18 Nm).

➡ **Before installing the timing belt, inspect it for cracks, wear or other damage and replace, if necessary. Do not allow the timing belt to come into contact with gasoline, oil or coolant. Do not twist or turn the belt inside out.**

21. Be sure the engine is at TDC for the No. 1 cylinder. Check that the camshaft sprocket marks line up with the index marks on the upper steel belt cover and that the crankshaft sprocket aligns with the index mark on the oil pump housing.

➡ **The timing belt has three yellow lines. Each line aligns with the index marks.**

22. Install the timing belt over the crankshaft and camshaft sprockets. The lettering on the belt **KOA** should be readable from the rear of the engine (top of the lettering to the front of the engine). Be sure the yellow lines are aligned with the index marks on the sprockets.
23. Release the timing belt tensioner idler pulley locknut. Leave the nut loose. Raise and safely support the vehicle on jackstands.
24. Install the center timing belt cover. Be sure the CKP sensor wiring and grommet are installed and routed properly. Tighten the mounting bolts to 60–90 inch lbs. (7–11 Nm).
25. Install the lower timing belt cover. Tighten the bolts to 60–90 inch lbs. (7–11 Nm).
26. Install the crankshaft damper and pulley using Installer T88T-6701-A, or equivalent. Install the retaining bolt and tighten to 113–126 ft. lbs. (152–172 Nm).
27. Rotate the crankshaft two revolutions in the clockwise direction until the yellow mark on the damper aligns with the **0** degree mark on the lower timing belt cover.
28. Remove the plastic door in the lower timing belt cover. Tighten the tensioner locknut to 25–37 ft. lbs. (33–51 Nm) and install the plastic door.
29. Rotate the crankshaft 60 degrees more in the clockwise direction until the white mark on the damper aligns with the **0** degree mark on the lower timing belt cover.
30. Lower the vehicle. Be sure the index marks on the camshaft sprockets align with the marks on the rear metal timing belt cover.
31. Route the CKP sensor wiring and connect it to the engine wiring harness.
32. Install the upper timing belt cover. Tighten the bolts to 60–90 inch lbs. (7–11 Nm).
33. Install the water pump/power steering tensioner pulley. Tighten the nut to 11–17 ft. lbs. (15–23 Nm).
34. Install the alternator/air conditioning tensioner pulley and bracket assembly. Tighten the bolts to 11–17 ft. lbs. (15–23 Nm).
35. Install the water pump/power steering and alternator/air conditioning drive belts, and set the tension. Tighten the idler pulley nut to 25–36 ft. lbs. (34–50 Nm).
36. Install the intake manifold crossover tube. Tighten the bolts to 11–17 ft. lbs. (15–23 Nm).
37. Install the engine roll damper. Install the battery. Connect the wiring to the ignition module.
38. Connect both battery cables, negative cable last. Raise and safely support the vehicle on jack stands.
39. Install the splash-shield. Install the right front wheel and tire assembly.
40. Lower the vehicle. Run the engine and check for proper operation.

3.2L (VIN P) SHO ENGINE

1990–94 Models

1. Disconnect the battery cables.
2. Remove the battery.
3. Remove the right-hand engine roll damper.
4. Disconnect the wiring to the ignition module.
5. Remove the intake manifold crossover tube bolts. Loosen the intake manifold tube hose clamps. Remove the intake manifold crossover tube.
6. Rotate accessory drive belt tensioner clockwise to relieve tension. Remove the belt.
7. Disconnect surge tank fitting.
8. Remove bolts retaining upper and lower idler pulleys to engine and remove pulleys.
9. Using strap wrench D85L-6000-A or equivalent, to hold power steering pump pulley, remove nut, washer and remove power steering pulley.

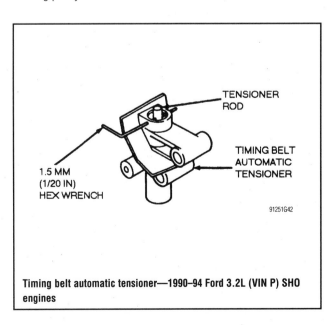

Timing belt automatic tensioner—1990–94 Ford 3.2L (VIN P) SHO engines

60 TIMING BELTS—FORD MOTOR COMPANY

10. Remove retaining bolt from belt tensioner and remove tensioner.
11. Remove the upper and center timing belt cover.
12. Disconnect the crankshaft sensor connectors.
13. Place the transaxle selector in **N**.
14. Rotate the crankshaft until the No. 1 cylinder piston is at TDC on the compression stroke. Make sure the white mark on the crankshaft damper aligns with the **0** degree index mark on the lower timing belt cover and the marks on the intake camshaft sprockets align with the index marks on the metal timing belt cover.
15. Raise the vehicle and support safely.
16. Remove the right front wheel and tire assembly.
17. Loosen the fender splash shield and place it aside.
18. Using a suitable puller, remove the crankshaft damper.
19. Remove the lower timing belt cover and belt guide.
20. Remove the upper timing belt tensioner bolt.
21. Slowly loosen the lower timing belt tension bolt and remove the tensioner.
22. Lower the vehicle and remove the timing belt.

To install:

➡ Before installing the timing belt, inspect it for cracks, wear or other damage and replace, if necessary. Do not allow the timing belt to come into contact with gasoline, oil, water, coolant or steam. Do not twist or turn the belt inside out.

23. Slowly compress timing belt tensioner in a soft jawed vice until hole in tensioner housing aligns with hole in tensioner rod.

※※ **CAUTION**

Use caution when compressing timing belt tensioner in vice to insure that tensioner does not slip from vice.

24. Insert a ¹⁄₂₀ inch (1.5mm) hex wrench through holes.
25. Release tension from vice.
26. If a new belt is being installed, loosen timing belt idler bolt.
27. Make sure the engine is at TDC on the No. 1 cylinder. Check that the camshaft sprocket marks line up with the index marks on the upper steel belt cover and that the crankshaft sprocket aligns with the index mark on the oil pump housing.

➡ The timing belt has 3 yellow lines. Each line aligns with the index marks.

28. Install the timing belt over the crankshaft and camshaft sprockets. The lettering on the belt **KOB** should be readable from the rear of the engine; top of the lettering to the front of the engine. Make sure the yellow lines are aligned with the index marks on the sprockets.

※※ **WARNING**

Do not install timing belt tensioner with rod extended.

29. Install timing belt tensioner on the cylinder block while pushing timing belt idler toward belt. Tighten tensioner bolts to 12–17 ft. lbs. (16–23 Nm).
30. Install grommets between timing belt tensioner and oil pump.
31. Remove ¹⁄₂₀ inch (1.5mm) hex wrench from timing belt tensioner.
32. If a new belt is being installed, perform the following steps:

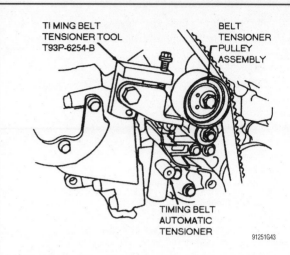

Timing belt tensioner pulley adjustment—1990–94 Ford 3.2L (VIN P) SHO engines

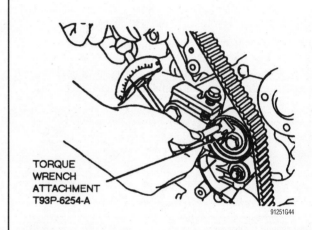

Timing belt tensioner pulley torque adjustment—1990–94 Ford 3.2L (VIN P) SHO engines

a. Position timing belt tensioner tool T93P-6254-B or equivalent, using power steering pump bracket holes.
b. Hand-tighten timing belt idler bolt.
c. Using torque wrench with attachment T93P-6254-A, rotate clockwise to 4.3 inch lbs. (0.5 Nm).
d. Tighten timing belt tensioner bolts to 27–37 ft. lbs. (36–50 Nm), then remove timing belt tensioning tool.

33. Raise the vehicle and support safely.
34. Install the belt guide and lower timing belt cover. Tighten the retaining bolts to 12–17 ft. lbs. (16–23 Nm).
35. Using a suitable tool, install the crankshaft damper. Tighten the damper attaching bolt to 113–126 ft. lbs. (152–172 Nm).
36. Rotate the crankshaft 2 revolutions in the clockwise direction until the yellow mark on the damper aligns with the **0** degree mark on the lower timing belt cover.
37. Lower the vehicle.
38. Make sure the index marks on the camshaft sprockets align with the marks on the rear metal timing belt cover.

39. Route the crankshaft sensor wiring and connect with the engine wiring harness.
40. Install the center and upper timing belt covers. Tighten the bolts to 12–17 ft. lbs. (16–23 Nm).
41. Install the water pump pulley. Tighten the nut to 12–17 ft. lbs. (16–23 Nm).
42. Install the single accessory drive belt while rotating accessory drive belt tensioner clockwise.
43. Install surge tank fitting.
44. Install the intake manifold crossover tube. Tighten the bolts to 11–17 ft. lbs. (15–23 Nm).
45. Install the engine roll damper and the battery.
46. Connect the wiring to the ignition module.
47. Connect the battery cables.
48. Raise the vehicle and support safely.
49. Install the splash shield and the right front wheel and tire assembly.
50. Lower the vehicle.

1995–99 Models

1. Disconnect the battery cables, negative cable first.
2. Remove the battery.
3. Remove the right engine roll damper.
4. Disconnect the wiring to the ignition module.
5. Remove the intake manifold crossover tube bolts. Loosen the intake manifold tube hose clamps. Remove the intake manifold crossover tube.
6. Rotate the accessory drive belt tensioner clockwise to relieve belt tension. Remove the accessory drive belt.
7. Disconnect the surge tank fitting.
8. Remove the bolts retaining the upper and lower idler pulleys to the engine and remove the pulleys.
9. Using Strap Wrench D85L-6000-A or equivalent, hold the power steering pump pulley. Remove the retaining nut and washer. Remove the power steering pulley.
10. Remove the retaining bolt from the belt tensioner and remove the tensioner.
11. Remove the upper and center timing belt covers.
12. Disengage the Crankshaft Position (CKP) sensor electrical connector.
13. Place the transaxle selector in **N** (Neutral).
14. Rotate the crankshaft until the piston for No. 1 cylinder is at TDC on its compression stroke. Be sure the white mark on the crankshaft damper aligns with the **0** degree index mark on the lower timing belt cover and the marks on the intake camshaft sprockets align with the index marks on the metal timing belt cover.
15. Raise and safely support the vehicle on jack stands.
16. Remove the right front wheel and tire assembly.
17. Loosen the fender splash-shield and place it aside.
18. Remove the crankshaft pulley and damper using Puller T67L-3600-A with the appropriate adapters, or equivalent.
19. Remove the lower timing belt cover and belt guide.
20. Remove the upper timing belt tensioner bolt.
21. Slowly loosen the lower timing belt tension bolt and remove the tensioner.
22. Lower the vehicle.
23. Remove the timing belt. If the belt is to be reused, use crayon to mark an arrow on the belt to indicate the direction of rotation, for installation reference.

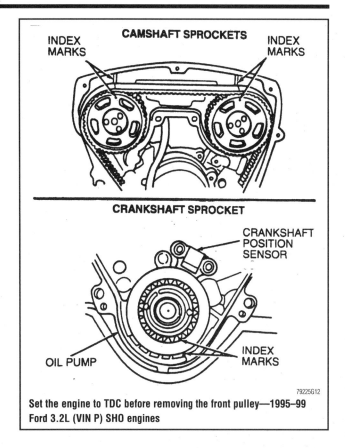

Set the engine to TDC before removing the front pulley—1995–99 Ford 3.2L (VIN P) SHO engines

24. If removing one or more camshaft sprockets, remove two retaining bolts securing each camshaft sprocket and remove the sprocket noting the location of the dowel pin.
25. If removing the crankshaft sprocket, install Puller T67L-3600-A, or equivalent and pull the crankshaft sprocket off the crankshaft using care not to damage the pulse wheel.

To install:
26. If the crankshaft sprocket was removed, install the crankshaft sprocket by aligning the keyway and pushing the sprocket on by hand.
27. If one or more camshaft sprockets was removed, install each camshaft sprocket by aligning the timing marks on the camshaft sprocket with the camshaft using the dowel pin as a guide. Install two retaining bolts and tighten to 10–13 ft. lbs. (14–18 Nm).

➡ **Before installing the timing belt, inspect it for cracks, wear or other damage and replace as necessary. Do not allow the timing belt to come into contact with gasoline, oil or coolant. Do not twist or turn the belt inside out.**

28. Slowly compress the timing belt tensioner in a soft-jawed vise until the hole in the tensioner housing aligns with the hole in the tensioner rod.

✸✸ CAUTION

Use care when compressing the timing belt tensioner in the vise to insure that the tensioner does not slip from the vise.

29. Insert a 1/20 in. (1.5mm) hex wrench through the holes in the tensioner.
30. Release the tension from the vise.

62 TIMING BELTS—FORD MOTOR COMPANY, GEO/CHEVROLET

31. If a new timing belt is being installed, loosen the timing belt idler bolt.

32. Ensure that the No. 1 cylinder is at TDC on its compression stroke. Check that the camshaft sprocket marks line up with the index marks on the upper steel belt cover and that the crankshaft sprocket aligns with the index mark on the oil pump housing.

➡ **The timing belt has three yellow lines. Each line aligns with the index marks.**

33. Install the timing belt over the crankshaft and camshaft sprockets. The lettering on the belt **KOB** should be readable from the rear of the engine (top of the lettering to the front of the engine). Be sure the yellow lines are aligned with the index marks on the sprockets.

✷✷ **WARNING**

Do not install the timing belt tensioner with the tensioner rod extended.

34. Install the timing belt tensioner on the cylinder block while pushing the timing belt idler toward the timing belt. Install and tighten the tensioner bolts to 12–17 ft. lbs. (16–23 Nm).

35. Install the grommets between the timing belt tensioner and the oil pump.

36. Remove the hex wrench from the timing belt tensioner.

37. If a new timing belt is being installed, perform the following steps:
   a. Remove the hex wrench from the timing belt tensioner, if installed.
   b. Mount Timing Belt Tensioner tool T93P-6254-B or equivalent, using the holes in the power steering pump support.
   c. Hand-tighten the timing belt idler pulley bolt.
   d. Using an in. pound torque wrench with attachment T93P-6254-A or equivalent, rotate the timing belt tensioner clockwise to 4.3 inch lbs. (0.5 Nm).
   e. Tighten the timing belt idler pulley bolt to 27–37 ft. lbs. (36–50 Nm). Remove both timing belt tensioning tools.

38. Raise and safely support the vehicle securely on jackstands.

39. Install the belt guide and lower timing belt cover. Tighten the retaining bolts to 12–17 ft. lbs. (16–23 Nm).

40. Using a suitable tool, install the crankshaft damper. Tighten the damper attaching bolt to 113–126 ft. lbs. (152–172 Nm).

41. Rotate the crankshaft two revolutions clockwise until the yellow mark on the damper aligns with the **0** degree mark on the lower timing belt cover.

42. Lower the vehicle.

43. Ensure that the index marks on the camshaft sprockets align with the marks on the rear metal timing belt cover.

44. Route the CKP sensor wiring and connect with the engine wiring harness.

45. Install the center and upper timing belt covers. Tighten the bolts to 12–17 ft. lbs. (16–23 Nm).

46. Install the steering pump pulley. Tighten the nut to 12–17 ft. lbs. (16–23 Nm).

47. Install the accessory drive belt while rotating the accessory drive belt tensioner clockwise.

48. Install the surge tank fitting.

49. Install the intake manifold crossover tube. Tighten the bolts to 11–17 ft. lbs. (15–23 Nm).

50. Install the engine roll damper.

51. Install the battery.

52. Connect the wiring to the ignition module.

53. Connect both battery cables, negative cable last.

54. Raise and safely support the vehicle.

55. Install the splash-shield.

56. Install the right front wheel and tire assembly.

57. Lower the vehicle.

58. Run the engine and check for leaks and proper engine operation.

Geo/Chevrolet

➡ **During these procedures, identify all components removed from the engine so that they may be reinstalled in their original positions. If discarding the old components so that new components can be installed, identifying the old items is not necessary.**

1.0L (VIN 6) & 1.3L (VIN 9) ENGINES

➡ **Timing belts must always be completely free of dirt, grease, fluids and lubricants. This includes the sprockets and contact surfaces on which the belt rides. The belt must never be crimped, twisted or bent. Never use tools to pry or wedge the belt.**

1. Disconnect the negative battery cable.
2. Raise and safely support the vehicle.
3. Remove the clips and right side splash-shield.
4. Remove the lower alternator cover plate.
5. Remove the alternator drive belt, if equipped, the A/C drive belt.
6. Remove the water pump pulley.

➡ **It is not necessary to remove the crankshaft timing sprocket bolt (center bolt) to remove the crankshaft pulley.**

7. Remove the crankshaft pulley bolts and crankshaft pulley.
8. Remove the retaining bolts and nut from the timing belt outside cover.
9. Remove the timing belt outside cover.

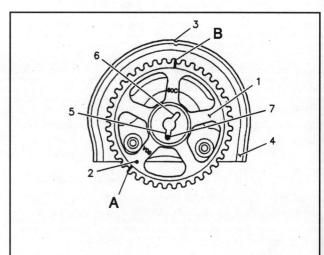

1 CAMSHAFT TIMING PULLEY	5 SLOT NO. 1
2 TIMING MARK	6 SLOT NO. 2
3 "V" MARK	7 PULLEY PIN
4 BELT INSIDE COVER	

79225G16

Upper timing pulley position—Geo/Chevrolet 1.0L (VIN 6) and 1.3L (VIN 9) engines

GEO/CHEVROLET—TIMING BELTS

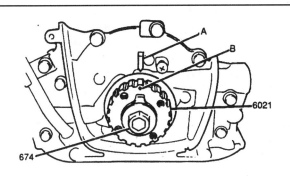

- A ARROW MARK ON OIL PUMP CASE
- B PUNCH MARK ON CRANKSHAFT TIMING GEAR
- 674 CRANKSHAFT PULLEY TIMING GEAR BOLT
- 6021 CRANKSHAFT TIMING GEAR

Crankshaft timing mark—Geo/Chevrolet 1.0L (VIN 6) and 1.3L (VIN 9) engines

10. Turn the crankshaft to align the timing marks. The mark on the crankshaft sprocket should align with the arrow mark on the oil pump housing. The mark on the camshaft sprocket should align with the **V** mark on the timing belt inner cover or cylinder head cover.

11. If the timing belt is to be reused, mark the direction of rotation on the belt.

12. Remove the timing belt tensioner, tensioner plate, tensioner spring, spring damper and timing belt.

➡ **Never turn the camshaft or crankshaft independently after the timing belt has been removed. Interference may occur between the pistons and valves, and parts may be damaged.**

13. Inspect the timing belt for wear or cracks, and replace as necessary. Check the tensioner for smooth rotation.

14. If the timing belt sprockets are to be removed, proceed as follows:

    a. Using a 0.39 in. (10mm) rod inserted into the camshaft, hold the camshaft and remove the retaining bolt and camshaft sprocket.

✴✴ WARNING

Be careful not to damage the cylinder head or cylinder head cover mating surfaces. Place a clean shop cloth between the rod and cylinder head. Do not bump the rod hard against the cylinder head when loosening the bolt.

   b. Raise and safely support the vehicle on jack stands.
   c. If equipped with a manual transaxle, lock the crankshaft in position by inserting a suitable flat-bladed tool into the hole in the bottom of the bell housing to engage the flywheel teeth.
   d. If equipped with an automatic transaxle, lock the crankshaft in position by inserting a suitable flat-bladed tool between the flywheel teeth and against the engine block.
   e. Remove the crankshaft sprocket bolt and crankshaft sprocket.

To install:

15. If the timing belt sprockets were removed, proceed as follows:
    a. Install the crankshaft sprocket, aligning the keyway. Lock the crankshaft in place and tighten the crankshaft sprocket bolt to 81 ft. lbs. (110 Nm).
    b. Lower the vehicle.

    c. Install the camshaft sprocket and retaining bolt. Lock the camshaft in place using the rod, and tighten the bolt to 44 ft. lbs. (60 Nm). Remove the locking rod.

16. Install the tensioner plate to the tensioner.
17. Insert the lug of the tensioner plate into the hole of the tensioner.
18. Install the tensioner, tensioner plate and spring. Do not fully tighten the tensioner bolt and stud at this time.
19. Move the tensioner plate in a counterclockwise direction. This should cause the tensioner to move in the same direction. If it does not, remove the tensioner and tensioner plate, and reinsert the tensioner plate lug in the timing plate tensioner hole.
20. Check that the camshaft timing marks are aligned. If not, align the two marks by turning the camshaft.
21. Check that the punch mark on the crankshaft timing belt sprocket is aligned with the arrow mark on the oil pump case. If not, align the two marks by turning the crankshaft.
22. With the timing marks aligned, install the timing belt on the two sprockets. If the old belt is being reused, be sure to install it running in the same direction of original rotation.
23. Install the tensioner spring and spring damper. Turn the timing belt two rotations clockwise after installing the tensioner spring and damper to remove any belt slack. Tighten the tensioner stud to 8 ft. lbs. (11 Nm), then the tensioner bolt to 20 ft. lbs. (27 Nm).

➡ **Confirm that both sets of timing marks are aligned properly.**

24. Using a new seal, install the timing belt cover and tighten the bolts and nut to 97 inch lbs. (11 Nm).
25. Install the crankshaft pulley. Fit the keyway on the pulley to the crankshaft timing belt sprocket and tighten the bolts to 8–12 ft. lbs. (11–16 Nm).
26. Install the alternator drive belt, if equipped, A/C drive belt.
27. Install the lower alternator cover plate. Tighten the bolts to 89 inch lbs. (10 Nm).
28. Install the right side splash-shield and clips.
29. Lower the vehicle and connect the negative battery cable.
30. Run the engine. Check for leaks.

1.6L (VIN 6) & 1.8L (VIN 8) ENGINES

Car Models

1990 MODELS

1. Disconnect the negative battery cable.
2. Elevate the vehicle and safely support it.
3. Remove the right splash shield under the vehicle.
4. Lower the vehicle. Remove the wiring harness from the upper timing belt cover.
5. Depending on equipment, loosen the air conditioner compressor, the power steering pump and the alternator on their adjusting bolts. Remove the drive belts.
6. Remove the crankshaft pulley. The use of a counter holding tool such as J–8614–01 or similar is highly recommended.
7. Remove the valve cover.
8. Remove the windshield washer reservoir.
9. Elevate and safely support the vehicle.
10. Support the engine either from above (Tool 28467–A or chain hoist) or below (floor jack and wood block) and remove the through bolt at the right engine mount.

64 TIMING BELTS—GEO/CHEVROLET

11. Remove the protectors on the mount nuts and studs for the center and rear transaxle mounts.
12. Remove the 2 rear transaxle mount-to-main crossmember nuts. Remove the 2 center transaxle mount-to-center crossmember nuts.
13. Carefully elevate the engine enough to gain access to the water pump pulley.
14. Remove the water pump pulley. Lower the engine to its normal position.
15. Remove the 4 bolts and the lower timing cover. Remove the center timing cover and its bolt, then the upper cover with its 4 bolts.
16. Rotate the crankshaft clockwise to the TDC compression position for No. 1 cylinder.
17. Loosen the timing belt idler pulley to relieve the tension on the belt, move the pulley away from the belt and temporarily tighten the bolt to hold it in the loose position.
18. Make matchmarks on the belt and both pulleys showing the exact placement of the belt. Mark an arrow on the belt showing its direction of rotation.
19. Carefully slip the timing belt off the pulleys.

➡ Do not disturb the position of the camshafts or the crankshaft during removal.

20. Remove the idler pulley bolt, pulley and return spring.
21. Use an adjustable wrench mounted on the flats of the camshaft to hold the camshaft from moving. Loosen the center bolt in the camshaft timing pulley and remove the pulley.
22. Check the timing belt carefully for any signs of cracking or deterioration. Pay particular attention to the area where each tooth or cog attaches to the backing of the belt. If the belt shows signs of damage, check the contact faces of the pulleys for possible burrs or scratches.
23. Check the idler pulley by holding it and spinning it. It should rotate freely and quietly. Any sign of grinding or abnormal noise indicates replacement of the pulley.
24. Check the free length of the tension spring. Correct length is 1.5 in. (38.5mm) measured at the inside faces of the hooks. A spring which has stretched during use will not apply the correct tension to the pulley; replace the spring.

To install:
25. Test the tension of the spring, look for 8.4 lbs. of tension at 2.0 in. (50mm) of length. If in doubt, replace the spring.
26. Reinstall the camshaft timing belt pulley, making sure the pulley fits properly on the shaft and that the timing marks align correctly. Tighten the center bolt to 43 ft. lbs. (58 Nm).

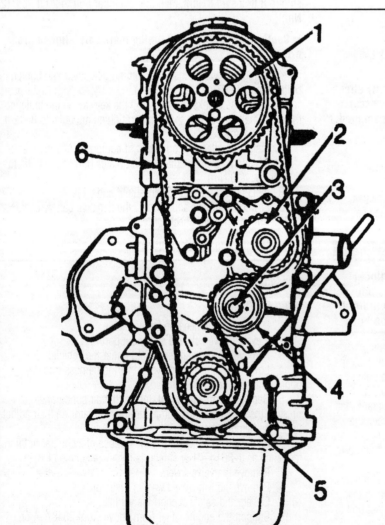

1. Camshaft timing pulley
2. Water pump timing pulley
3. Bolt
4. Tension pulley
5. Crankshaft timing pulley
6. Timing belt

View of the timing belt assembly—1990 Geo/Chevrolet 1.6L (VIN 6) and 1.8L (VIN 8) engines

GEO/CHEVROLET—TIMING BELTS

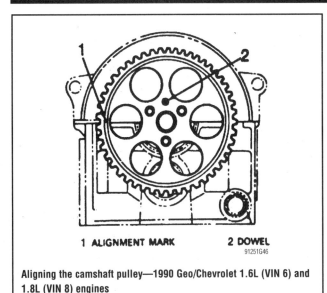

Aligning the camshaft pulley—1990 Geo/Chevrolet 1.6L (VIN 6) and 1.8L (VIN 8) engines

27. Before reinstalling the belt, double check that the crank and camshafts are exactly in their correct positions. The alignment mark on the end of the camshaft bearing cap should show through the small hole in the camshaft pulley and the small mark on the crankshaft timing belt pulley should align with the mark on the oil pump.
28. Reinstall the timing belt idler pulley and the tension spring. Pry the pulley to the left as far as it will go and temporarily tighten the retaining bolt. This will hold the pulley in its loosest position.
29. Install the timing belt, observing the matchmarks made earlier. Make sure the belt is fully and squarely seated on the upper and lower pulleys.
30. Loosen the retaining bolt for the timing belt idler pulley and allow it to tension the belt.
31. Temporarily install the crankshaft pulley bolt and turn the crank clockwise 2 full revolutions from TDC to TDC. Insure that each timing mark realigns exactly.
32. Tighten the timing belt idler pulley retaining bolt to 27 ft. lbs. (37 Nm).
33. Measure the timing belt deflection tool No. 23600–B or similar, looking for 0.20–0.24 in. (5–6mm) of deflection at 4.4 pounds of pressure. If the deflection is not correct, readjust the idler pulley by repeating Steps 15 through 18.
34. Remove the bolt from the end of the crankshaft.
35. Install the timing belt guide onto the crankshaft and install the lower timing belt cover.
36. When reinstalling, make certain that the gaskets and their mating surfaces are clean and free from dirt and oil. The gasket itself must be free of cuts and deformations and must fit securely in the grooves of the covers.
37. Install the covers and the bolts; tighten the bolts to 3.6 ft. lbs. (5 Nm).
38. Elevate the engine and install the water pump pulley.
39. Lower the engine to its normal position. Install the through bolt in the right engine mount and tighten it to 64 ft. lbs. (87 Nm) with the bolt secure, the engine lifting apparatus may be removed.
40. Install the valve cover.
41. Install the crankshaft pulley and tighten its bolt to 87 ft. lbs. (118 Nm).
42. Reinstall the air conditioning compressor, the power steering pump and the alternator. Install their belts and adjust them to the correct tension.
43. Reconnect the wiring harness to the upper timing belt cover.
44. Raise the vehicle and safely support it.
45. Install the 2 nuts on the center transaxle mount and the rear transaxle mount. Tighten all the nuts to 45 ft. lbs. (61 Nm).
46. Install the protectors on the nuts and studs.
47. Install the splash shield under the vehicle.
48. Lower the vehicle to the ground.
49. Install the windshield washer reservoir and connect the negative battery cable.

1991–99 MODELS

1. Disconnect the negative battery cable.
2. Remove the windshield washer reservoir from the engine compartment.
3. If equipped with cruise control, proceed as follows:
   a. Remove the cruise control actuator cover.
   b. Disconnect the cruise control harnesses.
   c. Disconnect the control cable.
   d. Remove the bolts and actuator from the vehicle.
4. Raise and safely support the vehicle on jack stands.
5. Remove the right front wheel.
6. Remove the bolts and plastic clips and the right front wheel housing.
7. Remove the alternator/water pump drive belt.
8. Lower the vehicle.
9. If equipped with A/C, proceed as follows:
   a. Remove the A/C compressor drive belt.
   b. Disconnect the compressor harness.
   c. Remove the bolts and compressor, without disconnecting the refrigerant lines. Suspend the compressor out of the way.
   d. Remove the compressor mounting bracket.
10. Remove the power steering pump drive belt.
11. Disconnect the wiring from the alternator and oil pressure switch.
12. Remove the engine wiring harness cover.
13. Remove the wiring harness from the cylinder head cover.
14. Disconnect the ignition wires from the spark plugs, then remove the spark plugs.
15. Remove the PCV hoses from the valve cover.
16. Remove the cap nuts, the seal washers and the cylinder head cover with the gasket.
17. Turn the crankshaft to align the timing mark on the crankshaft pulley at **0**, setting the piston in the No. 1 cylinder at Top Dead Center (TDC) on the compression stroke. Check that the valve lash adjusters on the No. 1 cylinder are loose. If not, turn the crankshaft pulley one complete revolution (360 degrees).
18. Remove the engine ground wire from the right fender apron.
19. Install a suitable support under the engine and remove the engine mount.
20. Remove the water pump pulley.
21. Remove the crankshaft pulley using a suitable puller.
22. Remove the 9 retaining bolts and the timing belt covers.
23. Slide the timing belt guide from the crankshaft.
24. Be sure the timing belt sprockets are properly aligned.

✴✴ WARNING

Do not turn the crankshaft or camshaft independently after removal of the timing belt; binding or damage to engine components could result. If the timing belt is to be reused, mark the belt with an arrow showing the direction of engine revolution.

66 TIMING BELTS—GEO/CHEVROLET

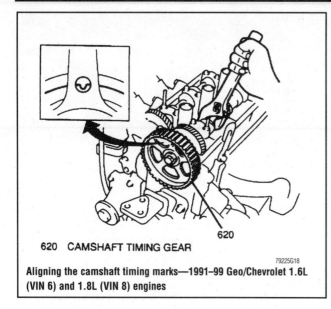

Aligning the camshaft timing marks—1991–99 Geo/Chevrolet 1.6L (VIN 6) and 1.8L (VIN 8) engines

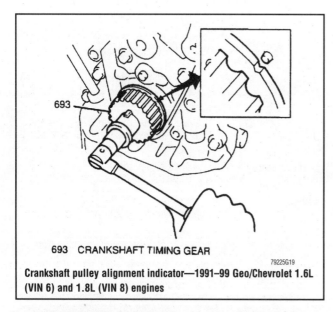

Crankshaft pulley alignment indicator—1991–99 Geo/Chevrolet 1.6L (VIN 6) and 1.8L (VIN 8) engines

25. Remove the timing belt tensioner bolt, tensioner and tension spring.
26. Remove the timing belt from the sprockets. Inspect the timing belt for cracked or damaged teeth. Replace as necessary.

※ WARNING
Do not bend, twist or turn the timing belt.

27. If the camshaft sprocket is to be removed, hold the camshaft stationary using a wrench positioned on the hexagon cast into the camshaft, and remove the sprocket retaining bolt and sprocket.

※ WARNING
Be careful not to damage the cylinder head when holding the camshaft in place.

28. If the crankshaft sprocket is to be removed, pry it from the crankshaft using two flat-bladed prybars.

To install:

29. Align the camshaft key with the groove on the sprocket and slide the sprocket on. Hold the camshaft with the wrench at the hexagonal portion of the camshaft, and tighten the camshaft timing sprocket bolt to 43 ft. lbs. (59 Nm).
30. Be sure the sprocket is still properly aligned.
31. Install the crankshaft timing sprocket. Align the crankshaft key with the groove on the sprocket and slide it on.
32. Reinstall the timing belt tensioner and the tension spring. Pry the tensioner to the left as far as it will go and temporarily tighten the retaining bolt.
33. Install the timing belt. If installing the old belt, observe the matchmarks made during removal.
34. Loosen the retaining bolt for the timing belt tensioner and allow it to tension the belt.
35. Temporarily install the crankshaft pulley bolt and turn the crankshaft clockwise two full revolutions. Be sure each timing mark realigns exactly.
36. Tighten the timing belt tensioner bolt to 27 ft. lbs. (37 Nm).
37. Measure the timing belt deflection. Correct deflection should be 0.20–0.24 in. (5–6mm) at 4 lbs. (20 Nm) of pressure. If the deflection is not correct, adjust it with the timing belt tensioner.
38. Install the timing belt guide, with the cup side facing outward.
39. Install the timing belt covers, installing the bottom one first. Tighten the 9 cover bolts to 62 inch lbs. (7 Nm).
40. Install the crankshaft pulley after aligning the pulley key with the slot on the pulley. Hold the pulley with tool J-8614–01 or equivalent, and tighten the pulley bolt to 87 ft. lbs. (118 Nm).
41. Temporarily install the water pump pulley.
42. Raise and safely support the vehicle on jack stands.
43. Install the engine mount.
44. Install or connect the remaining components.
45. If equipped with A/C, proceed as follows:
    a. Install the compressor mounting bracket and tighten the bolts to 35 ft. lbs. (47 Nm).
    b. Install the compressor and tighten the bolts to 18 ft. lbs. (25 Nm).
    c. Engage the compressor wiring connector.
    d. Install the compressor drive belt and adjust the tension.
46. If equipped with cruise control, proceed as follows:
    a. Install the cruise control actuator and tighten the bolts to 89 inch lbs. (10 Nm).
    b. Connect the cruise control cable.
    c. Install the cruise control actuator cover.
47. Connect the negative battery cable.
48. Start the engine and check vehicle operation.

Truck and SUV Models

The 1.6L 16-valve engine is known as an interference motor, because it is fabricated with such close tolerances between the pistons and valves that, if the timing belt is incorrectly positioned, jumps teeth on one of the sprockets or breaks, the valve and pistons will come into contact. This can cause severe internal engine damage

➡**Do not rotate the crankshaft counterclockwise or attempt to rotate the crankshaft by turning the camshaft sprocket.**

1. Remove the timing belt cover.
2. If the timing belt is not already marked with a directional arrow, use white paint, a grease pencil or correction fluid to do so.

GEO/CHEVROLET—TIMING BELTS

3. Rotate the crankshaft clockwise until the timing mark on the camshaft sprocket and the V mark on the timing belt inside cover are aligned, and the punch mark on the crankshaft sprocket is aligned with the mark on the engine.

✼✼ WARNING

Do not rotate the crankshaft or camshaft once the timing belt is removed, because the valves and pistons can come into contact, which may cause internal engine damage.

4. Disconnect one end of the tensioner spring. Loosen the timing belt tensioner bolt and stud, then, using your finger, press the tensioner plate up and remove the timing belt from the crankshaft and camshaft sprockets.
5. Remove the timing belt tensioner, tensioner plate and spring from the engine.
6. Install Suzuki Tool 09917–68220, or equivalent, onto the camshaft sprocket to hold the camshaft from rotating. Loosen the camshaft sprocket retaining bolt, then pull the camshaft sprocket off of the end of the camshaft.
7. Remove the crankshaft timing belt sprocket by loosening the center bolt, while preventing the crankshaft from rotating. To hold the crankshaft from turning, you can use Suzuki Tool 09927–56010 (or equivalent), or a large prybar inserted in the transmission housing slot and the flywheel teeth. Pull the sprocket off of the end of the crankshaft. Be sure to retain the crankshaft sprocket key and belt guide for assembly.
8. If necessary, remove the timing belt inside cover from the cylinder head.

To install:
9. If necessary, install the timing belt inside cover.
10. Slide the timing belt guide on the crankshaft so that the concave side faces the oil pump, then install the sprocket key in the groove in the crankshaft.
11. Slide the pulley onto the crankshaft, and install the center retaining bolt. Tighten the center bolt to 80 ft. lbs. (110 Nm). To hold the crankshaft from turning, you can use Suzuki Tool 09927–56010 (or equivalent), or a large prybar inserted in the transmission housing slot and the flywheel teeth.
12. Install the timing belt camshaft sprocket, ensuring that the slot in the sprocket engages the camshaft (pulley) pin; this ensures that the sprocket is properly positioned on the end of the camshaft. Secure the camshaft with the holding tool used during removal, then tighten the sprocket bolt to 44 ft. lbs. (60 Nm).
13. Assemble the timing belt tensioner plate and the tensioner, making sure that the lug of the tensioner plate engages the tensioner.

✼✼ WARNING

If any binding is felt when adjusting the timing belt tension by turning the crankshaft, STOP turning the engine, because the pistons may be hitting the valves.

14. Install the timing belt tensioner, tensioner plate and spring on the engine. Tighten the mounting bolt and stud only finger-tight at this time. Ensure that when the tensioner is moved in a counterclockwise direction, the tensioner moves in the same direction. If the tensioner does not move, remove it and the tensioner plate to reassemble them properly.
15. Loosen all rocker arm valve lash locknuts and adjusting screws. This will permit movement of the camshaft without any rocker arm associated drag, which is essential for proper timing belt tensioning. If the camshaft does not rotate freely (free of rocker arm drag), the belt will not be properly tensioned.
16. Rotate the camshaft sprocket clockwise until the timing mark on the sprocket and the V mark on the timing belt inside cover are aligned.
17. Using a wrench, or socket and breaker bar, on the crankshaft sprocket center bolt, turn the crankshaft clockwise until the punch mark on the sprocket is aligned with the arrow mark on the oil pump.
18. With the camshaft and crankshaft marks properly aligned, push the tensioner up with your finger and install the timing belt on the two sprockets, ensuring that the drive side of the belt is free of all slack. Release your finger from the tensioner. Be sure to install the timing belt so that the directional arrow is pointing in the appropriate direction.

➡ **In this position, the No. 4 cylinder is at Top Dead Center (TDC) on the compression stroke.**

19. Rotate the crankshaft clockwise two full revolutions, then tighten the tensioner stud to 97 inch lbs. (11 Nm). Then, tighten the tensioner bolt to 18 ft. lbs. (24 Nm).
20. Ensure that all four timing marks are still aligned as before; if they are not, remove the timing belt, and install and tension it again.
21. Install the timing belt cover and all related components.

1.6L (VIN U) ENGINE

➡ **Do not rotate the crankshaft counterclockwise or attempt to rotate the crankshaft by turning the camshaft sprocket.**

1. Remove the timing belt cover.
2. Remove rocker arm cover.
3. If the timing belt is not already marked with a directional arrow, use white paint, a grease pencil or correction fluid to do so.
4. Disconnect one end of the tensioner spring. Loosen the timing belt tensioner bolt and stud, then, using your finger, press the tensioner plate up and remove the timing belt from the crankshaft and camshaft sprockets.
5. Remove the timing belt tensioner, tensioner plate and spring from the engine.
6. Insert a metal rod through the hole in the camshaft to lock the camshaft from rotating. Loosen the camshaft sprocket retaining bolt, then pull the camshaft sprocket off of the end of the camshaft.

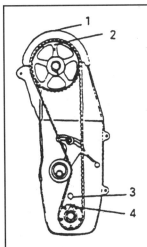

1. "V" mark on cylinder head cover
2. Timing mark by "E" on camshaft timing belt pulley
3. Arrow mark on oil pump case
4. Punch mark on crankshaft timing belt pulley

Rotate the crankshaft clockwise until the camshaft and crankshaft timing marks are aligned—Geo/Chevrolet 1.6L 16-valve engine

TIMING BELTS—GEO/CHEVROLET, General Motors Corp.

7. Remove the crankshaft timing belt sprocket by loosening the center bolt, while preventing the crankshaft from rotating. To hold the crankshaft from turning, you can use Suzuki Tool 09927–56010 (or equivalent), or a large prybar inserted in the transmission housing slot and the flywheel teeth. Pull the sprocket off of the end of the crankshaft. Be sure to retain the crankshaft sprocket key and belt guide for assembly.

8. If necessary, remove the timing belt inside cover from the cylinder head.

To install:

9. If necessary, install the timing belt inside cover.

10. Slide the timing belt guide on the crankshaft so that the concave side faces the oil pump, then install the sprocket key in the groove in the crankshaft.

11. Slide the pulley onto the crankshaft, and install the center retaining bolt. Tighten the center bolt to 58–65 ft. lbs. (80–90 Nm). To hold the crankshaft from turning, you can use Suzuki Tool 09927–56010 (or equivalent), or a large prybar inserted in the transmission housing slot and the flywheel teeth.

12. Install the timing belt camshaft sprocket, ensuring that the slot in the sprocket engages the camshaft (pulley) pin; this ensures that the sprocket is properly positioned on the end of the camshaft. Secure the camshaft with the metal rod used during removal, then tighten the sprocket bolt to 41–46 ft. lbs. (56–64 Nm).

13. Assemble the timing belt tensioner plate and the tensioner, making sure that the lug of the tensioner plate engages the tensioner.

14. Install the timing belt tensioner, tensioner plate and spring on the engine. Tighten the mounting bolt and stud only finger-tight at this time. Ensure that when the tensioner is moved in a counterclockwise direction, the tensioner moves in the same direction. If the tensioner does not move, remove it and the tensioner plate to reassemble them properly.

15. Loosen all rocker arm valve lash locknuts and adjusting screws. This will permit movement of the camshaft without any rocker arm associated drag, which is essential for proper timing belt tensioning. If the camshaft does not rotate freely (free of rocker arm drag), the belt will not be properly tensioned.

✲✲ WARNING

If any binding is felt when adjusting the timing belt tension by turning the crankshaft, STOP turning the engine, because the pistons may be hitting the valves.

16. Rotate the camshaft sprocket clockwise until the timing mark on the sprocket and the V mark on the timing belt inside cover are aligned.

17. Using a 17mm wrench, or socket and breaker bar, on the crankshaft sprocket center bolt, turn the crankshaft clockwise until the punch mark on the sprocket is aligned with the arrow mark on the oil pump.

18. With the camshaft and crankshaft marks properly aligned, push the tensioner up with your finger and install the timing belt on the two sprockets, ensuring that the drive side of the belt is free of all slack. Release your finger from the tensioner. Be sure to install the timing belt so that the directional arrow is pointing in the appropriate direction.

➡ **In this position, the No. 4 cylinder is at Top Dead Center (TDC) on the compression stroke.**

19. Rotate the crankshaft clockwise two full revolutions, then tighten the tensioner stud to 80–106 inch lbs. (9–12 Nm). Then, tighten the tensioner bolt to 18–21 ft. lbs. (24–30 Nm).

20. Ensure that all four timing marks are still aligned as before; if they are not, remove the timing belt, and install and tension it again.

21. Install the timing belt cover and all related components.

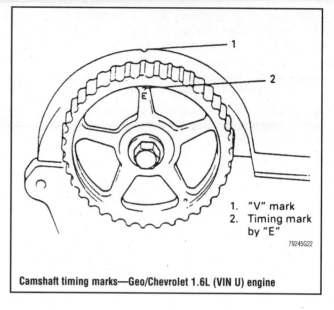

1. "V" mark
2. Timing mark by "E"

Camshaft timing marks—Geo/Chevrolet 1.6L (VIN U) engine

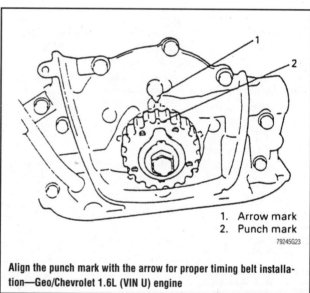

1. Arrow mark
2. Punch mark

Align the punch mark with the arrow for proper timing belt installation—Geo/Chevrolet 1.6L (VIN U) engine

General Motors Corporation

2.0L ENGINE

1990–94 Models

1. Disconnect the negative battery cable and drain the engine cooling system to a level below the water pump.

2. For the VIN H engine, remove the coolant reservoir tank.

3. Remove the serpentine belt from the engine. For the VIN H engine, remove the fuel vapor pipe assembly.

4. Loosen the drive belt tensioner bolt and the tensioner will swing downward. If necessary, remove the bolt and tensioner from the engine.

5. Remove the cover attaching bolts, then remove the timing belt cover from the engine.

6. Turn the engine to align the timing marks on the gears with the marks on the timing belt rear cover.

7. Loosen the water pump bolts and release timing belt tension with tension adjusting tool J–33039 or equivalent. If equipped, remove the A/C drive belt.

GENERAL MOTORS CORP.—TIMING BELTS

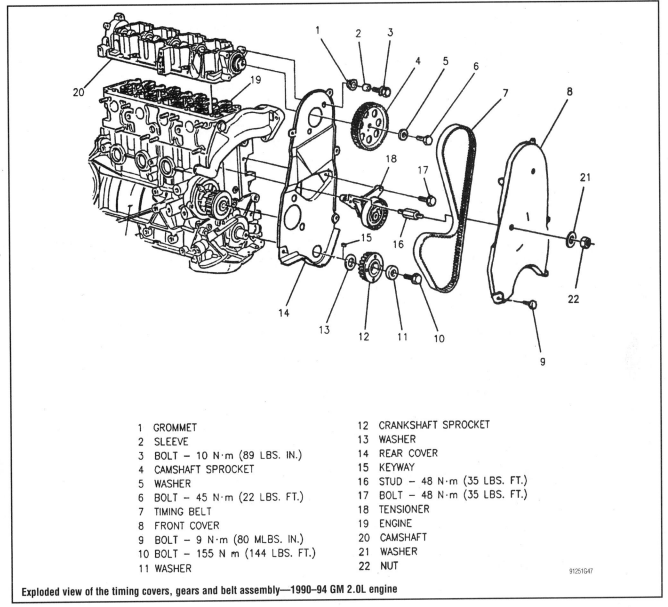

1	GROMMET	12	CRANKSHAFT SPROCKET
2	SLEEVE	13	WASHER
3	BOLT – 10 N·m (89 LBS. IN.)	14	REAR COVER
4	CAMSHAFT SPROCKET	15	KEYWAY
5	WASHER	16	STUD – 48 N·m (35 LBS. FT.)
6	BOLT – 45 N·m (22 LBS. FT.)	17	BOLT – 48 N·m (35 LBS. FT.)
7	TIMING BELT	18	TENSIONER
8	FRONT COVER	19	ENGINE
9	BOLT – 9 N·m (80 MLBS. IN.)	20	CAMSHAFT
10	BOLT – 155 N·m (144 LBS. FT.)	21	WASHER
11	WASHER	22	NUT

Exploded view of the timing covers, gears and belt assembly—1990–94 GM 2.0L engine

8. Raise and support the vehicle safely, then remove the right splash shield.

9. Remove the pulley retaining bolt, then remove the pulley from the end of the crankshaft.

10. Lower the vehicle and remove the timing belt. If necessary, remove the retainers and the belt tensioner from the engine.

To install:

11. If removed, install the timing belt tensioner and tighten the retainers to 18 ft. lbs. (25 Nm).

12. If necessary, turn the crankshaft and/or the camshaft gears clockwise to align the timing marks on the gears with the timing marks on the rear cover.

13. Install the timing belt, making sure the portion between the camshaft gear and crankcase gear is in tension.

14. Using tool J–33039 or equivalent, turn the water pump eccentric clockwise until the tensioner contacts the high torque stop. Tighten the water pump screws slightly.

15. Turn the engine clockwise 720 degrees using the crankshaft gear bolt in order to fully seat the belt into the gear teeth.

16. Turn the water pump eccentric counterclockwise until the hole in the tensioner arm is aligned with the hole in the base. In order for the holes to properly align, the engine must be at or near room temperature of 68°F (20°C).

17. Tighten the water pump screws to 18 ft. lbs. (25 Nm) while checking that the tensioner holes remain as adjusted in the prior step.

18. Raise and support the vehicle safely, then install the crankshaft pulley. Coat the pulley retainer bolt threads with Loctite® 242, or equivalent threadlock. Install the retainer and tighten to 20 ft. lbs. (27 Nm) except for the VIN H engine which should be tightened to 15 ft. lbs. (21 Nm).

19. Install the right splash shield and lower the vehicle.

20. Install the timing belt cover and attaching bolts. Tighten to 89 inch lbs. (10 Nm) for 1990–91 vehicles, or to 62 inch lbs. (7 Nm) for 1992–94 vehicles.

21. Pivot or install the timing belt tensioner into position and tighten the retaining bolt to 40 ft. lbs. (54 Nm) for 1990–91 vehicles, or to 35 ft. lbs. (48 Nm) for 1992–94 vehicles.

TIMING BELTS—GENERAL MOTORS CORP.

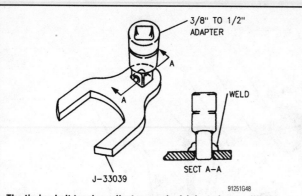

The timing belt tension adjuster may be fabricated or modified to suit the engine—1990-94 GM 2.0L engine

22. Install the serpentine belt.
23. Install fuel vapor pipes.
24. Install the serpentine belt and, if equipped, the A/C drive belt. For the VIN H engine, install the coolant reservoir.
25. Connect the negative battery cable and properly fill the engine cooling system.

3.0L (VIN R) ENGINE

➡ The steps in this procedure are critical in preventing catastrophic engine damage, adhering to this sequence is imperative. There are special tools needed to perform this procedure. It is a good idea to read this procedure several times before attempting to perform this job. This is an interference engine.

TIMING BELT INSTALLATION AND ADJUSTMENT TABLE

Step	Action	Value	Yes	No
1	Install the timing belt and align marks on the belt with the marks on the camshaft gears and the crankshaft gear. Check the timing belt deflection between the idler pulley for camshafts 3 & 4 and camshaft number 4. Is the timing belt installed, the marks aligned and the timing belt deflection adjusted?	1 cm (0.4 in) maximum	Go to Step 2	—
2	Set the initial timing belt tension at the timing belt tensioner. Is the initial timing belt tension set?	—	Go to Step 3	—
3	Rotate the engine two complete revolutions and secure the crankshaft at Top Dead Center (TDC) with the J 42069-10. Has the engine been rotated and the crankshaft secured to TDC?	—	Go to Step 4	—
4	Starting with camshafts 3 and 4, check the alignment of the marks on the camshaft gears with the marks on the J 42069-20 checking gauge. Do the marks on the camshaft gears align exactly with the marks on J 42069-20?	—	Go to Step 5	Go to Step 6
5	Check the alignment of the marks on camshafts gears 1 and 2 with the marks on the J 42069-20 checking gauge. Do the marks on the camshaft gears align exactly with the marks on J 42069-20?	—	Go to Step 14	Go to Step 10
6	Do the camshaft gear marks line up to the left (BTDC) of the marks on the J 42069-20 checking gauge?	—	Go to Step 8	Go to Step 7
7	Do the camshaft gear marks line up to the right (ATDC) of the marks on the J 42069-20 checking gauge?	—	Go to Step 9	—
8	Turn the idler pulley eccentric, for camshafts 3 and 4, counterclockwise until the marks on the camshaft gear align exactly with the marks on J 42069-20. Rotate the engine two complete revolutions, lock the crankshaft at TDC with J 42069-10 and recheck the alignment of the camshaft gear marks to the marks on J 42069-20. Do the marks on the camshaft gears align exactly with the marks on J 42069-20?	—	Go to Step 5	Go to Step 6
9	Turn the idler pulley eccentric, for camshafts 3 and 4, clockwise until the marks on the camshaft gear align exactly with the marks on J 42069-20. Rotate the engine two complete revolutions, lock the crankshaft at TDC with J 42069-10 and recheck the alignment of the camshaft gear marks to the marks on J 42069-20. Do the marks on the camshaft gears align exactly with the marks on J 42069-20?	—	Go to Step 5	Go to Step 6

Timing belt installation and adjustment table—GM 3.0L (VIN R) engine

GENERAL MOTORS CORP.—TIMING BELTS

Step	Action	Value	Yes	No
10	Do the camshaft gear marks line up to the left (BTDC) of the marks on the J 42069-20 checking gauge?	—	Go to Step 12	Go to Step 11
11	Do the camshaft gear marks line up to the right (ATDC) of the marks on the J 42069-20 checking gauge?	—	Go to Step 13	—
12	Turn the idler pulley eccentric, for camshafts 1 and 2, counterclockwise until the marks on the camshaft gear align exactly with the marks on J 42069-20. Rotate the engine two complete revolutions, lock the crankshaft at TDC with J 42069-10 and recheck the alignment of the camshaft gear marks to the marks on J 42069-20. Do the marks on the camshaft gears align exactly with the marks on J 42069-20?	—	Go to Step 14	Go to Step 10
13	Turn the idler pulley eccentric, for camshafts 1 and 2, clockwise until the marks on the camshaft gear align exactly with the marks on J 42069-20. Rotate the engine two complete revolutions, lock the crankshaft at TDC with J 42069-10 and recheck the alignment of the camshaft gear marks to the marks on J 42069-20. Do the marks on the camshaft gears align exactly with the marks on J 42069-20?	—	Go to Step 14	Go to Step 10
14	Set the final timing belt tension at the timing belt tensioner. Is the final timing belt tension set?	—	Go to Step 15	—
15	Again, rotate the engine two complete revolutions and lock the crankshaft at TDC. Do a final inspection of the camshaft gear marks' relationship to the J 42069-20 marks. The marks must align exactly. Do the marks on the camshaft gears align exactly with the marks on the J 42069-20?	—	Go to Step 16	Go to Step 2
16	Remove all checking tools and ensure all idler pulleys and the tensioner locking nut are tightened to specifications. Continue with re-assembly of the engine.	—	—	—

Timing belt installation and adjustment table (continued)—GM 3.0L (VIN R) engine

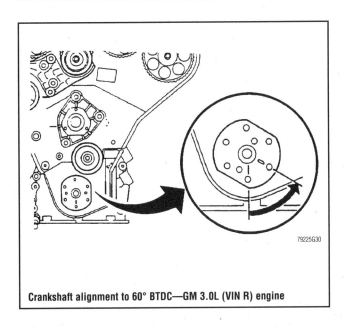

Crankshaft alignment to 60° BTDC—GM 3.0L (VIN R) engine

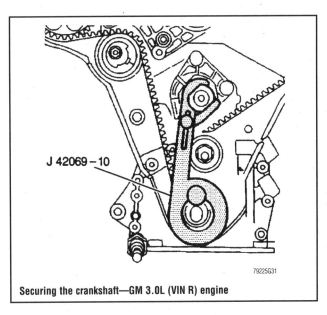

Securing the crankshaft—GM 3.0L (VIN R) engine

72 TIMING BELTS—GENERAL MOTORS CORP.

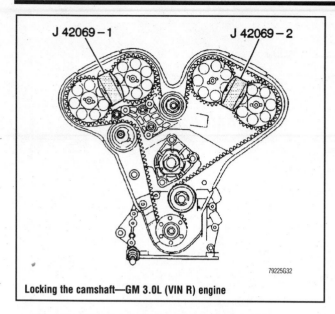

Locking the camshaft—GM 3.0L (VIN R) engine

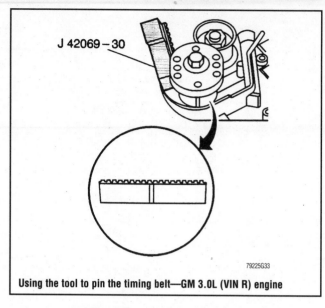

Using the tool to pin the timing belt—GM 3.0L (VIN R) engine

Always turn the crankshaft in the direction of rotation (clockwise), never against engine rotation. Never remove the timing belt without first setting the camshaft gears and crankshaft drive gear to TDC and locking them in place with tool J42069 (or equivalent).

The "Timing Belt Installation and Adjustment Table" provides an overview of the steps needed to properly install and adjust the timing belt. Use this table as a reference, not as a substitution, for the steps in this procedure.

1. Disconnect the negative battery cable.
2. Remove the resonance chamber.
3. Remove the front timing belt cover.
4. Remove the harmonic balancer from the crankshaft.
5. Rotate the crankshaft clockwise to 60°BTDC.
6. Install J42069–10 (or equivalent) to the crankshaft drive gear with knurled bolt.
7. Turn the engine clockwise, with J42098 (or equivalent), until the lever of J42069–10 (or equivalent) firmly contacts the water pump pulley flange. Secure the lever to the water pump.

➡**Be sure the engine is not 180°off. The camshaft marks must line up with the rear timing cover.**

8. Lock the camshaft gears, using J42069–1 (or equivalent) and J42069–2 (or equivalent). It may be necessary to loosen the relevant idler pulley to lock the gears, then tighten the idler pulley bolt to 30 ft. lb. (40 Nm).
9. Loosen the timing belt tensioner and remove the belt.

❊❊❊ WARNING

With the belt removed, do not rotate the camshaft or crankshaft, or remove the locking tools, because the pistons may contact the valves and cause internal engine damage.

To install:

10. Remove J42069–10 (or equivalent).
11. Raise and safely support the vehicle securely on jackstands.
12. Install the timing belt, starting at the crankshaft gear and aligning the double dash (TDC) mark on the belt with the oil pump and belt drive gear.
13. Using J42069–30 (or equivalent), secure the belt to prevent the splines from jumping.
14. Lower the vehicle.
15. Route the belt between the idler pulley for camshafts 3 and 4, then between the gears for 3 and 4.
16. If the dash marks on the belt do not line up with the camshaft and rear timing cover marks, loosen the idler pulley, or move the cam gears slightly, with the camshaft gears still locked in place, until the timing belt can be properly installed. Temporally tighten the idler pulley locking bolt; the locking bolt will be tightened to specification after final adjustments are made.

➡**The timing belt deflection must be no more than 0.4 in (10mm) between camshaft gear 4 and the idler pulley. To adjust the deflection, rotate the timing belt idler pulley for camshafts 3 and 4 counterclockwise with tool J42069–40 (or equivalent).**

17. Route the belt between the idler pulley for camshafts 1 and 2, then between the gears for 1and 2.
18. If the dash marks on the belt do not line up with the camshaft and rear timing cover marks, loosen the idler pulley, or move the cam gears slightly, with the camshaft gears still locked in place, until the timing belt can be installed. Temporally tighten the idler pulley; the locking bolt will be tightened to specification after final adjustments are made.
19. Apply tension to the belt to keep it from slipping off the gears by turning the timing belt idler pulley for camshafts 1and 2 counterclockwise with J42069–40 (or equivalent). Temporally tighten the idler pulley; the locking bolt will be tightened to specification after final adjustments are made.
20. Complete the routing of the timing belt through the belt tensioner.
21. Apply initial tension by turning the timing belt tensioner counterclockwise, with a 5mm Allen wrench, until the marks are set as shown in the initial timing belt adjustment illustration.
22. Tighten the tensioner locking nut to 15 ft. lbs. (20 Nm).
23. Ensure that the alignment marks are at their specific reference points.
24. Remove J42069–30, J42069–1 and J42069–2.

GENERAL MOTORS CORP.—TIMING BELTS

25. Rotate the engine two revolutions clockwise, stopping at 60° BTDC.
26. Install J42069–10 to the crankshaft gear with the knurled bolt.
27. Turn the engine clockwise, with J42098 or equivalent, until the lever of J42069–10 or equivalent firmly contacts the water pump pulley flange. Secure the lever to the water pump.

➡ **The alignment marks on the timing belt will no longer match the marks on the camshaft gears after one or more revolutions. The marks on the camshaft gears must line up with the notches on the rear timing cover, and the crankshaft drive gear and the oil pump housing should match up to their mark.**

28. If timing belt adjustment is necessary, first adjust camshafts 3 and 4.
29. Check the alignment of camshafts 3 and 4, 1 and 2 with gauge J42069–20.
30. If alignment is OK, then set final belt tension as follows:
    a. Loosen the timing belt tensioner locking nut.
    b. Adjust the timing belt tensioner by turning the eccentric cam, with a 5mm hex wrench, until the marks are set.
31. Tighten the tensioner locking nut to 15 ft. lb. (20 Nm).
32. Remove J42069–10 and J42069–20.
33. Rotate the engine clockwise two revolutions, stopping at 60° BTDC.
34. If further adjustment is needed, then repeat the applicable steps.
35. Tighten the idler pulley bolts for the camshafts to 30 ft lb. (40 Nm).
36. Be sure all tools are removed from the engine.
37. Install the harmonic balancer and tighten the bolts to 15 ft. lb. (20 Nm).
38. Install the timing belt covers.
39. Install the resonance chamber.
40. Connect the negative battery cable, and reprogram applicable accessories.

3.4L (VIN X) ENGINE

The 3.4L (VIN X) engine uses a timing chain and camshaft timing belts.

1. Disconnect the negative battery cable.
2. Disconnect and remove the power steering pump from the pump mounting bracket.
3. Remove the left, right and center timing belt covers.
4. Rotate the engine clockwise to align the timing marks, TDC on the No.1 exhaust stroke, on the camshaft sprockets and intermediate shaft.
5. Loosely clamp the two camshaft sprockets on each side of the engine together using clamping pliers or the equivalent. Secure the belt to the right side cam sprocket with a C-clamp and a wide pad on the belt.

➡ **When clamping the sprockets no deflection should be noticed. If any deflection is noticed, loosen the clamping devices. DO NOT mar the camshaft sprockets with the clamping device.**

6. Remove the tensioner side plate retaining bolts from the tensioner and remove the side plate from the actuator and base.
7. Rotate the actuator assembly around the arm pivot and out of the base. Removal of the tensioner from the base allows it to extend to its maximum travel.
8. Set the actuator aside on a table in a vertical position to allow the oil to drain into the boot end. The tensioner should be allowed to sit for 5 minutes prior to refilling with oil.
9. Reset the timing belt actuator as follows:
    a. Straighten out a paper clip or a piece of stiff wire 0.032 in. (0.75mm) diameter to a minimum straight length of 1.85 in. (47mm). Form a double loop in the remaining end.
    b. Remove the rubber end plug from the rear of the tensioner assembly. This will aid in allowing the oil in the tensioner to escape.
    c. Hold the tensioner in your hand with the rubber boot end of the tensioner pointing down.
    d. DO NOT remove the vent plug. Push the paper clip through the center hole in the vent plug and into the pilot hole.
    e. Insert a small screwdriver into the screw slot inside the end of the tensioner.
    f. Retract the tensioner by rotating the tensioner plunger in a clockwise direction while pushing the rod tip against a table top.
    g. Align the screw slot to align with the vent hole, and push the straight section of the wire into the screw slot to retain the plunger in the retracted position.
    h. If tensioner oil has been lost, fill the tensioner with SAE 5W30 Mobil 1®. Fill the tensioner to the bottom of the plug. The tensioner **MUST** be fully retracted before being filled with oil.

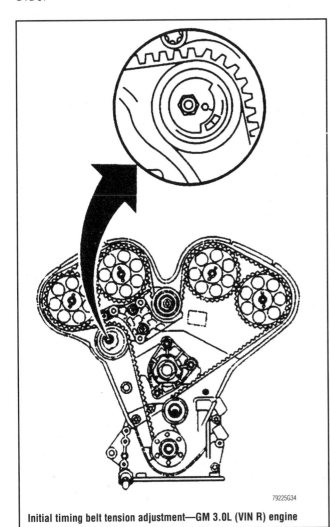

Initial timing belt tension adjustment—GM 3.0L (VIN R) engine

74 TIMING BELTS—GENERAL MOTORS CORP.

10. If the belt is being reused, mark the direction of rotation on the belt.
11. Remove the timing belt tensioner pulley mounting bolt and pulley.
12. Remove the timing belt after first removing the C-clamp retaining the belt to the right side camshaft sprocket.
13. Remove the Torx® head bolts securing the idler pulleys, if the idlers need to be replaced.
14. Remove the intermediate shaft sprocket using the following procedure:
    a. Use a suitable tool to hold the engine from turning.
    b. Remove the intermediate shaft sprocket mounting bolt and washer.
    c. Using J-38616, or an equivalent sprocket puller, remove the sprocket from the intermediate shaft.
15. If the camshaft sprockets need to be removed, proceed as follows:
    a. Remove the camshaft carrier cover (s).
    b. Remove the camshaft sprocket clamping pliers.
    c. Rotate the camshaft being serviced so the flats on the camshaft are face up.
    d. Install a camshaft hold-down tool, J-38613 or the equivalent, and tighten to 22 ft. lbs. (30 Nm).
    e. Remove the camshaft sprocket mounting bolt and washer while holding the camshaft from turning with J-38613 and J-38614, or the equivalent
    f. Using J-38616, or an equivalent sprocket puller remove the sprocket from the camshaft.
    g. Repeat for each camshaft sprocket as necessary.
16. Drain the cooling system.
17. Disconnect the lower radiator hose from water pump inlet pipe.
18. If equipped with a manual transaxle, disconnect the front AIR hose at the AIR pipe.
19. Disconnect the heater hose at the front cover.
20. Remove the heater pipe bracket mounting bolts at the frame.
21. Raise and safely support the vehicle on jack stands. Remove the right front tire and wheel assembly and the right inner fender splash-shield.
22. Remove the crankshaft pulley mounting bolts and remove the pulley from the damper.
23. Remove the crankshaft damper as follows:
    a. While holding the crankshaft from turning using a suitable tool, remove the damper mounting bolt and washer.
    b. Install tool J-24420-B, or the equivalent, and remove the damper from the crankshaft.
24. Place an oil catch pan under the oil filter and remove the oil filter.
25. Remove the A/C compressor mounting bracket bolts.
26. Remove the lower front cover bolts.
27. On automatic transaxle vehicles, remove the halfshaft following the recommended procedure.
28. Remove the rear alternator bracket.
29. Disconnect and remove the starter following the recommended procedure.
30. Lower the vehicle.
31. Remove the intermediate shaft drive belt sprocket retaining bolt and remove the intermediate shaft drive belt sprocket using J38616, or an equivalent puller.
32. Remove the upper alternator mounting bolts.
33. Remove the forward light relay center screws and position the relay center aside.
34. Disconnect the oil cooler hose from the front cover.
35. Remove the water pump pulley.
36. Remove the upper front cover bolts and remove the front cover.
37. Mark the intermediate shaft sprocket, chain link, crankshaft sprocket and cylinder block for assembly reference. The marks should be made with paint so they won't be lost when the components are removed.
38. Retract the timing chain tensioner shoe as follows:
    a. Insert J-33875 or an equivalent, on both sides of the tensioner.
    b. Pull on the through-pin in the tensioner arm to retract the spring located in the tensioner arm.
    c. While compressing the spring, use a suitable tool, a cotter pin or nail, and insert the pin in the hole in the tensioner assembly to hold the tensioner compressed. The tool used must be strong enough to hold the tensioner compressed.

➡ **The timing chain, crankshaft sprocket and intermediate shaft sprocket will be removed at the same time. If, when removing the assembly, the intermediate shaft sprocket does not easily come off the intermediate shaft, rotate the crankshaft back and forth to loosen the intermediate shaft sprocket.**

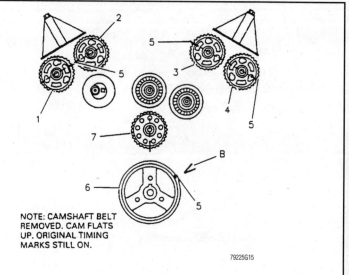

A LOCATION OF TIMING MARKS WITH CAM HOLD DOWN TOOLS J 38613 INSTALLED (#4 TDC COMPRESSION STROKE)
B FRONT COVER TIMING MARK
C LOCATION OF TIMING MARKS WITH DRIVE BELT INSTALLED
D LOCATION WHERE CAM HOLD DOWN TOOLS ARE INSTALLED
1 RH EXHAUST CAMSHAFT SPROCKET
2 RH INTAKE CAMSHAFT SPROCKET
3 LH INTAKE CAMSHAFT SPROCKET
4 LH EXHAUST CAMSHAFT SPROCKET
5 PERMANENT MARKS PAINTED DOTS REMOVE PREVIOUS MARKS IF TIMING IS BEING CHANGED AND MARKS AGAIN IN THESE LOCATIONS
6 CRANKSHAFT BALANCER
7 INTERMEDIATE SHAFT SPROCKET

NOTE: CAMSHAFT BELT REMOVED. CAM FLATS UP. ORIGINAL TIMING MARKS STILL ON.

Timing marks with hold-down tool in place—GM 3.4L (VIN X) engines

GENERAL MOTORS CORP.—TIMING BELTS

39. Install a suitable puller, J-38611 and J-8433 or their equivalents.
40. Tighten the bolt on the puller and slowly pull the crankshaft sprocket off the crankshaft. Be sure the intermediate shaft sprocket is moving along with the crankshaft sprocket.
41. Remove the timing chain and sprockets.
42. Remove the tensioner mounting bolts and remove the tensioner assembly.

To install:

43. Install the tensioner assembly and tensioner assembly mounting bolts finger-tight first. Tighten the bolt in the slotted hole first to 18 ft. lbs. (25 Nm), then tighten the remainder of the bolts to 18 ft. lbs. (25 Nm).
44. Check to ensure that the crankshaft key is fully seated in the crankshaft cutout and the tensioner assembly is fully retracted.
45. Assemble the timing chain, intermediate shaft sprocket and crankshaft sprocket on a work bench. The timing marks made should be in alignment. The large chamfer and counterbore of the crankshaft sprocket are installed facing toward the engine and the intermediate shaft spline sockets are installed facing away from the engine.
46. Install the sprocket and chain assembly onto the engine. As the sprockets are installed, parallel alignment must be maintained.
47. The crankshaft sprocket will have to be pressed on the final 0.31 in. (8mm). This can be done using J-38612, or an equivalent puller.
48. Ensure timing was maintained.
49. Remove the retaining pin from tensioner. Clean all gasket surfaces completely.
50. Apply GM Sealer 1052080 or equivalent, to the lower edges of the sealing surface of the front cover. Install a new gasket on the front cover.
51. Install the front cover on the engine. Apply thread sealant to the large bolts and tighten the bolts enough to pull the front cover against the engine block.
52. Install the water pump pulley.
53. Connect the oil cooler hose to the front cover.
54. Position the forward light relay center and install the mounting screws.
55. Install the upper alternator mounting bolt and tighten it to 22 ft. lbs. (30 Nm).
56. Install the intermediate shaft drive belt sprocket. The sprocket must lock into the intermediate shaft timing chain sprocket. Install the mounting bolt and washer. While holding the engine from turning, tighten the bolt to 95 ft. lbs. (130 Nm).
57. Raise and safely support vehicle on jack stands.
58. Connect and install the starter. Tighten the starter mounting bolts to 32 ft. lbs. (43 Nm).
59. Install the halfshaft following the recommended procedure.
60. Install the rear alternator bracket. Tighten the mounting bolt to 22 ft. lbs. (30 Nm) and the mounting stud to 41 ft. lbs. (55 Nm) and the lower bolt to 61 ft. lbs. (83 Nm).
61. Install the lower front cover bolts and tighten the small bolts to 18 ft. lbs. (25 Nm).
62. Install the A/C compressor mounting bolts and tighten the mounting bolts to 37 ft. lbs. (50 Nm).
63. Install a new oil filter.
64. Install the crankshaft damper as follows:
    a. Coat the seal contact area on the damper with clean engine oil.
    b. Line up the notch inside the damper with the crankshaft key and slide the damper on until the key is started into the notch.
    c. Using J-29113, or an equivalent puller, press the damper into position on the crankshaft.
    d. Install the crankshaft pulley and pulley mounting bolts. Tighten the pulley mounting bolts to 37 ft. lbs. (50 Nm).
    e. Install the crankshaft damper mounting bolt and washer, and tighten to 78 ft. lbs. (105 Nm).
65. Install the right side inner fender splash-shield.
66. Install the tire and wheel assembly.
67. Lower the vehicle.
68. Tighten the upper front cover small bolts to 18 ft. lbs. (25 Nm) and the large bolts to 35 ft. lbs. (47 Nm).
69. Connect the heater hose to the front cover.
70. Install the heater pipe bracket mounting bolts.
71. If equipped with a manual transaxle, connect the front AIR hose to the AIR pipe.
72. Connect the lower radiator hose to the water pump.
73. Install the right side cooling fan. Install the upper radiator support.
74. Position the torque strut mounting bracket and install the mounting bolts and tighten to 52 ft. lbs. (70 Nm).
75. Install the front engine lift hook and tighten the mounting bolt to 52 ft. lbs. (70 Nm).
76. To install the camshaft sprockets, proceed as follows:
    a. Wipe the camshaft noses with clean engine oil.
    b. Install the camshaft sprocket onto the nose of the camshaft.
    c. Install the lockring and shim ring.
    d. Install, but DO NOT tighten, the camshaft sprocket mounting bolts at this time.
77. Install the intermediate shaft sprocket as follows:
    a. Lubricate the seal contact area on the intermediate shaft sprocket with clean engine oil.
    b. Slide the sprocket through the intermediate shaft sprocket seal and engage the locking tangs into the sockets of the chain sprocket.
    c. Lightly lubricate the shaft seal and place it in position on the end of the intermediate shaft.
    d. Install the intermediate shaft sprocket mounting bolt and washer. Tighten the bolt to 96 ft. lbs. (130 Nm) while holding the crankshaft from turning.
78. Install the timing belt idler pulleys and tighten the Torx®bolts to 37 ft. lbs. (50 Nm).
79. Install the actuator assembly and side plate. Tighten the actuator mounting bracket bolts to 37 ft. lbs. (50 Nm).
80. Install the belt, taking note of direction of rotation if the old belt was used.
81. Install the tensioner pulley to the mounting base. Tighten the bolt to 37 ft. lbs. (50 Nm).
82. Rotate the tensioner pulley counterclockwise into the belt using the cast square lug on the body and engage the ball end of the actuator into the socket on the pulley arm.
83. Remove the tensioner lockpin allowing the tensioner shaft to extend and the pulley to move into the belt.
84. Rotate the tensioner pulley counterclockwise, applying 14 ft. lbs. (18 Nm) of torque.
85. Rotate the engine clockwise three times to seat the belt. Align the crankshaft reference marks during the final rotation to TDC. Do not allow the crankshaft to spring back or reverse its direction of rotation.

➡**The timing flats on the camshafts should be 180 degrees apart from the left side to the right side. Both camshafts on the same side should be the same.**

86. To perform the camshaft timing procedure, proceed as follows:

76 TIMING BELTS—GENERAL MOTORS CORP., HONDA

 a. Rotate the camshaft flats up on the right side camshafts and install a camshaft hold-down tool, J-38613 or the equivalent, and tighten to 22 ft. lbs. (30 Nm).
 b. Seat the lockring on the right exhaust and intake camshaft sprockets by threading in the mounting bolt and washer.
 c. Hold the sprocket from turning using tool J-38614, or equivalent.

➡ **Running torque of the bolts before seating should be 55 ft. lbs. (75 Nm).**

 d. If less torque is required, replace the shim ring and lockring.
 e. If more torque is required, replace the shim ring and lockring and inspect the bolts for burring.
 f. Seating of the lockring is accomplished when the edge is flush with the sprocket hub.
 g. With the lockring seated tighten the bolt to final torque of 81 ft. lbs. (110 Nm).
 h. Remove J-38613, or equivalent.
 i. Rotate the engine clockwise one full revolution, or any number of odd revolutions. DO NOT rotate the engine backward.
 j. Be sure the timing mark on the damper lines up with the mark on the front cover.
 k. Repeat Substeps a through i for the left side.

87. Install the camshaft carrier covers.
88. Install timing the belt left, right and center covers and retaining bolts.
89. Connect the negative battery cable.
90. Start the engine and verify proper operation and engine performance.

Honda

1.5L & 1.6L ENGINES

1990–94 Models

1. Disconnect the negative battery cable.
2. Bring the piston in No. 1 cylinder to TDC on the compression stroke.
3. Raise and safely support the vehicle.

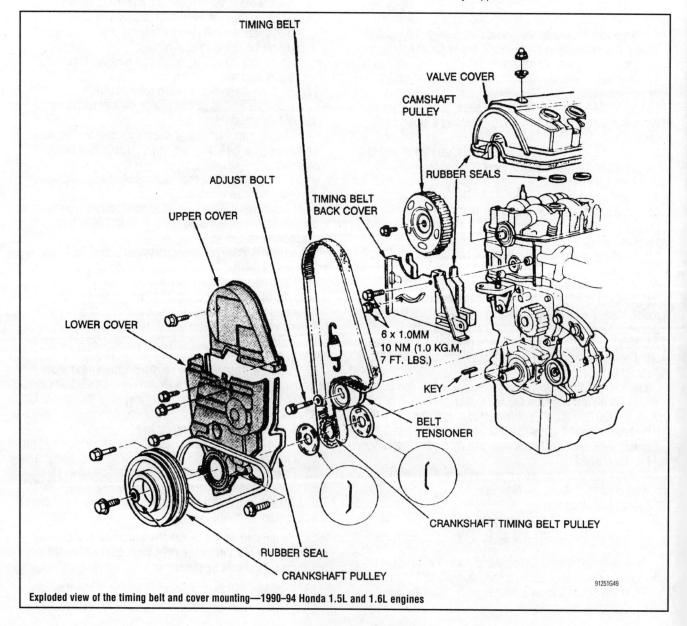

Exploded view of the timing belt and cover mounting—1990–94 Honda 1.5L and 1.6L engines

HONDA—TIMING BELTS

4. Remove the left front wheel and tire assembly.
5. Remove the left front wheelwell splash shield.
6. If equipped, remove the power steering belt and pump.
7. If equipped with air conditioning, remove the adjust pulley with bracket and the belt.
8. Remove the power steering bracket, loosen the alternator adjust bolt and through bolt and remove the alternator belt.
9. Use a suitable device to support the engine. Remove the engine side support bolts and nut and then remove the side mount rubber.
10. Remove the valve cover, the crankshaft pulley bolt and the crankshaft pulley.
11. Remove the timing belt upper and lower cover.
12. Mark the direction of timing belt rotation. On 1990–92 Accord and 1992–94 Prelude, mark the direction of timing balancer belt rotation.
13. On all except Accord and 1992–94 Prelude, loosen the adjusting bolt and remove the timing belt. On Accord and 1992–94 Prelude, push the timing balancer belt tensioner and the timing belt tensioner to remove tension on the belts, then reinstall and tighten the adjusting nut. Remove the timing balancer belt and the timing belt.

To install:

14. Align the camshaft sprocket(s) and crankshaft sprocket as follows:

    a. Make sure the **UP** mark on the camshaft sprocket(s) is at the top most position. The sprocket timing marks should be aligned with the cylinder head upper surface.

    b. On Accord and Prelude, remove the timing inspection hole cover at the rear of the engine block. Make sure the TDC mark on the flywheel, indicated by a white painted mark, is aligned with the pointer in the inspection hole.

    c. On Civic, Civic del Sol and CRX, temporarily reinstall the lower timing belt cover and crankshaft pulley. Make sure the TDC mark on the crankshaft pulley, indicated by a white painted mark, is aligned with the pointer on the timing cover. Remove the lower timing belt cover and crankshaft pulley.

15. Install the timing belt. If the old timing belt is reused, install the belt in the same rotational direction, as indicated by the mark that was made during removal.

16. On 1990–92 Accord and 1992–94 Prelude, align the timing belt balancer pulleys and install the balancer belt as follows:

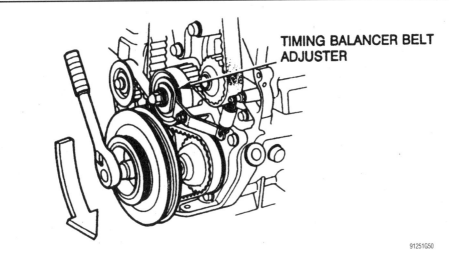

Timing balancer belt adjustment—1992–94 Honda Prelude

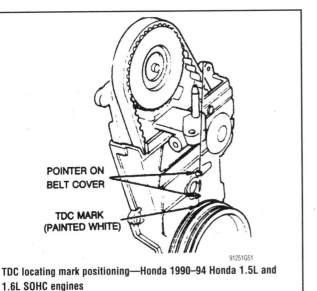

TDC locating mark positioning—Honda 1990–94 Honda 1.5L and 1.6L SOHC engines

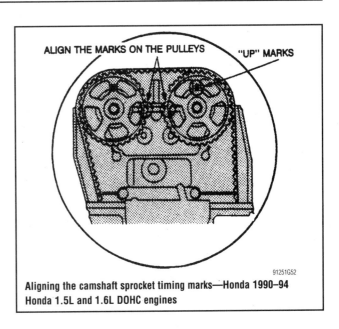

Aligning the camshaft sprocket timing marks—Honda 1990–94 Honda 1.5L and 1.6L DOHC engines

78 TIMING BELTS—HONDA

a. The timing belt balancer drive pulley should already be at TDC, if the timing belt is installed correctly.

b. Align the groove on the front timing balancer belt driven pulley with the pointer on the oil pump body.

c. Remove the bolt from the maintenance hole on the cylinder block, next to the rear balancer shaft. Align the rear timing balancer belt driven pulley using a 6 x 100mm bolt, or equivalent. Mark a line at 3 in. (74mm) length of the bolt. Align the pulley by inserting the bolt through the maintenance hole. The bolt must be inserted to a depth where the line is flush with the maintenance hole surface.

d. Install the timing balancer belt. If the old balancer belt is reused, install the belt in the same rotational direction, as indicated by the mark that was made during removal. After the balancer belt is installed, remove the rear balancer belt driven pulley alignment bolt and reinstall the original bolt in the maintenance hole. Tighten the bolt to 22 ft. lbs. (30 Nm).

17. Installation of the remaining components is the reverse of the removal procedure. Be sure to properly adjust the timing belt tension. On 1990–92 Accord, the balancer belt tension is automatically adjusted when the timing belt tension is adjusted.

18. Tighten the crankshaft pulley bolt as follows:
- 1990–91 Prelude and 1989 Accord—108 ft. lbs. (150 Nm)
- 1990–92 Civic and CRX—119 ft. lbs. (165 Nm)
- 1990–92 Accord and 1992–94 Prelude—159 ft. lbs. (220 Nm)

1995–97 Models

1. Rotate the crankshaft to set the engine at Top Dead Center (TDC) on the compression stroke for the No. 1 piston. The white mark on the crankshaft pulley should align with the pointers on the timing cover. Once the engine is in this position, it must not be disturbed.

2. Remove all necessary components for access to the cylinder head and timing belt covers. Cover the rocker arm and shaft assemblies with a towel or sheet of plastic to keep out dust and foreign objects.

3. Remove the timing belt covers.

4. Loosen the timing belt adjusting bolt 180 degrees (½ turn). Push the tensioner pulley down to release the belt tension. After releasing the tension, retighten the tensioner pulley bolt until snug.

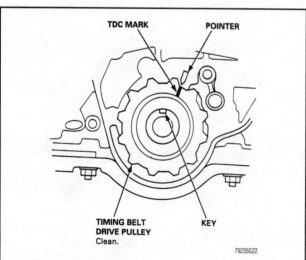

TDC alignment mark locations for the crankshaft sprocket—1995–97 Honda 1.5L and 1.6L engines

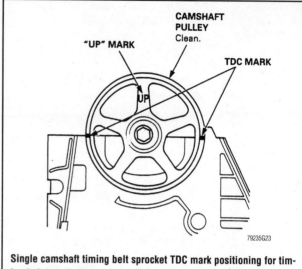

Single camshaft timing belt sprocket TDC mark positioning for timing belt installation—1995–97 Honda 1.5L and 1.6L engines

➡Do not remove the tensioner pulley unless it is to be replaced.

5. Remove the timing belt. Mark the direction of the belt's rotation if it is to be reinstalled.

To install:

➡Inspect the water pump when replacing the timing belt; the manufacturer recommends replacing the water pump at the timing belt's service interval. Replace the timing belt if it shows any signs of wear, or if it is contaminated with oil or coolant.

6. Verify that the timing is set at TDC on the compression stroke for the No. 1 cylinder.
- The groove in the crankshaft sprocket must line up with the pointer on the oil pump.
- On 1.6L (B16A2 and B16A3) engines, the TDC marks on the camshaft sprockets must line up with the pointer located between the sprockets. The TDC marks will also be in line with the upper surface of the head.
- On other engines, the TDC mark on the camshaft sprocket must line up with the pointer on the back cover.
- The **UP** mark on the camshaft sprocket must point up.

7. Install the timing belt onto the crankshaft sprocket, then around the adjusting pulley and water pump sprocket, and finally over the camshaft sprocket.

8. Loosen the adjusting pulley bolt 180 degrees (½ turn). Then, tighten the adjusting bolt to 40 ft. lbs. (55 Nm) on 1.6L (B16A2 and B16A3) engines or to 33 ft. lbs. (45 Nm) on all other engines.

9. Install the lower timing belt cover and the crankshaft pulley. Apply a light coat of fresh oil to the pulley bolt threads, then tighten it to 134 ft. lbs. (181 Nm).

10. Rotate the crankshaft five or six turns counterclockwise to position the belt on the sprockets.

11. Adjust the timing belt tension, as follows:

a. Set the No. 1 piston at TDC on the compression stroke for the No. 1 cylinder.

b. Loosen the adjusting pulley bolt 180 degrees (½ turn).

c. Rotate the crankshaft counterclockwise so that the camshaft sprocket moves three teeth from the TDC/compression position.

HONDA—TIMING BELTS

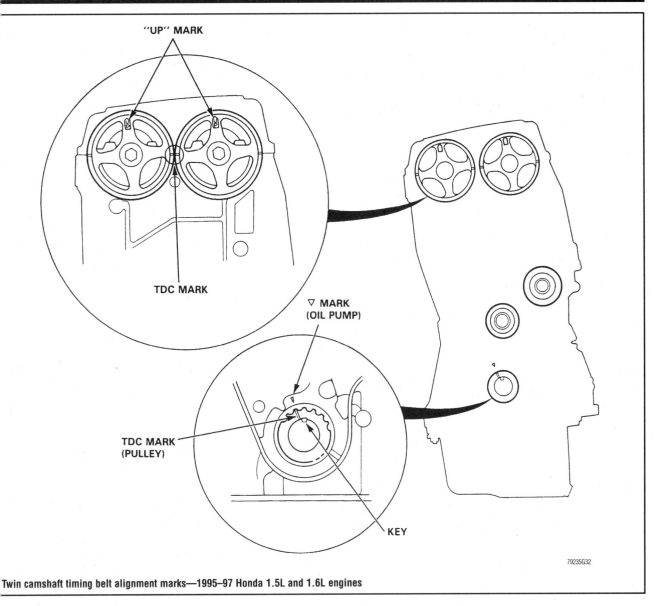

Twin camshaft timing belt alignment marks—1995–97 Honda 1.5L and 1.6L engines

d. Tighten the adjusting bolt to 33 ft. lbs. (45 Nm).
e. Tighten the crankshaft pulley to 134 ft. lbs. (181 Nm).

12. Verify that the crankshaft and camshaft sprockets will align properly at the TDC/compression position. If the camshaft pulley is not at TDC/compression, remove the timing belt, adjust the sprocket positions, and reinstall the belt.

13. Install the upper timing and cylinder head covers, and all other applicable components. When reattaching the side engine mount, tighten the support nuts to 54 ft. lbs. (75 Nm).

2.0L (B20A5), 2.1L (B21A1), 2.2L (F22B4) & 2.3L (H23A1) ENGINES

1990–94 Models

1. Disconnect the negative battery cable.
2. Bring the piston in No. 1 cylinder to TDC on the compression stroke.
3. Raise and safely support the vehicle.
4. Remove the left front wheel and tire assembly.
5. Remove the left front wheelwell splash shield.
6. If equipped, remove the power steering belt and pump.
7. If equipped with air conditioning, remove the adjust pulley with bracket and the belt.
8. Remove the power steering bracket, loosen the alternator adjust bolt and through bolt and remove the alternator belt.
9. Use a suitable device to support the engine. Remove the engine side support bolts and nut and then remove the side mount rubber.
10. Remove the valve cover, the crankshaft pulley bolt and the crankshaft pulley.
11. Remove the timing belt upper and lower cover.
12. Mark the direction of timing belt rotation. On 1990–92 Accord and 1992–94 Prelude, mark the direction of timing balancer belt rotation.
13. On all except Accord and 1992–94 Prelude, loosen the adjusting bolt and remove the timing belt. On Accord and 1992–94 Prelude, push the timing balancer belt tensioner and the timing belt tensioner to remove tension on the belts, then reinstall and tighten the adjusting nut. Remove the timing balancer belt and the timing belt.

To install:

14. Align the camshaft sprocket(s) and crankshaft sprocket as follows:

80 TIMING BELTS—HONDA

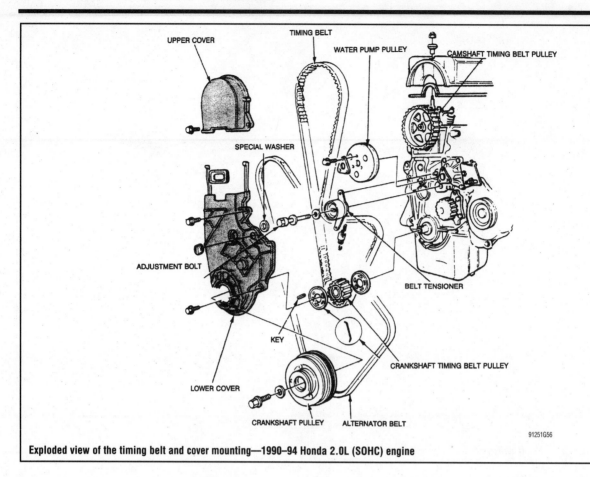

Exploded view of the timing belt and cover mounting—1990–94 Honda 2.0L (SOHC) engine

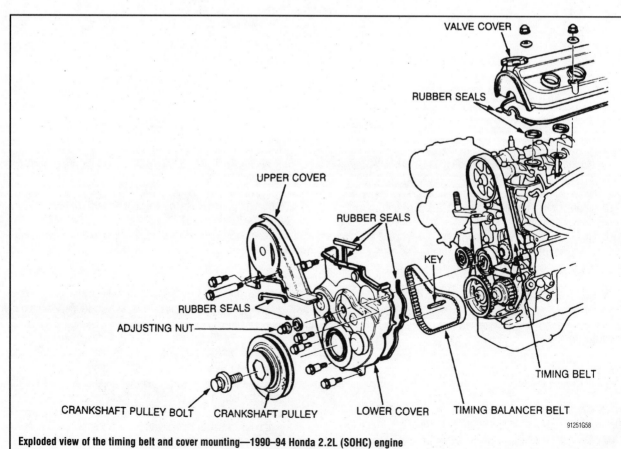

Exploded view of the timing belt and cover mounting—1990–94 Honda 2.2L (SOHC) engine

HONDA—TIMING BELTS

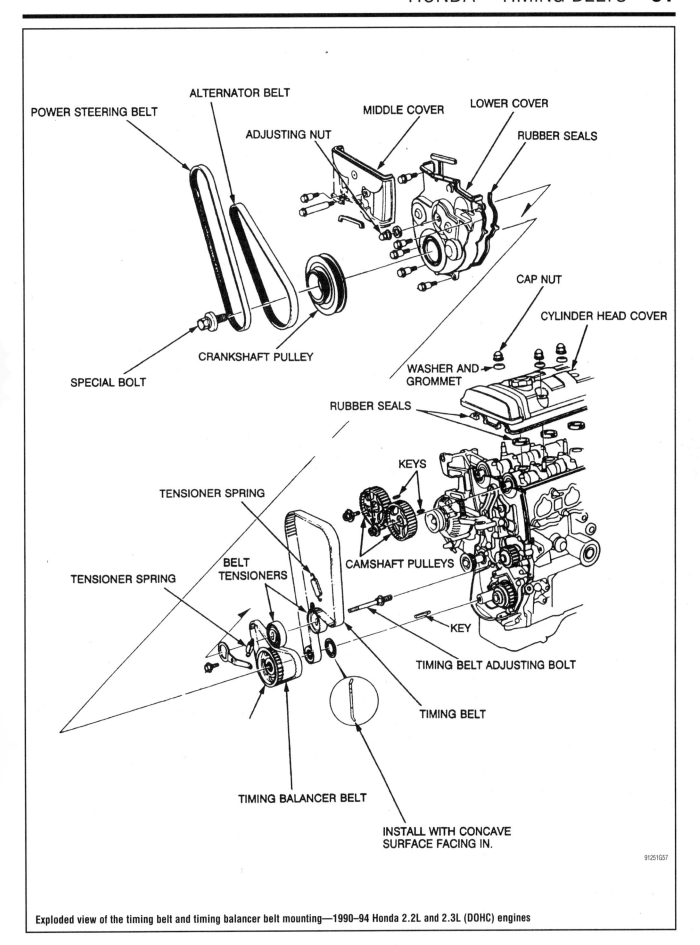

Exploded view of the timing belt and timing balancer belt mounting—1990–94 Honda 2.2L and 2.3L (DOHC) engines

82 TIMING BELTS—HONDA

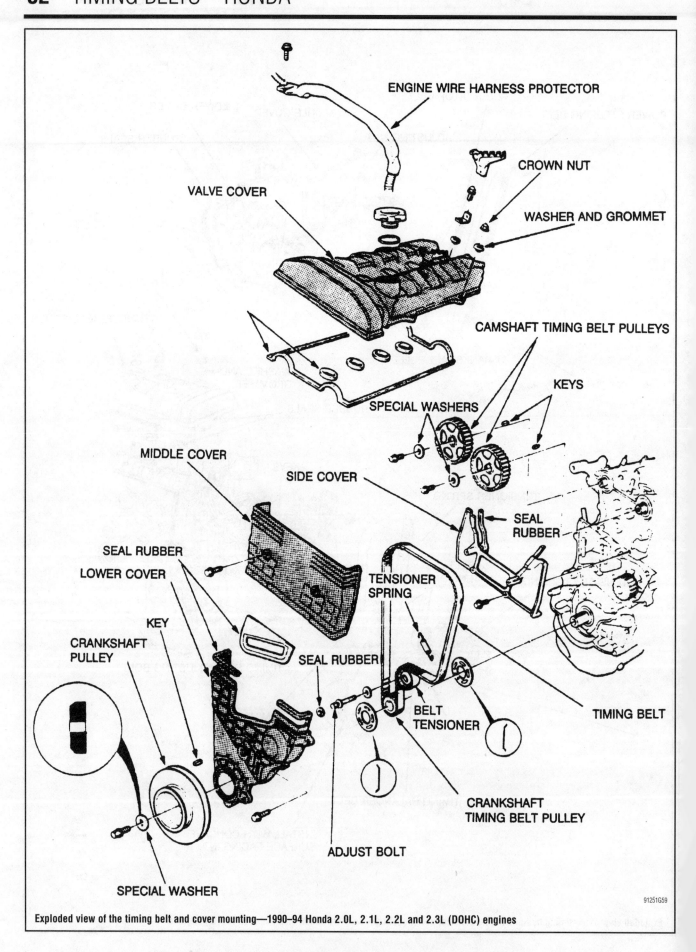

Exploded view of the timing belt and cover mounting—1990–94 Honda 2.0L, 2.1L, 2.2L and 2.3L (DOHC) engines

HONDA—TIMING BELTS

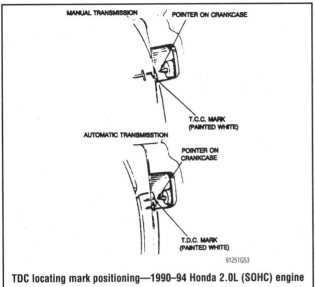

TDC locating mark positioning—1990–94 Honda 2.0L (SOHC) engine

a. Make sure the **UP** mark on the camshaft sprocket(s) is at the top most position. The sprocket timing marks should be aligned with the cylinder head upper surface.

b. On Accord and Prelude, remove the timing inspection hole cover at the rear of the engine block. Make sure the TDC mark on the flywheel, indicated by a white painted mark, is aligned with the pointer in the inspection hole.

c. On Civic, Civic del Sol and CRX, temporarily reinstall the lower timing belt cover and crankshaft pulley. Make sure the TDC mark on the crankshaft pulley, indicated by a white painted mark, is aligned with the pointer on the timing cover. Remove the lower timing belt cover and crankshaft pulley.

15. Install the timing belt. If the old timing belt is reused, install the belt in the same rotational direction, as indicated by the mark that was made during removal.

16. On 1990–92 Accord and 1992–94 Prelude, align the timing belt balancer pulleys and install the balancer belt as follows:

a. The timing belt balancer drive pulley should already be at TDC, if the timing belt is installed correctly.

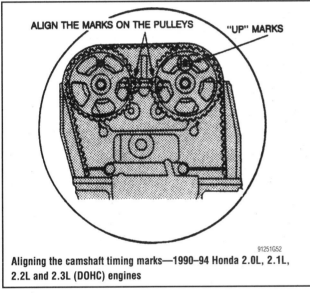

Aligning the camshaft timing marks—1990–94 Honda 2.0L, 2.1L, 2.2L and 2.3L (DOHC) engines

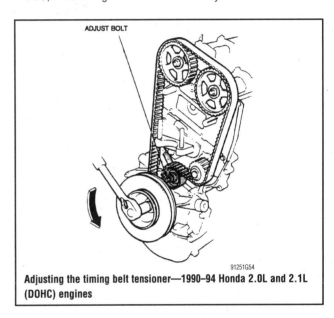

Adjusting the timing belt tensioner—1990–94 Honda 2.0L and 2.1L (DOHC) engines

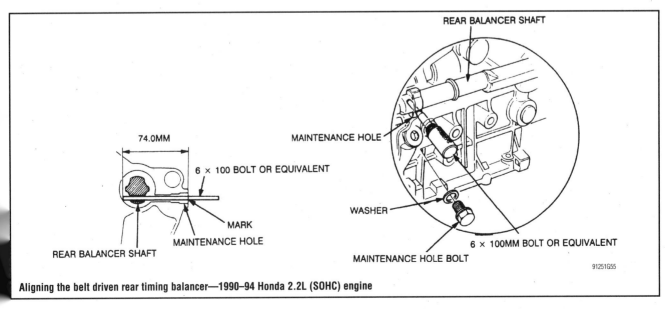

Aligning the belt driven rear timing balancer—1990–94 Honda 2.2L (SOHC) engine

84 TIMING BELTS—HONDA

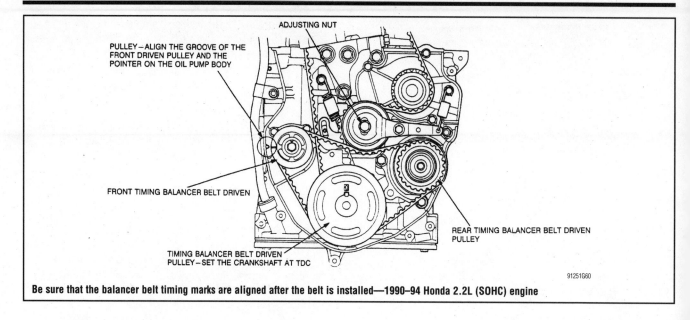

Be sure that the balancer belt timing marks are aligned after the belt is installed—1990–94 Honda 2.2L (SOHC) engine

b. Align the groove on the front timing balancer belt driven pulley with the pointer on the oil pump body.

c. Remove the bolt from the maintenance hole on the cylinder block, next to the rear balancer shaft. Align the rear timing balancer belt driven pulley using a 6 x 100mm bolt, or equivalent. Mark a line at 3 in. (74mm) length of the bolt. Align the pulley by inserting the bolt through the maintenance hole. The bolt must be inserted to a depth where the line is flush with the maintenance hole surface.

d. Install the timing balancer belt. If the old balancer belt is reused, install the belt in the same rotational direction, as indicated by the mark that was made during removal. After the balancer belt is installed, remove the rear balancer belt driven pulley alignment bolt and reinstall the original bolt in the maintenance hole. Tighten the bolt to 22 ft. lbs. (30 Nm).

17. Installation of the remaining components is the reverse of the removal procedure. Be sure to properly adjust the timing belt tension. On 1990–92 Accord, the balancer belt tension is automatically adjusted when the timing belt tension is adjusted.

18. Tighten the crankshaft pulley bolt as follows:
- 1990–91 Prelude and 1989 Accord—108 ft. lbs. (150 Nm)
- 1990–92 Accord and 1992–94 Prelude—159 ft. lbs. (220 Nm)

2.0L (B20B4) ENGINE

1. Disconnect the negative battery cable.
2. Position crankshaft so that No. 1 piston is at TDC.
3. Remove the splash guard.
4. Remove the accessory drive belts.
5. If equipped, remove the cruise control actuator.
6. Place a piece of wood between the oil pan and the jack, support the engine with a jack.
7. Remove upper engine bracket.
8. Remove the valve cover.
9. Remove the timing belt covers.
10. Loosen the adjusting bolt 180°. Release the tension from the belt by pushing on the tensioner, then retighting the adjusting bolt.
11. Remove the belt.

To install:

12. Be sure the timing marks are properly aligned.

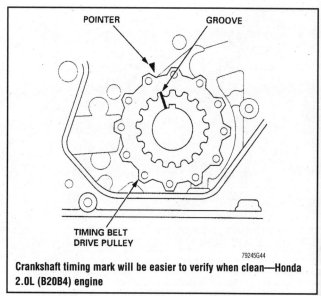

Crankshaft timing mark will be easier to verify when clean—Honda 2.0L (B20B4) engine

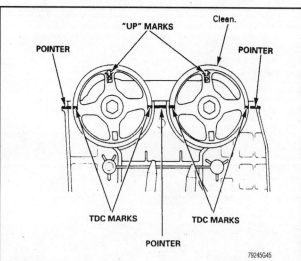

Intake and exhaust camshaft timing marks properly aligned at TDC—Honda 2.0L (B20B4) engine

HONDA—TIMING BELTS

13. Install the timing belt on the pulleys following this sequence:
    a. Crankshaft pulley.
    b. Adjusting pulley.
    c. Water pump pulley.
    d. Exhaust camshaft pulley.
    e. Intake camshaft pulley.
14. Loosen and retighten the adjusting bolt to allow tension to be applied to the belt.
15. Install the lower and middle timing covers.
16. Install the crankshaft pulley and tighten the bolt to 130 ft. lbs. (177 Nm).

✳✳ WARNING

If any binding is felt when adjusting the timing belt tension by turning the crankshaft, STOP turning the engine, because the pistons may be hitting the valves.

17. Rotate the crankshaft about five or six times counterclockwise to seat the timing belt.
18. Position the No. 1 piston to TDC.
19. Loosen the adjusting bolt ½ turn.
20. Rotate the crankshaft counterclockwise three teeth on the camshaft pulley.
21. Tighten the adjusting bolt to 40 ft. lbs. (54Nm).
22. Retighten the crankshaft pulley bolt to 130 ft. lbs. (177 Nm).
23. Install valve cover.
24. Install the engine mounting bracket, then remove the jack.
25. If removed, install the cruise control actuator.
26. Install the accessory drive belts.
27. Install the splash guard.
28. Connect the negative battery cable.
29. Check the engine operation and road test.

2.2L (F22A1) ENGINES

1. Turn the crankshaft to align the timing belt matchmarks and set cylinder No.1 to Top Dead Center on the compression stroke. Once in this position, the engine must NOT be turned or disturbed.
2. Remove all necessary components to gain access to the cylinder head and timing belt covers.
3. Remove the cylinder head and timing belt covers.

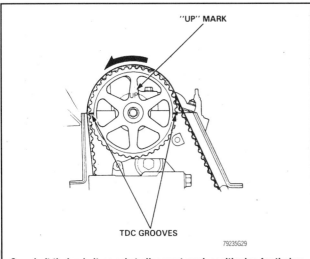

Camshaft timing belt sprocket alignment mark positioning for timing belt installation—Honda 2.2L (F22A1) engines

4. There are two belts in this system; the one running to the camshaft pulley is the timing belt. The other, shorter one drives the balance shafts and is referred to as the balancer belt or timing balancer belt. Lock the timing belt adjuster in position by installing one of the lower timing belt cover bolts to the adjuster arm.
5. Loosen the timing belt and balancer shafts tensioner adjuster nut, but do not loosen the nut more than one turn. Push the tensioner for the balancer belt away from the belt to relieve the tension. Hold the tensioner and tighten the adjusting nut to hold the tensioner in place.
6. Carefully remove the balancer belt. Do not crimp or bend the belt; protect it from contact with oil or coolant. Slide the belt off the pulleys.
7. Remove the balancer belt drive sprocket from the crankshaft.
8. Loosen the lockbolt installed to the timing belt adjuster and loosen the adjusting nut. Push the timing belt adjuster to remove the tension on the timing belt, then tighten the adjuster nut.
9. Remove the timing belt. Do not crimp or bend the belt; protect it from contact with oil or coolant. Slide the belt off the pulleys.
10. If defective, remove the belt tensioners by performing the following:
    a. Remove the springs from the balancer belt and the timing belt tensioners.
    b. Remove the adjusting nut.
    c. Remove the bolt from the balancer belt adjuster lever, then remove the lever and the tensioner pulley.
    d. Remove the lockbolt from the timing belt tensioner lever, then remove the tensioner pulley and lever from the engine.
11. This is an excellent time to check or replace the water pump. Even if the timing belt is only being replaced as part of a good maintenance schedule, consider replacing the pump at the same time.

To install:

12. If the water pump is to be replaced, install a new O-ring and make certain it is properly seated. Install the water pump and retaining bolts. Tighten the mounting bolts to 106 inch lbs. (12 Nm).
13. If the tensioners were removed perform the following to install them:
    a. Install the timing belt tensioner lever and tensioner pulley.

➡**The tensioner lever must be properly positioned on it's pivot pin located on the oil pump. Be sure that the timing belt lever and tensioner moves freely and does not bind.**

    b. Install the lockbolt to the timing belt tensioner, do not tighten the lockbolt at this time.
    c. Install the balancer belt pulley and adjuster lever.
    d. Install the adjusting nut and the bolt to the balancer belt adjuster lever. Do not tighten the adjuster nut or bolt at this time.

➡**Be sure that the balancer lever and tensioner moves freely and does not bind.**

    e. Install the springs to the tensioners.
    f. Move the timing belt tensioner its full deflection and tighten the lockbolt.
    g. Move the balancer its full deflection and tighten the adjusting nut.
14. The crankshaft timing pointer must be perfectly aligned with the white mark on the flywheel or flex-plate; the camshaft pulley must be aligned so that the word UP is at the top of the pulley and the marks on the edge of the pulley are aligned with the surfaces of the head.
15. Install the timing belt over the pulleys and tensioners.

86 TIMING BELTS—HONDA

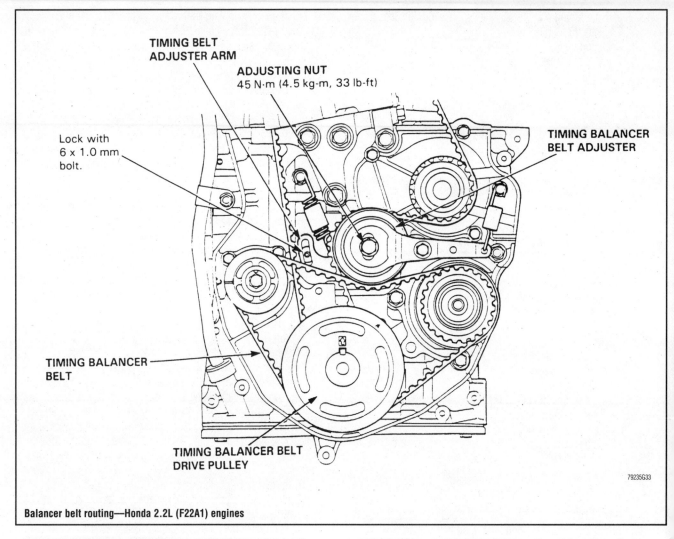

Balancer belt routing—Honda 2.2L (F22A1) engines

16. Loosen the bolt used to lock the timing belt tensioner. Loosen, then tighten the timing belt adjusting nut.

17. Turn the crankshaft counterclockwise until the cam pulley has moved 3 teeth; this creates tension on the timing belt. Loosen, then tighten the adjusting nut and tighten it to 33 ft. lbs. (45 Nm). Tighten the bolt used to lock the timing belt tensioner.

18. Realign the timing belt marks, then install the balancer belt drive sprocket on the crankshaft.

19. Align the front balancer pulley; the face of the front timing balancer pulley has a mark which must be aligned with the notch on the oil pump body. This pulley is the one at 10 o'clock to the crank pulley when viewed from the pulley end.

20. Align the rear timing balancer pulley (2 o'clock from the crank pulley) using a 6 x 100mm bolt or rod. Mark the bolt or rod at a point 2.9 in. (74mm) from the end. Remove the bolt from the maintenance hole on the side of the block; insert the bolt or rod into the hole. Align the 2.9 in. (74mm) mark with the face of the hole. This pin will hold the shaft in place during installation.

21. Install the balancer belt. Once the belts are in place, be sure that all the engine alignment marks are still correct. If not, remove the belts, realign the engine and reinstall the belts. Once the belts are properly installed, slowly loosen the adjusting nut, allowing the tensioner to move against the belt. Remove the pin from the maintenance hole and reinstall the bolt and washer.

22. Turn the crankshaft one full turn, then tighten the adjuster nut to 33 ft. lbs. (45 Nm). Remove the bolt used to lock the timing belt tensioner.

23. Install the lower cover, ensuring the rubber seals are in place. Install a new seal around the adjusting nut, DO NOT loosen the adjusting nut.

24. Install the key on the crankshaft and install the crankshaft pulley. Apply oil to the bolt threads and tighten it to 181 ft. lbs. (250 Nm).

25. Install the upper timing belt cover and all applicable components. When installing the side engine mount, tighten the bolt and nut attaching the mount to the engine to 40 ft. lbs. (55 Nm) and the through-bolt and nut to 47 ft. lbs. (65 Nm).

2.2L (F22B1 & F22B2) ENGINES

1. Remove the cylinder head (valve) and upper timing belt covers.

2. Turn the engine to align the timing marks and set cylinder No.1 to TDC. The white mark on the crankshaft sprocket should align with the pointer on the timing belt cover. The words **UP** embossed on the camshaft sprocket should be aligned in the upward position. The marks on the edge of the sprocket should be aligned with the cylinder head or the back cover upper edge. Once in this position, the engine must NOT be turned or disturbed.

HONDA—TIMING BELTS

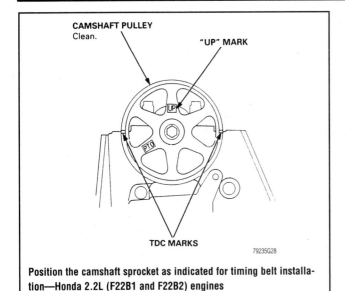

Position the camshaft sprocket as indicated for timing belt installation—Honda 2.2L (F22B1 and F22B2) engines

3. Remove all necessary components for access to the lower timing belt cover, then remove the cover.
4. There are two belts in this system; the one running to the camshaft sprocket is the timing belt. The other, shorter one drives the balance shaft and is referred to as the balancer shaft belt or timing balancer belt. Lock the timing belt adjuster in position, by installing one of the lower timing belt cover bolts to the adjuster arm.
5. Loosen the timing belt and balancer shafts tensioner adjuster nut, do not loosen the nut more than one turn. Push the tensioner for the balancer belt away from the belt to relieve the tension. Hold the tensioner and tighten the adjusting nut to hold the tensioner in place.
6. Carefully remove the balancer belt. Do not crimp or bend the belt; protect it from contact with oil or coolant.
7. Remove the balancer belt sprocket from the crankshaft.
8. Loosen the lockbolt installed to the timing belt adjuster and loosen the adjusting nut. Push the timing belt adjuster to remove the tension on the timing belt, then tighten the adjuster nut.
9. Remove the timing belt by sliding it off the sprockets. Do not crimp or bend the belt; protect it from contact with oil or coolant.

10. If defective, remove the belt tensioners by performing the following:
   a. Remove the springs from the balancer belt and the timing belt tensioners.
   b. Remove the adjusting nut from the belt tensioners.
   c. Remove the bolt from the balancer belt adjuster lever, then remove the lever and the tensioner pulley.
   d. Remove the lockbolt from the timing belt tensioner lever, then remove the tensioner pulley and lever from the engine.
11. This is an excellent time to check or replace the water pump. Even if the timing belt is only being replaced as part of a good maintenance schedule, consider replacing the pump at the same time.

To install:
12. If the water pump is to be replaced, install a new O-ring and make certain it is properly seated. Install the water pump and tighten the mounting bolts to 106 inch lbs. (12 Nm).
13. If the tensioners were removed perform the following to install them:
   a. Install the timing belt tensioner lever and the tensioner pulley.
   b. Install the balancer belt pulley and adjuster lever.
   c. Install the adjusting nut and the bolt to the balancer belt adjuster lever.
   d. Install the springs to the tensioners.
   e. Install the lockbolt to the timing belt tensioner, then move it it's full deflection and tighten the lockbolt.
   f. Move the balancer it's full deflection and tighten the adjusting nut to hold it's position.
14. The pointer on the crankshaft sprocket should be aligned with the pointer on the oil pump; the camshaft sprocket must be aligned so that the word UP is at the top of the sprocket and the marks on the edge of the sprocket are aligned with the surfaces of the head or the back cover upper edge.
15. Install the timing belt on the sprockets in the following sequence: crankshaft sprocket, tensioner sprocket, water pump sprocket, camshaft sprocket.
16. Check the timing marks to be sure that they did not move.
17. Loosen, then retighten the timing belt adjusting nut; this will apply the proper amount of tension to the timing belt.
18. Install the timing balancer belt drive sprocket and the lower timing belt cover.
19. Install the crankshaft pulley and bolt, tighten the bolt to 181 ft. lbs. (245 Nm). Rotate the crankshaft sprocket five or six turns to position the timing belt on the sprockets.
20. Set the No. 1 cylinder to TDC and loosen the timing belt adjusting nut one turn. Turn the crankshaft counterclockwise until the cam sprocket has moved 3 teeth; this creates tension on the timing belt.
21. Tighten the timing belt adjusting nut.
22. Set the crankshaft sprocket and the camshaft sprocket to TDC. If the sprockets do not align, remove the belt to realign the marks, then install the belt.
23. Remove the crankshaft pulley and the lower cover.
24. With the timing marks aligned, lock the timing belt adjuster in place with one of the lower cover mounting bolts.
25. Loosen the adjusting nut and ensure the timing balancer belt adjuster moves freely.
26. Align the rear timing balancer sprocket using a 6 x 100mm bolt or rod. Mark the bolt or rod at a point 2.9 in. (74mm) from the end. Remove the bolt from the maintenance hole on the side of the block; insert the bolt/rod into the hole and align the 2.9 in. (74mm) mark with the face of the hole. This will hold the shaft in place during installation.

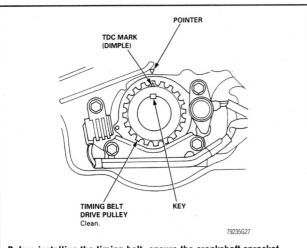

Before installing the timing belt, ensure the crankshaft sprocket marks are properly aligned—Honda 2.2L (F22B1 and F22B2) engines

TIMING BELTS—HONDA

27. Align the groove on the front balancer shaft sprocket with the pointer on the oil pump.
28. Install the balancer belt. Once the belts are in place, be sure that all the engine alignment marks are still correct. If not, remove the belts, realign the engine and reinstall the belts. Once the belts are properly installed, slowly loosen the adjusting nut, allowing the tensioner to move against the belt. Remove the bolt from the maintenance hole and reinstall the bolt and washer.
29. Install the crankshaft pulley, then turn the crankshaft sprocket 1 turn counterclockwise and tighten the timing belt adjusting nut to 33 ft. lbs. (45 Nm).
30. Remove the crankshaft pulley and the bolt locking the timing belt adjuster in place.
31. Install the lower and upper timing belt covers, and all applicable components. When installing the crankshaft pulley, coat the threads and seating face of the pulley bolt with engine oil, then install and tighten the bolt to 181 ft. lbs. (250 Nm).
32. Install the cylinder head cover gasket cover to the groove of the cylinder head cover. Before installing the gasket thoroughly clean the seal and the groove. Seat the recesses for the camshaft first, then work it into the groove around the outside edges. Be sure the gasket is seated securely in the corners of the recesses.
33. Apply liquid gasket to the four corners of the recesses of the cylinder head cover gasket. Do not install the parts if 5 minutes or more have elapsed since applying liquid gasket. After assembly, wait at least 20 minutes before filling the engine with oil.
34. Install the cylinder head (valve) cover and all other applicable components.

2.2L (F22B6) & 2.3L (F23A7) ENGINES

➡ The radio may contain a coded theft protection circuit. Always make note of your code number before disconnecting the battery.

1. Disconnect the negative and positive battery cables.
2. Remove the valve cover.
3. Remove the upper timing belt cover.
4. Turn the engine to align the timing marks and set cylinder No.1 to TDC for the compression stroke. The white mark on the crankshaft pulley should align with the pointer on the timing belt cover. The words **UP** embossed on the camshaft pulley should be aligned in the upward position. The marks on the edge of the pulley should be aligned with the cylinder head or the back cover upper edge. Once in this position, the engine must NOT be turned or disturbed.
5. Remove the splash shield from below the engine.
6. Remove the wheel well splash shield.
7. Loosen and remove the power steering pump belt. Remove the power steering pump.
8. Loosen the adjusting and mounting bolts for the alternator and remove the drive belt.
9. Support the engine with a floor jack cushioned with a piece of wood.
10. Remove the through-bolt for the side engine mount and remove the mount.
11. Remove the crankshaft pulley bolt and remove the crankshaft pulley. Use a crank pulley holder (part No. 07MAB-PY3010A) and holder handle (part No. 07JAB-001020A),or there equivalents, to hold the crankshaft pulley in place while removing the bolt.
12. Remove the lower timing belt cover.
13. Remove the balancer shaft belt and its drive pulley.
14. Insert a suitable tool into the maintenance hole in the front balancer shaft. Unbolt and remove the balancer driven pulley.

➡ For servicing the balance shafts, front refers to the side of the engine facing the radiator. Rear refers to the side of the engine facing the firewall.

15. Remove the timing belt.
16. If equipped with a TDC sensor assembly at the crankshaft sprocket, unbolt the assembly and move it to the side before removing the sprocket.
17. Remove the key and the spacers to remove the crankshaft timing sprocket.
18. Unbolt and remove the camshaft timing sprocket.

To install:

19. Install the camshaft timing sprocket so that the **UP** mark is up and the TDC marks are parallel to the cylinder head gasket surface. Install the key and tighten the bolt to 27 ft. lbs. (37 Nm).
20. Install the crankshaft timing sprocket so that the TDC mark aligns with the pointer on the oil pump. Install the spacers with their

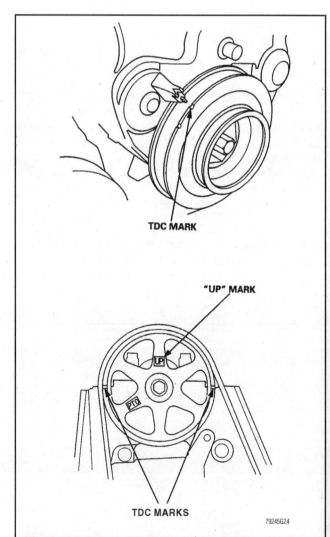

Align the camshaft, crankshaft and engine marks before removing the timing belt and pulleys—Honda 2.2L (F22B6) and 2.3L (F23A7) engines

HONDA—TIMING BELTS

concave surfaces facing in. Install the key. Install the TDC sensor assembly back into position before installing the timing belt.

21. Install and tension the timing belt.
22. Rotate the crankshaft counterclockwise five to six turns to be sure the belt is properly seated.
23. Set the No. 1 piston at TDC for its compression stroke.

※※ WARNING

If any binding is felt when adjusting the timing belt tension by turning the crankshaft, STOP turning the engine, because the pistons may be hitting the valves.

24. Rotate the crankshaft counterclockwise so that the camshaft pulley moves only three teeth beyond its TDC mark.
25. Tighten the tensioner adjusting nut to 33 ft. lbs. (45 Nm).
26. Tighten the crankshaft pulley bolt to 181 ft. lbs. (245 Nm).
27. Install the balancer shaft belt drive pulley.
28. Align the groove on the pulley edge to the pointer on the balancer gear case.
29. Check the alignment of the pointer on the balancer pulley to the pointer on the oil pump.
30. Install and tension the balancer shaft belt.
31. Be sure the timing belts have been tensioned correctly and that all TDC and alignment marks are in their proper positions.
32. Install the lower timing cover and the crankshaft pulley. Apply engine oil to the pulley bolt threads and washer surface. Install the pulley bolt and tighten it to 181 ft. lbs. (245 Nm).
33. Install the upper timing cover and the valve cover. Be sure the seals are properly seated.
34. Install the side engine mount. Tighten the through-bolt to 47 ft. lbs. (64 Nm). Tighten the mount nut and bolt to 40 ft. lbs. (55 Nm) each.
35. Remove the floor jack.
36. Install and tension the alternator belt.
37. Install the power steering pump and tension its belt.
38. Install the splash shields.
39. Reconnect the positive and negative battery cables. Enter the radio security code.
40. Check engine operation.

2.2L (H22A1) ENGINE

1. Disconnect the negative battery cable.
2. Turn the crankshaft so the No. 1 piston is at top dead center. The No. 1 piston is at top dead center when the pointer on the block aligns with the white painted mark on the driveplate.
3. Remove the cylinder head and upper timing belt covers.
4. Ensure the words UP embossed on the camshaft pulleys are aligned in the upward position.
5. Support the engine with a floor jack below the center of the center beam. Tension the jack so that it is just supporting the beam but not lifting it. Remove the 2 rear bolts from the center beam to allow the engine to drop down for clearance to remove the lower cover.
6. Remove and discard the rubber seal from the timing belt adjuster. Do not loosen the adjusting nut.
7. Remove the lock pin from the maintenance bolt.
8. Remove the lower timing belt cover.
9. There are two belts in this system; the one running to the camshaft pulley is the timing belt. The other, shorter one drives the balance shafts and is referred to as the balancer belt or timing balancer belt.

10. Loosen the balancer shafts tensioner adjusting nut, do not loosen the nut more than one turn. Push the tensioner for the balancer belt away from the belt to relieve the tension. Hold the tensioner and tighten the adjusting nut to hold the tensioner in place.
11. Carefully remove the balancer belt by sliding it off of the pulleys. Do not crimp or bend the belt; protect it from contact with oil or coolant.
12. Remove the balancer belt drive sprocket from the crankshaft.
13. Remove the bolts attaching the Crankshaft Position/Top Dead Center (CKP/TDC) sensor and remove the sensor.
14. Remove the timing belt by sliding it off of the pulleys. Do not crimp or bend the belt; protect it from contact with oil or coolant.
15. If defective, remove the two bolts mounting the timing belt auto-tensioner and remove the tensioner from the vehicle.
16. If defective, remove the balancer belt tensioner by performing the following:
    a. Remove the spring from the balancer belt tensioner.
    b. Remove the adjusting nut.
    c. Remove the bolt from the balancer belt adjuster lever, then remove the lever and the tensioner pulley.
17. This is an excellent time to check or replace the water pump. Even if the timing belt is only being replaced as part of a good maintenance schedule, consider replacing the pump at the same time.

To install:

18. If the water pump is to be replaced, install a new O-ring and make certain it is properly seated. Install the water pump and retaining bolts. Tighten the mounting bolts to 106 inch lbs. (12 Nm).

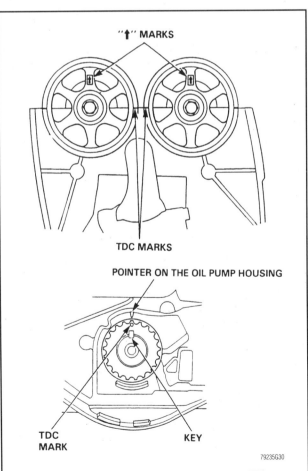

Camshaft and crankshaft alignment mark positioning for TDC—Honda 2.2L (H22A1) and 2.3L (H23A1) engines

TIMING BELTS—HONDA

19. If the balancer tensioner was removed, perform the following to install it:
    a. Install the balancer belt pulley and adjuster lever.
    b. Install the adjusting nut and the bolt to the balancer belt adjuster lever. Do not tighten the adjuster nut or bolt at this time.

➡ **Be sure that the balancer lever and tensioner moves freely and does not bind.**

    c. Install the spring to the tensioner.
    d. Move the balancer its full deflection and tighten the adjusting nut.
20. Hold the auto-tensioner with the maintenance bolt pointing up. Remove the maintenance bolt and discard the gasket.

➡ **Handle the tensioner carefully so the oil inside does not spill or leak. Replenish the auto-tensioner with oil if any spills or leaks out. The auto-tensioner total capacity is 1/4 oz. (8 ml).**

21. Clamp the mounting boss of the auto-tensioner in a vise. Use pieces of wood or a cloth to protect the mounting boss.

✷✷ WARNING

Do not clamp the housing of the auto-tensioner, component damage may occur.

22. Insert a flat-bladed prytool into the maintenance hole. Place the stopper (part No. 14540-P13-003) on the auto-tensioner while turning the prytool clockwise to compress the tensioner. Take care not to damage the threads or the gasket contact surface with the prytool.
23. Remove the prytool and install the maintenance bolt with a new gasket. Tighten the maintenance bolt to 71 inch lbs. (8 Nm).
24. Be sure no oil is leaking from the maintenance bolt and install the auto-tensioner to the engine. Tighten the auto-tensioner mounting bolts to 16 ft. lbs. (22 Nm).
25. The pointer on the crankshaft pulley should be aligned with the pointer on the oil pump; the camshaft pulley must be aligned so that the word **UP** is at the top of the pulley and the marks on the edge of the pulley are aligned with the surfaces of the head.
26. Install the timing belt.
27. Remove the stopper from the timing belt adjuster.
28. Install the CKP/TDC sensors and tighten the bolts to 106 inch lbs. (12 Nm). Connect the CKP/TDC sensors connector.
29. Install the balancer belt drive sprocket to the crankshaft.
30. Align the groove on the front balancer shaft pulley with the pointer on the oil pump.
31. Align the rear timing balancer pulley using a 6 x 100mm bolt or rod. Mark the bolt or rod at a point 2.9 in. (74mm) from the end. Remove the bolt from the maintenance hole on the side of the block; insert the bolt/rod into the hole and align the 2.9 in. (74mm) mark with the face of the hole. This pin will hold the shaft in place during installation.
32. Ensure the timing balancer belt adjuster moves freely.
33. Install the balancer belt. Once the belts are in place, be sure that all the engine alignment marks are still correct. If not, remove the belts, realign the engine and reinstall the belts. Once the belts are properly installed, slowly loosen the adjusting nut, allowing the tensioner to move against the belt. Remove the pin from the maintenance hole and reinstall the bolt and washer.
34. Turn the crankshaft pulley one full turn, then tighten the adjusting nut to 33 ft. lbs. (45 Nm).

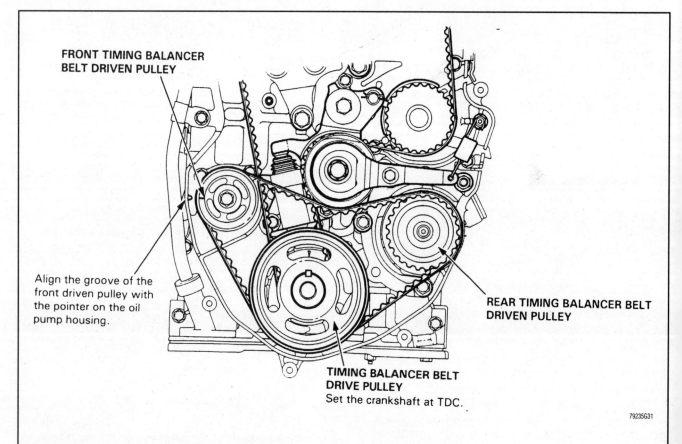

View of the balancer belt and the related timing marks—Honda 2.2L (H22A1) and 2.3L (H23A1) engines

HONDA—TIMING BELTS

✱✱ WARNING
Do not apply extra pressure to the pulleys or tensioners while performing the adjustment.

35. Install the timing belt and cylinder head covers, and any other applicable components. When installing the side engine mount, tighten the bolt and nut attaching the mount to the engine to 40 ft. lbs. (55 Nm), and the through-bolt and nut to 47 ft. lbs. (65 Nm).

2.6L ENGINE

1. Disconnect the negative battery cable.
2. Loosen and remove the engine accessory drive belts.
3. Remove the cooling fan assembly and the water pump pulley.
4. Drain the fluid from the power steering reservoir.
5. Unbolt and remove the power steering pump. Unbolt the hydraulic line brackets from the upper timing cover and move the pump out of the work area without disconnecting the hydraulic lines.
6. Disconnect and remove the starter motor if a flywheel holder (part No. J–38674 or equivalent) is to be used.
7. Remove the upper timing belt cover.
8. Rotate the crankshaft to set the engine at TDC/compression for the No. 1 cylinder. The arrow mark on the camshaft sprocket will be aligned with the mark on the rear timing cover.
9. Remove the crankshaft pulley.
10. Remove the lower timing belt cover.
11. Verify that the engine is set at TDC/compression for the No. 1 cylinder. The notch on the crankshaft sprocket will be aligned with the pointer on the oil seal retainer.
12. Release and remove the tensioner spring to release the timing belt's tension.
13. Remove the timing belt.
14. Unbolt the tensioner pulley bracket from the engine's front cover.
15. If necessary, unbolt and remove the camshaft sprockets. Use a puller to remove the crankshaft pulley if necessary. Don't lose the crankshaft sprocket key.

To install:

16. If removed, install the camshaft and crankshaft sprockets. Align the camshaft and crankshaft timing marks and be sure to install any keys. Tighten the camshaft sprocket bolt to 43 ft. lbs. (59 Nm).
17. Install the tensioner assembly. Tighten the tensioner mounting bolt to 14 ft. lbs. (19 Nm) and the cap bolt to 9 ft. lbs. (13 Nm).
18. Be sure the crankshaft and the camshaft sprockets are aligned with their timing marks. Install the timing belt onto the sprockets using the following sequence: first around the crankshaft sprocket; second around the oil pump sprocket; third around the camshaft sprocket.
19. Loosen the tensioner mounting bolt. This will allow the tensioner spring to apply pressure to the timing belt.
20. After the spring has pulled the timing belt as far as possible, temporarily tighten the tensioner mounting bolt to 14 ft. lbs. (19 Nm).

➡**Remove the flywheel holder before rotating the crankshaft. Reinstall the holder to tighten the crankshaft pulley bolt.**

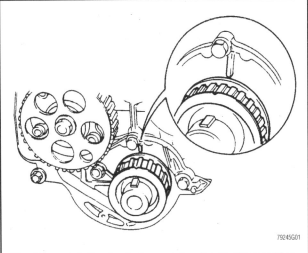

Align the crankshaft pulley timing mark the with oil retainer setting mark—Honda Passport 2.6L engine

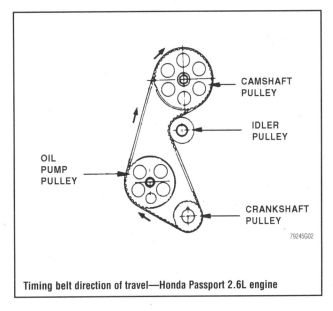

Timing belt direction of travel—Honda Passport 2.6L engine

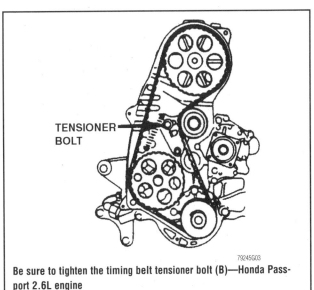

Be sure to tighten the timing belt tensioner bolt (B)—Honda Passport 2.6L engine

92 TIMING BELTS—HONDA

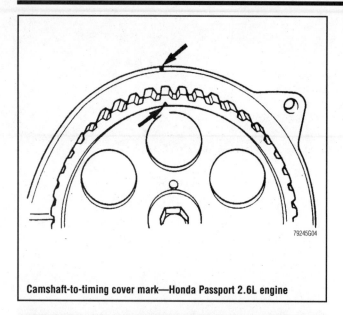

Camshaft-to-timing cover mark—Honda Passport 2.6L engine

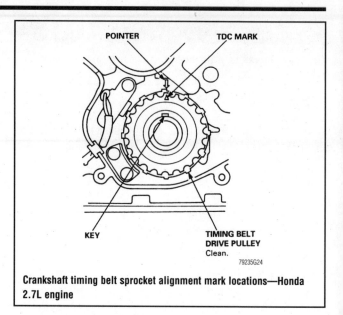

Crankshaft timing belt sprocket alignment mark locations—Honda 2.7L engine

✼✼✼ WARNING

If any binding is felt when adjusting the timing belt tension by turning the crankshaft, STOP turning the engine, because the pistons may be hitting the valves.

21. Rotate the crankshaft counterclockwise two complete revolutions to check the rotation of the belt and the alignment of the timing marks. Listen for any rubbing noises which may mean the belt is binding.
22. Loosen the tensioner pulley bolt to allow the spring to adjust the correct tension. Then, retighten the tensioner pulley bolt to 14 ft. lbs. (19 Nm).
23. Install the lower timing cover and the crankshaft pulley.
24. Tighten the crankshaft pulley bolt to 87 ft. lbs. (118 Nm). Tighten the small pulley bolts to 6 ft. lbs. (8 Nm).
25. Install the upper timing cover.
26. Install the starter if it was removed. Tighten the bolts to 30 ft. lbs. (40 Nm).
27. Install the power steering pump. If the hydraulic lines were disconnected, refill and bleed the power steering system.
28. Install the water pump pulley and tighten its nut to 20 ft. lbs. (26 Nm).
29. Install the cooling fan assembly.
30. Install and adjust the accessory drive belts.
31. Connect the negative battery cable.

2.7L ENGINE

1. Turn the engine to align the timing marks and set cylinder No.1 to TDC. The white mark on the crankshaft pulley should align with the pointer on the timing belt cover. Remove the inspection caps on the upper timing belt covers to check the alignment of the timing marks. The pointers for the camshafts should align with the green marks on the camshaft sprockets.
2. Remove all necessary components for access to the timing belt covers, then remove the covers.

➡ **Do not use the covers to store removed items.**

3. Loosen the timing belt adjuster bolt 180 degrees (½ turn). Push the tensioner to remove the tension from the timing belt, then retighten the adjusting bolt.

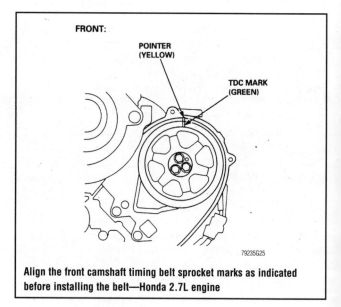

Align the front camshaft timing belt sprocket marks as indicated before installing the belt—Honda 2.7L engine

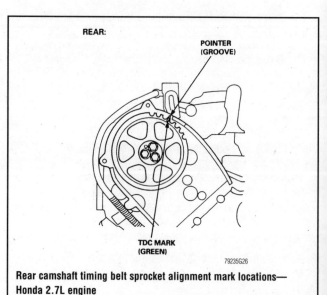

Rear camshaft timing belt sprocket alignment mark locations—Honda 2.7L engine

HONDA—TIMING BELTS

4. Remove the timing belt. Do not crimp or bend the belt; protect it from contact with oil or coolant. Slide the belt off the sprockets.
5. Remove the bolts attaching the camshaft sprockets to the camshafts, then remove the sprockets.
6. If the timing belt tensioner is defective, remove the spring from the timing belt tensioner. Remove the tensioner pulley adjusting bolt and the adjuster assembly from the engine.

➡ **This is an excellent time to check or replace the water pump. Even if the timing belt is only being replaced as part of a good maintenance schedule, consider replacing the pump at the same time.**

To install:

7. If the water pump is to be replaced, install a new O-ring and make certain it is properly seated. Install the water pump and retaining bolts. Tighten the mounting bolts to 16 ft. lbs. (22 Nm).
8. If removed, install the tensioner pulley and the adjusting bolt, be sure the tensioner is properly positioned on its pivot pin. Install the spring to the tensioner, then push the tensioner to it's full deflection and tighten the adjusting bolt.
9. Set the timing belt drive sprocket so that the No. 1 piston is at top dead center (TDC). Align the TDC mark on the tooth of the timing belt drive sprocket with the pointer on the oil pump.
10. Set the camshaft sprockets so that the No. 1 piston is at TDC. Align the TDC marks (green mark) on the camshaft sprockets to the pointers on the back covers.
11. Install the timing belt onto the sprockets in the following sequence: crankshaft sprocket, tensioner pulley, front camshaft sprocket, water pump pulley, rear camshaft sprocket.
12. Loosen, then retighten the timing belt adjuster bolt to tension the timing belt.
13. Install the lower timing belt cover.
14. Install the crankshaft sprocket and the crankshaft pulley bolt. Tighten the bolt to 181 ft. lbs. (245 Nm) with the aid of the crank pulley holder.
15. Rotate the crankshaft five or six turns clockwise so that the timing belt positions on the sprockets.
16. Set cylinder No. 1 to TDC by aligning the timing marks. If the timing marks do not align, remove the timing belt, then adjust the components and reinstall the timing belt.
17. Loosen the timing belt adjusting bolt 180 degrees (½ turn) and retighten the adjusting bolt. Tighten the adjusting bolt to 31 ft. lbs. (42 Nm).
18. Install the upper timing belt cover and all other applicable components. When installing the side engine mount to the engine, use three new attaching bolts. Tighten the new bolts to 40 ft. lbs. (54 Nm).

3.2L ENGINE

1. Disconnect the negative battery cable.
2. Drain the engine coolant into a sealable container.
3. Remove the air cleaner assembly and intake air duct.
4. Disconnect the upper radiator hose from the coolant inlet.
5. Remove the upper fan shroud from the radiator.
6. Remove the four nuts retaining the cooling fan assembly. Remove the cooling fan from the fan pulley.
7. Loosen and remove the drive belts.
8. Remove the upper timing belt covers.
9. Remove the fan pulley assembly.
10. Rotate the crankshaft to align the camshaft timing marks with the pointer dots on the back covers. Verify that the pointer on the crankshaft aligns with the mark on the lower timing cover.

➡ **When the timing marks are aligned on 1995 vehicles, no pistons will be at TDC/compression. When the timing marks are aligned on 1995½–99 vehicles, the No. 2 piston is at TDC/compression.**

✱✱ WARNING

Align the camshaft and crankshaft sprockets with their alignment marks before removing the timing belt. Failure to align the belt and sprocket marks may result in valve damage.

11. Use tool No. J-8614–01, or a suitable pulley holding tool to remove the crankshaft pulley center bolt. Remove the crankshaft pulley.
12. If present, disconnect the two oil cooler hose bracket bolts on the timing cover. Move the oil cooler hoses and bracket off of the lower timing cover.

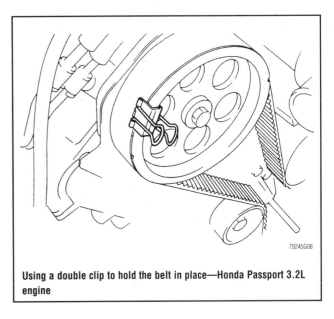

Using a double clip to hold the belt in place—Honda Passport 3.2L engine

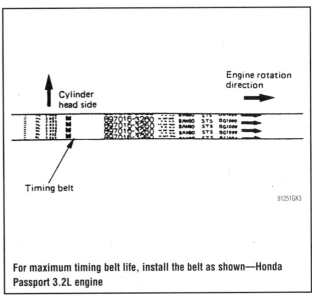

For maximum timing belt life, install the belt as shown—Honda Passport 3.2L engine

94 TIMING BELTS—HONDA

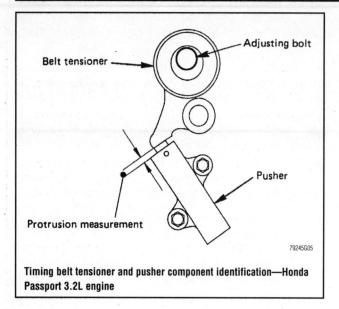

Timing belt tensioner and pusher component identification—Honda Passport 3.2L engine

13. Remove the lower timing belt cover.
14. Remove the pusher assembly (tensioner) from below the belt tensioner pulley. The pusher rod must always face upward to prevent oil leakage. Depress the pusher rod, and insert a wire pin into the hole to keep the pusher rod retracted.
15. Remove the timing belt.
16. Inspect the water pump and replace it if there is any doubt about its condition.

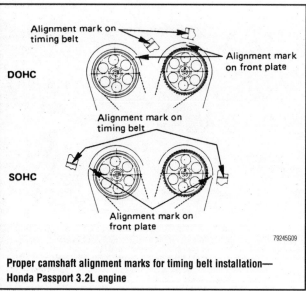

Proper camshaft alignment marks for timing belt installation—Honda Passport 3.2L engine

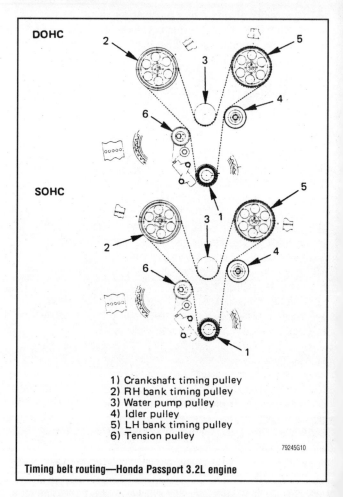

1) Crankshaft timing pulley
2) RH bank timing pulley
3) Water pump pulley
4) Idler pulley
5) LH bank timing pulley
6) Tension pulley

Timing belt routing—Honda Passport 3.2L engine

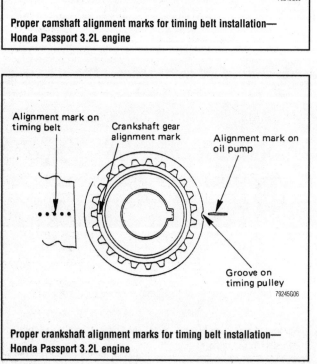

Proper crankshaft alignment marks for timing belt installation—Honda Passport 3.2L engine

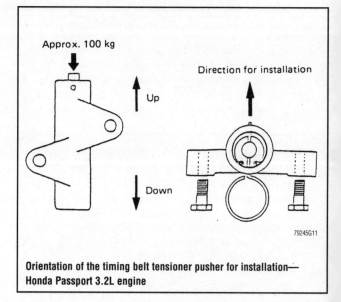

Orientation of the timing belt tensioner pusher for installation—Honda Passport 3.2L engine

HONDA, HYUNDAI—TIMING BELTS

17. Repair any oil or coolant leaks before installing a new timing belt. If the timing belt has been contaminated with oil or coolant, or is damaged, it must be replaced.

To install:

18. Verify that the sprocket timing marks are still aligned and that the groove and the keyway on the crankshaft timing sprocket align with the mark on the oil pump. The white pointers on the camshaft timing sprockets should align with the dots on the front plate.

19. Install the timing belt. Use clips to secure the belt onto each sprocket until the installation is complete. Align the dotted marks on the timing belt with the timing mark opposite the groove on the crankshaft sprocket.

➡ The arrows on the timing belt must follow the belt's direction of rotation. The manufacturer's trademark on the belt's spine should be readable left-to-right when the belt is installed.

20. Align the white line on the timing belt with the alignment mark on the right bank camshaft timing pulley. Secure the belt with a clip.

✳✳ WARNING

If any binding is felt when adjusting the timing belt tension by turning the crankshaft, STOP turning the engine, because the pistons may be hitting the valves.

21. Rotate the crankshaft counterclockwise to remove the slack between the crankshaft sprocket and the right camshaft timing belt sprocket.
22. Install the belt around the water pump pulley.
23. Install the belt on the idler pulley.
24. Align the white alignment mark on the timing belt with the alignment mark on the left bank camshaft timing belt sprocket.
25. Install the crankshaft pulley and tighten the center bolt by hand. Rotate the crankshaft pulley clockwise to give slack between the crankshaft timing belt pulley and the right bank camshaft timing belt pulley.
26. Insert a 1.4mm piece of wire through the hole in the pusher to hold the rod in. Install the pusher assembly while pushing the tension pulley toward the belt.
27. Pull the pin out from the pusher to release the rod.
28. Remove the clamps from the sprockets. Rotate the crankshaft pulley clockwise two turns. Measure the rod protrusion to ensure it is between 0.16–0.24 in. (4–6mm).
29. If the tensioner pulley bracket pivot bolt was removed, tighten it to 31 ft. lbs. (42 Nm).
30. Tighten the pusher bolts to 14 ft. lbs. (19 Nm).
31. Remove the crankshaft pulley. Install the lower and upper timing belt covers and tighten their bolts to 12 ft. lbs. (17 Nm).
32. Fit the oil cooler hose onto the timing cover and tighten its mounting bracket bolts to 16 ft. lbs. (22 Nm).
33. Install the crankshaft pulley and tighten the pulley bolt to 123 ft. lbs. (167 Nm).
34. Install fan pulley assembly and tighten the bolts to 16 ft. lbs. (22 Nm).
35. Install and adjust the accessory drive belts.
36. Install the cooling fan assembly and tighten the bolts to 6 ft. lbs. (8 Nm).
37. Install the upper fan shroud.
38. Install the air cleaner assembly and intake air duct.
39. Connect the upper radiator hose to the coolant inlet.
40. Refill and bleed the cooling system.
41. Connect the negative battery cable.

Hyundai

✳✳ CAUTION

Timing belt maintenance is extremely important. All Hyundai models use interference-type non-freewheeling engines. Should the timing belt break in these engines, the valves in the cylinder head will come in contact with the pistons, causing major engine damage. The recommended replacement interval for timing belts is 60,000 miles.

1.5L ENGINES

1990–94 Models

1. Disconnect the negative battery cable.
2. Remove the engine undercover.
3. Using the proper equipment, slightly raise the engine to take the weight off the side engine mount. Remove the engine mount bracket.
4. Remove the accessory drive belts, tension pulley brackets, water pump pulley and crankshaft pulley.
5. Remove all attaching screws and remove the upper and lower timing belt covers.

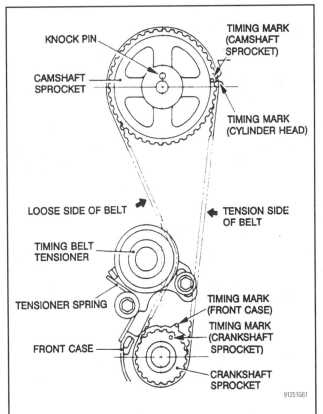

Timing belt component identification and timing mark alignment—1990–94 Hyundai 1.5L engine

TIMING BELTS—HYUNDAI

6. Rotate the crankshaft clockwise and align the timing marks so No. 1 piston will be at TDC of the compression stroke. Loosen the tensioning bolt, it runs in the slotted portion of the tensioner, and the pivot bolt on the timing belt tensioner. Move the tensioner as far as it will go toward the water pump. Tighten the adjusting bolt. Mark the timing belt with an arrow showing direction of rotation if the belt is to be reused.
7. Remove the timing belt.
8. Remove the camshaft sprocket as required.
9. Remove the crankshaft sprocket bolts and remove the crankshaft sprocket and flange, noting the direction of installation for each. Remove the timing belt tensioner.
10. Inspect the belt thoroughly. The back surface must be pliable and rough. If it is hard and glossy, the belt should be replaced. Any cracks in the belt backing or teeth or missing teeth mean the belt must be replaced. The canvas cover should be intact on all the teeth. If rubber is exposed anywhere, the belt should be replaced.
11. Inspect the tensioner for grease leaking from the grease seal and any roughness in rotation. Replace a tensioner for either defect.
12. The sprockets should be inspected and replaced, if there is any sign of damaged teeth or cracking. Do not immerse sprockets in solvent, as solvent that has soaked into the metal may cause deterioration of the timing belt later. Do not clean the tensioner in solvent either, as this may wash the grease out of the bearing.

To install:
13. Install the flange and crankshaft sprocket. The flange must go on first with the chamfered area outward. The sprocket is installed with the boss forward and the studs for the fan belt pulley outward. Install and tighten the crankshaft sprocket bolt to 51–72 ft. lbs. (69–98 Nm). Install the camshaft sprocket and bolt, torquing it to 47–54 ft. lbs. (64–74 Nm).
14. Align the timing marks of the camshaft sprocket. Check that the crankshaft timing marks are still in alignment (the locating pin on the front of the crankshaft sprocket is aligned with a mark on the front case).
15. Mount the tensioner, spring and spacer with the bottom end of the spring free. Then, install the bolts and tighten the adjusting bolt slightly with the tensioner moved as far as possible away from the water pump. Install the free end of the spring into the locating tang on the front case. Position the belt over the crankshaft sprocket and then over the camshaft sprocket. Slip the back of the belt over the tensioner wheel. Turn the camshaft sprocket in the opposite of its normal direction of rotation until the straight side of the belt is tight and make sure the timing marks align. If not, shift the belt 1 tooth at a time in the appropriate direction until this occurs.
16. Loosen the tensioner mounting bolts so the tensioner works, without the interference of any friction, under spring pressure. Make sure the belt follows the curve of the camshaft pulley so the teeth are engaged all the way around.
17. Correct the path of the belt, if necessary. Tighten the tensioner adjusting bolt to 15–18 ft. lbs. (20–26 Nm). Then, tighten the tensioner pivot bolt to the same figure. Bolts must be torqued, in order, or tension won't be correct.
18. Turn the crankshaft 1 turn clockwise until timing marks again align to seat the belt. Loosen both tensioner attaching bolts and let the tensioner position itself under spring tension as before. Tighten the bolts in order. Check belt tension by putting a finger on the water pump side of the tensioner wheel and pull the belt toward it. The belt should move toward the pump until the teeth are about ¼ of the way across the head of the tensioner adjusting bolt. Retension the belt, if necessary.
19. Install the timing belt covers.
20. Install the crankshaft pulley, making sure the pin on the crankshaft sprocket fits through the hole in the rear surface of the pulley. Install the bolts and tighten to 7.5–8.5 ft. lbs. (9–12 Nm).
21. Connect the negative battery cable.

1995–99 Models

1. Disconnect the negative battery cable.
2. Remove the engine undercover.
3. Using the proper equipment, slightly raise the engine to take the weight off the side engine mount. Remove the engine mount bracket.
4. Remove the accessory drive belts, tension pulley brackets, water pump pulley and crankshaft pulley.
5. Remove all attaching screws and remove the upper and lower timing belt covers.
6. Rotate the crankshaft clockwise and align the timing marks so No. 1 piston will be at TDC of the compression stroke.
7. Loosen the tensioning bolt and the pivot bolt on the timing belt tensioner. Move the tensioner as far as it will go toward the water pump. Tighten the adjusting bolt.
8. Mark the timing belt with an arrow showing direction of rotation.
9. Remove the timing belt.
10. If defective, remove the timing belt tensioner.

To install:
11. Align the timing marks of the camshaft sprocket and check that the crankshaft timing marks are still in alignment.
12. If removed, install the timing belt tensioner, spring and spacer with the bottom end of the spring free. Tighten the adjusting bolt slightly with the tensioner moved as far as possible away from the water pump.
13. Install the free end of the spring into the locating tang on the front case.
14. Position the timing belt over the crankshaft sprocket, then over the camshaft sprocket. Slip the back of the belt over the tensioner wheel.
15. Turn the camshaft sprocket in the opposite of its normal direction of rotation until the straight side of the belt is tight and be sure the timing marks align.

➡️**If the timing marks are not properly aligned, shift the belt 1 tooth at a time in the appropriate direction until they are lined up.**

16. Loosen the tensioner mounting bolts so the tensioner works, without the interference of any friction, under spring pressure. Be sure the belt follows the curve of the camshaft pulley so the teeth are engaged all the way around. Correct the path of the belt, if necessary.
17. Tighten the tensioner adjusting bolt, then the tensioner pivot bolt to 15–18 ft. lbs. (20–26 Nm).

➡️**Bolts must be tightened in the stated order or tension won't be correct.**

18. Turn the crankshaft 1 turn clockwise until timing marks again align to seat the belt.
19. Loosen both tensioner attaching bolts and let the tensioner position itself under spring tension. Retighten the bolts.
20. Check belt tension by putting a finger on the water pump side of the tensioner wheel and pull the belt toward the water pump. The belt should move toward the pump until the teeth are approximately ½ of the way across the head of the tensioner adjusting bolt. Re-tension the belt, if necessary.
21. Install the timing belt covers and all other related components.

HYUNDAI—TIMING BELTS

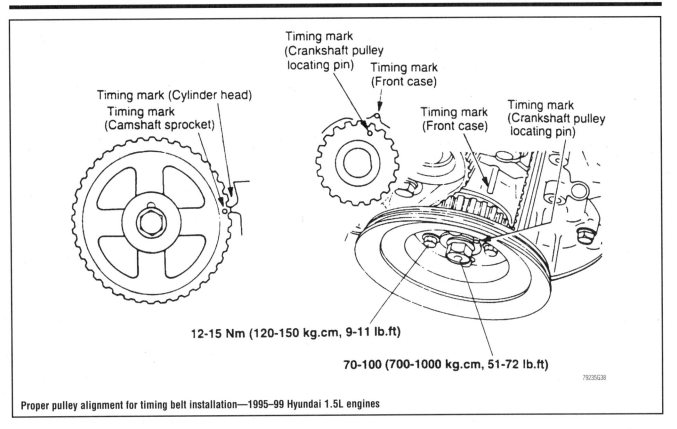

Proper pulley alignment for timing belt installation—1995–99 Hyundai 1.5L engines

1.6L & 2.0L (VIN P) ENGINES

1990–94 Models

1. Disconnect the negative battery cable.
2. Remove the engine undercover.
3. Using the proper equipment, slightly raise the engine to take the weight off the side engine mount. Remove the engine mount bracket.
4. Remove the accessory drive belts, tension pulley brackets, water pump pulley and crankshaft pulley.
5. Remove all attaching screws and remove the upper and lower timing belt covers.
6. Remove the timing belt upper and lower covers.
7. Rotate the crankshaft clockwise and align the timing marks so No. 1 piston will be at TDC of the compression stroke. At this time the timing marks on the camshaft sprocket and the upper surface of the cylinder head should coincide, and the dowel pin of the camshaft sprocket should be at the upper side.

➡ **Always rotate the crankshaft in a clockwise direction. Make a mark on the back of the timing belt indicating the direction of rotation so it may be reassembled in the same direction if it is to be reused.**

8. Remove the auto tensioner and remove the outermost timing belt.
9. Remove the timing belt tensioner pulley, tensioner arm, idler pulley, oil pump sprocket, special washer, flange and spacer.
10. Remove the silent shaft (inner) belt tensioner and remove the inner belt.

To install:

11. Align the timing marks on the crankshaft sprocket and the silent shaft sprocket. Fit the inner timing belt over the crankshaft and silent shaft sprocket. Ensure that there is no slack in the belt.

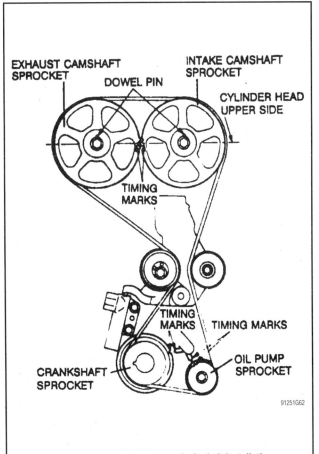

Correct alignment of engine timing marks for belt installation—1990–94 Hyundai 1.6L and 2.0L (VIN P) engines

12. While holding the inner timing belt tensioner with your fingers, adjust the timing belt tension by applying a force towards the center of the belt, until the tension side of the belt is taut. Tighten the tensioner bolt.

➡ When tightening the bolt of the tensioner, ensure that the tensioner pulley shaft does not rotate with the bolt. Allowing it to rotate with the bolt can cause excessive tension on the belt.

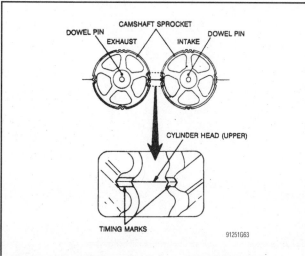

Alignment of camshaft sprocket timing marks—1990–94 Hyundai 1.6L and 2.0L (VIN P) engines

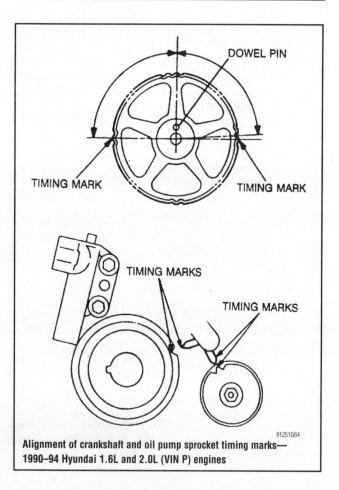

Alignment of crankshaft and oil pump sprocket timing marks—1990–94 Hyundai 1.6L and 2.0L (VIN P) engines

13. Check belt for proper tension by depressing the belt on its long side with your finger and noting the belt deflection. The desired reading is 0.20–0.28 in. (5–7mm). If tension is not correct, readjust and check belt deflection.

14. Install the flange, crankshaft and washer to the crankshaft. The flange on the crankshaft sprocket must be installed towards the inner timing belt sprocket. Tighten bolt to 80–94 ft. lbs. (110–130 Nm).

15. To install the oil pump sprocket, insert a Phillips screwdriver with a shaft 0.31 in. (8mm) in diameter into the plug hole in the left side of the cylinder block to hold the left silent shaft. Tighten the nut to 36–43 ft. lbs. (50–60 Nm).

16. Using a wrench, hold the camshaft hexagon found between journal No. 2 and 3 and tighten bolt to 58–72 ft. lbs. (80–100 Nm). If no hexagon is present between journal No. 2 and 3, hold the sprocket stationary with a spanner wrench while tightening the retainer bolt.

17. Carefully push the auto tensioner rod in until the set hole in the rod aligned up with the hole in the cylinder. Place a wire into the hole to retain the rod.

18. Install the tensioner pulley onto the tensioner arm. Locate the pinhole in the tensioner pulley shaft to the left of the center bolt. Then, tighten the center bolt finger-tight.

19. When installing the timing belt, turn the 2 camshaft sprockets so their dowel pins are located on top. Align the timing marks facing each other with the top surface of the cylinder head. When you let go of the exhaust camshaft sprocket, it will rotate 1 tooth in the counter-clockwise direction. This should be taken into account when installing the timing belts on the sprocket.

➡ Both camshaft sprockets are used for the intake and exhaust camshafts and are provided with 2 timing marks. When the sprocket is mounted on the exhaust camshaft, use the timing mark on the right with the dowel pin hole on top. For the intake camshaft sprocket, use the 1 on the left with the dowel pin hole on top.

20. Align the crankshaft sprocket and oil pump sprocket timing marks.

21. After alignment of the oil pump sprocket timing marks, remove the plug on the cylinder block and insert a Phillips screw driver with a shaft diameter of 0.31 in. (8mm) through the hole. If the shaft can be inserted 2.4 in. (70mm) deep, the silent shaft is in the correct position. If the shaft of the tool can only be inserted 0.8—1.0 in. (20–25mm) deep, turn the oil pump sprocket 1 turn and realign the marks. Reinsert the tool making sure it is inserted 2.4 in. (70mm) deep. Keep the tool inserted in hole for the remainder of this procedure.

➡ The above step assures that the oil pump socket is in correct orientation to the silent shafts. This step must not be skipped or a vibration may develop during engine operation.

22. Install the timing belt as follows:
   a. Install the timing belt around the intake camshaft sprocket and retain it with 2 spring clips or binder clips.
   b. Install the timing belt around the exhaust sprocket, aligning the timing marks with the cylinder head top surface using 2 wrenches. Retain the belt with 2 spring clips.
   c. Install the timing belt around the idler pulley, oil pump sprocket, crankshaft sprocket and the tensioner pulley. Remove the 2 spring clips.
   d. Lift upward on the tensioner pulley in a clockwise direction and tighten the center bolt. Make sure all timing marks are aligned.

e. Rotate the crankshaft ¼ turn counterclockwise. Then, turn in clockwise until the timing marks are aligned again.

23. To adjust the timing (outer) belt, turn the crankshaft ¼ turn counterclockwise, then turn it clockwise to move No. 1 cylinder to TDC.

24. Loosen the center bolt. Using tool MD998738 or equivalent and a torque wrench, apply a torque of 1.88–2.03 ft. lbs. (2.6–2.8 Nm). Tighten the center bolt.

25. Screw the special tool into the engine left support bracket until its end makes contact with the tensioner arm. At this point, screw the special tool in some more and remove the set wire attached to the auto tensioner, if the wire was not previously removed. Then remove the special tool.

26. Rotate the crankshaft 2 complete turns clockwise and let it sit for approximately 15 minutes. Then, measure the auto tensioner protrusion (the distance between the tensioner arm and auto tensioner body) to ensure that it is within 0.15–0.18 in. (3.8–4.5mm). If out of specification, repeat Step 1–4 until the specified value is obtained.

27. If the timing belt tension adjustment is being performed with the engine mounted in the vehicle, and clearance between the tensioner arm and the auto tensioner body cannot be measured, the following alternative method can be used:

   a. Screw in special tool MD998738 or equivalent, until its end makes contact with the tensioner arm.

   b. After the special tool makes contact with the arm, screw it in some more to retract the auto tensioner pushrod while counting the number of turns the tool makes until the tensioner arm is brought into contact with the auto tensioner body. Make sure the number of turns the special tool makes conforms with the standard value of 2½–3 turns.

   c. Install the rubber plug to the timing belt rear cover.

28. Install the timing belt covers and all related items.
29. Connect the negative battery cable.

1995–99 Models

1. Disconnect the negative battery cable.
2. Remove the engine undercover.
3. Using the proper equipment, slightly raise the engine to take the weight off the side engine mount. Remove the engine mount bracket.
4. Remove the accessory drive belts, tension pulley brackets, water pump pulley and crankshaft pulley.
5. Remove all attaching screws and remove the upper and lower timing belt covers.

➡ **Always rotate the crankshaft in a clockwise direction.**

6. Rotate the crankshaft clockwise and align the timing marks so No. 1 piston will be at TDC of the compression stroke. At this time the timing marks on the camshaft sprocket and the upper surface of the cylinder head should coincide, and the dowel pin of the camshaft sprocket should be at the upper side.
7. Remove the outer timing belt tensioner.
8. Mark the timing belts, indicating the direction of rotation.
9. Remove the outer timing belt.
10. Remove the camshaft sprockets.
11. Insert a prytool with a 0.32 in. (8mm) diameter shaft into the left side cylinder block plug hole. The prytool will hold the counterbalance shaft stable while removing the oil pump sprocket retaining nut.
12. Remove the oil pump sprocket.
13. Loosen the right counterbalance shaft sprocket bolt.
14. Remove the inner timing belt tensioner.
15. Remove the inner timing belt.

To install:

16. Install the counterbalance shaft sprocket and tighten the flange bolt finger-tight.
17. Align the timing mark on each sprocket with the corresponding timing mark on the front case.
18. Install the inner timing belt.

➡ **When installing the inner timing belt, ensure that the tension side has no slack.**

19. Install the inner timing belt tensioner with the center of the pulley on the left side of the mounting bolt and with the pulley flange facing the front of the engine.
20. Lift the inner timing belt tensioner to tighten the inner timing belt so that its tension side will be pulled tight.
21. Tighten the bolt to secure the inner tensioner.

➡ **When tightening the bolt of the tensioner, ensure that the tensioner pulley shaft does not rotate with the bolt. Allowing it to rotate with the bolt can cause excessive tension on the belt.**

22. Ensure the timing marks are in alignment.
23. Check the belt for proper tension by depressing the belt on its long side with your finger and noting the belt deflection. The desired deflection should be 0.20–0.28 in. (5–7mm).
24. Install the flange, crankshaft sprocket and washer on the crankshaft. The flange on the crankshaft sprocket must be installed towards the inner timing belt sprocket. Tighten the bolt to 80–94 ft. lbs. (110–130 Nm).
25. Insert a prytool with a 0.32 in. (8mm) diameter shaft into the left side cylinder block plug hole. The prytool will hold the counterbalance shaft stable while removing the oil pump sprocket retaining nut.
26. Install the oil pump sprocket and tighten the nut to 36–43 ft. lbs. (50–60 Nm).
27. Install the camshaft sprocket and tighten the bolt to 56–72 ft. lbs. (80–100 Nm).
28. Carefully push the auto-tensioner rod in until the set hole in the rod is aligned with the hole in the cylinder. Place a wire into the hole to retain the rod.
29. Install the outer timing belt tensioner.
30. Install the outer tensioner pulley onto the tensioner arm. Locate the pinhole in the tensioner pulley shaft to the left of the center bolt. Tighten the center bolt finger-tight.
31. Turn the two camshaft sprockets so their dowel pins are located on top. Align the timing marks facing each other with the top surface of the cylinder head.

➡ **Both camshaft sprockets are used for the intake and exhaust camshafts and are provided with two timing marks. When the sprocket is mounted on the exhaust camshaft, use the timing mark on the right with the dowel pin hole on top. For the intake camshaft sprocket, use the 1 on the left with the dowel pin hole on top.**

32. Align the crankshaft sprocket and oil pump sprocket timing marks.
33. Insert a prytool with a 0.32 in. (8mm) diameter shaft into the left side cylinder block plug hole. If the shaft can be inserted 2.4 in. (61mm), the silent shaft is in the correct position. If the shaft of the tool can only be inserted 0.8–1.0 in. (20–25mm) deep, turn the oil pump sprocket one full turn and realign the marks.

TIMING BELTS—HYUNDAI

➡ Keep the tool inserted in hole for the remainder of this procedure. The above step assures that the oil pump socket is in correct orientation to the silent shafts. This step must not be skipped or a vibration may develop during engine operation.

34. Install the timing belt around the tensioner pulley and crankshaft sprocket. Hold the belt with your left-hand.
35. Pulling the belt with your right-hand, install it around the oil pump sprocket.
36. Install the belt around the idler pulley and intake camshaft sprocket.
37. Turn the exhaust camshaft sprocket one tooth clockwise to align its timing mark with the cylinder head top surface. Pulling the belt with both hands, install it around the exhaust camshaft sprocket.
38. Gently raise the tensioner pulley so that the belt does not sag and temporarily tighten the center bolt.
39. Turn the crankshaft ¼ turn counterclockwise. Turn the crankshaft clockwise to move the No. 1 cylinder to TDC.
40. Loosen the center bolt and attach special tool (PN 09244–28100) or equivalent to a torque wrench. Apply a torque of 23–25 inch lbs. (2.6–2.8 Nm). Tighten the center bolt.
41. Screw the special tool (PN 09244–28000) or equivalent into the engine left support bracket until its end makes contact with the tensioner arm. At this point, screw the special tool in some more and remove the set wire attached to the auto-tensioner, if the wire was not previously removed. Remove the special tool.
42. Rotate the crankshaft 2 complete turns clockwise and let it sit for approximately 15 minutes. Then, measure the auto-tensioner protrusion (the distance between the tensioner arm and auto-tensioner body) to ensure that it is within 0.15–0.18 in. (3.8–4.5mm).
43. If the timing belt tension adjustment is being performed with the engine mounted in the vehicle, and clearance between the tensioner arm and the auto-tensioner body cannot be measured, the following alternative method can be used:
   a. Screw in special tool (PN 09244–28000) or equivalent, until its end makes contact with the tensioner arm.
   b. After the tool makes contact with the arm, screw it in some more to retract the auto-tensioner pushrod while counting the number of turns the tool makes until the tensioner arm is brought into contact with the auto-tensioner body. Be sure the number of turns the special tool makes conforms with the standard value of 2 ½ –3 turns.

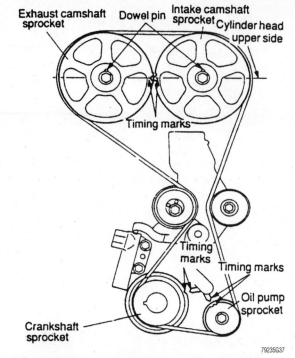

Timing belt sprocket alignment mark locations and positioning for belt removal and installation—1995–99 Hyundai 1.6L and 2.0L (VIN P) and 1995 1.8L engines

   c. Install the rubber plug to the timing belt rear cover.
44. Install the timing belt covers.

1.8L (VIN M) ENGINES

1992–94 Models

1. Disconnect the negative battery cable.
2. Remove the engine undercover.
3. Using the proper equipment, slightly raise the engine to take the weight off the side engine mount. Remove the engine mount bracket.
4. Remove the accessory drive belts, tension pulley brackets, water pump pulley and crankshaft pulley.
5. Remove all attaching screws and remove the upper and lower timing belt covers.
6. Remove the timing belt upper and lower covers.
7. Rotate the crankshaft clockwise and align the timing marks so No. 1 piston will be at TDC of the compression stroke. At this time the timing marks on the camshaft sprocket and the upper surface of the cylinder head should coincide, and the dowel pin of the camshaft sprocket should be at the upper side.

➡ Always rotate the crankshaft in a clockwise direction. Make a mark on the back of the timing belt indicating the direction of rotation so it may be reassembled in the same direction if it is to be reused.

8. Remove the auto tensioner and remove the outermost timing belt.
9. Remove the timing belt tensioner pulley, tensioner arm, idler pulley, oil pump sprocket, special washer, flange and spacer.
10. Remove the silent shaft (inner) belt tensioner and remove the inner belt.

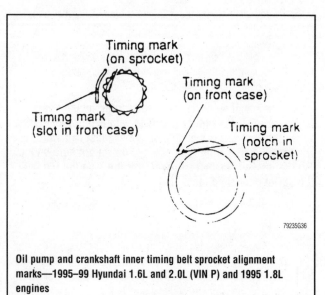

Oil pump and crankshaft inner timing belt sprocket alignment marks—1995–99 Hyundai 1.6L and 2.0L (VIN P) and 1995 1.8L engines

HYUNDAI—TIMING BELTS

To install:

11. Align the timing marks on the crankshaft sprocket and the silent shaft sprocket. Fit the inner timing belt over the crankshaft and silent shaft sprocket. Ensure that there is no slack in the belt.

12. While holding the inner timing belt tensioner with your fingers, adjust the timing belt tension by applying a force towards the center of the belt, until the tension side of the belt is taut. Tighten the tensioner bolt.

➡ **When tightening the bolt of the tensioner, ensure that the tensioner pulley shaft does not rotate with the bolt. Allowing it to rotate with the bolt can cause excessive tension on the belt.**

13. Check belt for proper tension by depressing the belt on its long side with your finger and noting the belt deflection. The desired reading is 0.20–0.28 in. (5–7mm). If tension is not correct, readjust and check belt deflection.

14. Install the flange, crankshaft and washer to the crankshaft. The flange on the crankshaft sprocket must be installed towards the inner timing belt sprocket. Tighten bolt to 80–94 ft. lbs. (110–130 Nm).

15. To install the oil pump sprocket, insert a Phillips screwdriver with a shaft 0.31 in. (8mm) in diameter into the plug hole in the left side of the cylinder block to hold the left silent shaft. Tighten the nut to 36–43 ft. lbs. (50–60 Nm).

16. Using a wrench, hold the camshaft hexagon found between journal No. 2 and 3 and tighten bolt to 58–72 ft. lbs. (80–100 Nm). If no hexagon is present between journal No. 2 and 3, hold the sprocket stationary with a spanner wrench while tightening the retainer bolt.

17. Carefully push the auto tensioner rod in until the set hole in the rod aligned up with the hole in the cylinder. Place a wire into the hole to retain the rod.

18. Install the tensioner pulley onto the tensioner arm. Locate the pinhole in the tensioner pulley shaft to the left of the center bolt. Then, tighten the center bolt finger-tight.

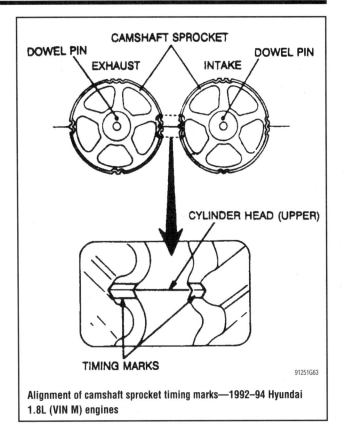

Alignment of camshaft sprocket timing marks—1992–94 Hyundai 1.8L (VIN M) engines

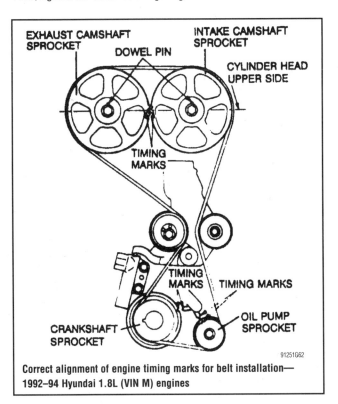

Correct alignment of engine timing marks for belt installation—1992–94 Hyundai 1.8L (VIN M) engines

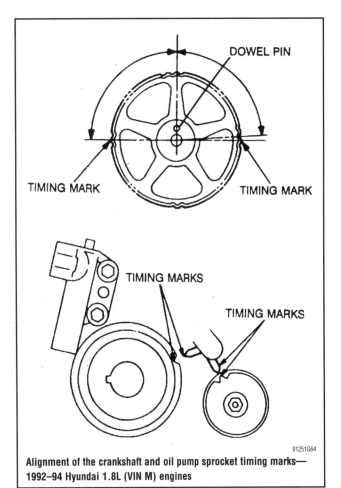

Alignment of the crankshaft and oil pump sprocket timing marks—1992–94 Hyundai 1.8L (VIN M) engines

TIMING BELTS—HYUNDAI

19. When installing the timing belt, turn the 2 camshaft sprockets so their dowel pins are located on top. Align the timing marks facing each other with the top surface of the cylinder head. When you let go of the exhaust camshaft sprocket, it will rotate 1 tooth in the counter-clockwise direction. This should be taken into account when installing the timing belts on the sprocket.

➡ **Both camshaft sprockets are used for the intake and exhaust camshafts and are provided with 2 timing marks. When the sprocket is mounted on the exhaust camshaft, use the timing mark on the right with the dowel pin hole on top. For the intake camshaft sprocket, use the 1 on the left with the dowel pin hole on top.**

20. Align the crankshaft sprocket and oil pump sprocket timing marks.
21. After alignment of the oil pump sprocket timing marks, remove the plug on the cylinder block and insert a Phillips screw driver with a shaft diameter of 0.31 in. (8mm) through the hole. If the shaft can be inserted 2.4 in. (70mm) deep, the silent shaft is in the correct position. If the shaft of the tool can only be inserted 0.8—1.0 in. (20—25mm) deep, turn the oil pump sprocket 1 turn and realign the marks. Reinsert the tool making sure it is inserted 2.4 in. (70mm) deep. Keep the tool inserted in hole for the remainder of this procedure.

➡ **The above step assures that the oil pump socket is in correct orientation to the silent shafts. This step must not be skipped or a vibration may develop during engine operation.**

22. Install the timing belt as follows:
    a. Install the timing belt around the intake camshaft sprocket and retain it with 2 spring clips or binder clips.
    b. Install the timing belt around the exhaust sprocket, aligning the timing marks with the cylinder head top surface using 2 wrenches. Retain the belt with 2 spring clips.
    c. Install the timing belt around the idler pulley, oil pump sprocket, crankshaft sprocket and the tensioner pulley. Remove the 2 spring clips.
    d. Lift upward on the tensioner pulley in a clockwise direction and tighten the center bolt. Make sure all timing marks are aligned.
    e. Rotate the crankshaft ¼ turn counterclockwise. Then, turn in clockwise until the timing marks are aligned again.
23. To adjust the timing (outer) belt, turn the crankshaft ¼ turn counterclockwise, then turn it clockwise to move No. 1 cylinder to TDC.
24. Loosen the center bolt. Using tool MD998738 or equivalent and a torque wrench, apply a torque of 1.88–2.03 ft. lbs. (2.6–2.8 Nm). Tighten the center bolt.
25. Screw the special tool into the engine left support bracket until its end makes contact with the tensioner arm. At this point, screw the special tool in some more and remove the set wire attached to the auto tensioner, if the wire was not previously removed. Then remove the special tool.
26. Rotate the crankshaft 2 complete turns clockwise and let it sit for approximately 15 minutes. Then, measure the auto tensioner protrusion (the distance between the tensioner arm and auto tensioner body) to ensure that it is within 0.15–0.18 in. (3.8–4.5mm). If out of specification, repeat Step 1–4 until the specified value is obtained.
27. If the timing belt tension adjustment is being performed with the engine mounted in the vehicle, and clearance between the tensioner arm and the auto tensioner body cannot be measured, the following alternative method can be used:
    a. Screw in special tool MD998738 or equivalent, until its end makes contact with the tensioner arm.
    b. After the special tool makes contact with the arm, screw it in some more to retract the auto tensioner pushrod while counting the number of turns the tool makes until the tensioner arm is brought into contact with the auto tensioner body. Make sure the number of turns the special tool makes conforms with the standard value of 2½–3 turns.
    c. Install the rubber plug to the timing belt rear cover.
28. Install the timing belt covers and all related items.
29. Connect the negative battery cable.

1995 Models

1. Disconnect the negative battery cable.
2. Remove the engine undercover.
3. Using the proper equipment, slightly raise the engine to take the weight off the side engine mount. Remove the engine mount bracket.
4. Remove the accessory drive belts, tension pulley brackets, water pump pulley and crankshaft pulley.
5. Remove all attaching screws and remove the upper and lower timing belt covers.

➡ **Always rotate the crankshaft in a clockwise direction.**

6. Rotate the crankshaft clockwise and align the timing marks so No. 1 piston will be at TDC of the compression stroke. At this time the timing marks on the camshaft sprocket and the upper surface of the cylinder head should coincide, and the dowel pin of the camshaft sprocket should be at the upper side.
7. Remove the outer timing belt tensioner.
8. Mark the timing belts, indicating the direction of rotation.
9. Remove the outer timing belt.
10. Remove the camshaft sprockets.
11. Insert a prytool with a 0.32 in. (8mm) diameter shaft into the left side cylinder block plug hole. The prytool will hold the counterbalance shaft stable while removing the oil pump sprocket retaining nut.
12. Remove the oil pump sprocket.
13. Loosen the right counterbalance shaft sprocket bolt.
14. Remove the inner timing belt tensioner.
15. Remove the inner timing belt.

To install:
16. Install the counterbalance shaft sprocket and tighten the flange bolt finger-tight.
17. Align the timing mark on each sprocket with the corresponding timing mark on the front case.
18. Install the inner timing belt.

➡ **When installing the inner timing belt, ensure that the tension side has no slack.**

19. Install the inner timing belt tensioner with the center of the pulley on the left side of the mounting bolt and with the pulley flange facing the front of the engine.
20. Lift the inner timing belt tensioner to tighten the inner timing belt so that its tension side will be pulled tight.
21. Tighten the bolt to secure the inner tensioner.

➡ **When tightening the bolt of the tensioner, ensure that the tensioner pulley shaft does not rotate with the bolt. Allowing it to rotate with the bolt can cause excessive tension on the belt.**

HYUNDAI—TIMING BELTS

22. Ensure the timing marks are in alignment.
23. Check the belt for proper tension by depressing the belt on its long side with your finger and noting the belt deflection. The desired deflection should be 0.20–0.28 in. (5–7mm).
24. Install the flange, crankshaft sprocket and washer on the crankshaft. The flange on the crankshaft sprocket must be installed towards the inner timing belt sprocket. Tighten the bolt to 80–94 ft. lbs. (110–130 Nm).
25. Insert a prytool with a 0.32 in. (8mm) diameter shaft into the left side cylinder block plug hole. The prytool will hold the counterbalance shaft stable while removing the oil pump sprocket retaining nut.
26. Install the oil pump sprocket and tighten the nut to 36–43 ft. lbs. (50–60 Nm).
27. Install the camshaft sprocket and tighten the bolt to 56–72 ft. lbs. (80–100 Nm).
28. Carefully push the auto-tensioner rod in until the set hole in the rod is aligned with the hole in the cylinder. Place a wire into the hole to retain the rod.
29. Install the outer timing belt tensioner.
30. Install the outer tensioner pulley onto the tensioner arm. Locate the pinhole in the tensioner pulley shaft to the left of the center bolt. Tighten the center bolt finger-tight.
31. Turn the two camshaft sprockets so their dowel pins are located on top. Align the timing marks facing each other with the top surface of the cylinder head.

➡ **Both camshaft sprockets are used for the intake and exhaust camshafts and are provided with two timing marks. When the sprocket is mounted on the exhaust camshaft, use the timing mark on the right with the dowel pin hole on top. For the intake camshaft sprocket, use the 1 on the left with the dowel pin hole on top.**

32. Align the crankshaft sprocket and oil pump sprocket timing marks.
33. Insert a prytool with a 0.32 in. (8mm) diameter shaft into the left side cylinder block plug hole. If the shaft can be inserted 2.4 in. (61mm), the silent shaft is in the correct position. If the shaft of the tool can only be inserted 0.8–1.0 in. (20–25mm) deep, turn the oil pump sprocket one full turn and realign the marks.

➡ **Keep the tool inserted in hole for the remainder of this procedure. The above step assures that the oil pump socket is in correct orientation to the silent shafts. This step must not be skipped or a vibration may develop during engine operation.**

34. Install the timing belt around the tensioner pulley and crankshaft sprocket. Hold the belt with your left-hand.
35. Pulling the belt with your right-hand, install it around the oil pump sprocket.
36. Install the belt around the idler pulley and intake camshaft sprocket.
37. Turn the exhaust camshaft sprocket one tooth clockwise to align its timing mark with the cylinder head top surface. Pulling the belt with both hands, install it around the exhaust camshaft sprocket.
38. Gently raise the tensioner pulley so that the belt does not sag and temporarily tighten the center bolt.
39. Turn the crankshaft ¼ turn counterclockwise. Turn the crankshaft clockwise to move the No. 1 cylinder to TDC.

40. Loosen the center bolt and attach special tool (PN 09244–28100) or equivalent to a torque wrench. Apply a torque of 23–25 inch lbs. (2.6–2.8 Nm). Tighten the center bolt.
41. Screw the special tool (PN 09244–28000) or equivalent into the engine left support bracket until its end makes contact with the tensioner arm. At this point, screw the special tool in some more and remove the set wire attached to the auto-tensioner, if the wire was not previously removed. Remove the special tool.
42. Rotate the crankshaft 2 complete turns clockwise and let it sit for approximately 15 minutes. Then, measure the auto-tensioner protrusion (the distance between the tensioner arm and auto-tensioner body) to ensure that it is within 0.15–0.18 in. (3.8–4.5mm).
43. If the timing belt tension adjustment is being performed with the engine mounted in the vehicle, and clearance between the tensioner arm and the auto-tensioner body cannot be measured, the following alternative method can be used:

   a. Screw in special tool (PN 09244–28000) or equivalent, until its end makes contact with the tensioner arm.

   b. After the tool makes contact with the arm, screw it in some more to retract the auto-tensioner pushrod while counting the number of turns the tool makes until the tensioner arm is brought into contact with the auto-tensioner body. Be sure the number of turns the special tool makes conforms with the standard value of 2 ½ –3 turns.

   c. Install the rubber plug to the timing belt rear cover.

44. Install the timing belt covers.

1996–99 Models

1. Disconnect the negative battery cable.
2. Remove the timing belt upper and lower covers.
3. Rotate the crankshaft clockwise and align the timing marks so No. 1 piston will be at TDC of the compression stroke. At this time, the timing marks on the camshaft sprocket and the upper surface of the cylinder head should coincide, and the dowel pin of the camshaft sprocket should be at the upper side.

➡ **Always rotate the crankshaft in a clockwise direction. Make a mark on the back of the timing belt indicating the direction of rotation so it may be reassembled in the same direction if it is to be reused.**

4. Remove the auto tensioner and remove the outermost timing belt.
5. Remove the timing belt tensioner pulley, tensioner arm, idler pulley, oil pump sprocket, special washer, flange and spacer.
6. Remove the silent shaft (inner) belt tensioner and remove the inner belt.

To install:
7. Align the timing marks on the crankshaft sprocket and the silent shaft sprocket. Fit the inner timing belt over the crankshaft and silent shaft sprocket. Ensure that there is no slack in the belt.
8. While holding the inner timing belt tensioner with your fingers, adjust the timing belt tension by applying a force towards the center of the belt, until the tension side of the belt is taut. Tighten the tensioner bolt.

➡ **When tightening the bolt of the tensioner, ensure that the tensioner pulley shaft does not rotate with the bolt. Allowing it to rotate with the bolt can cause excessive tension on the belt.**

TIMING BELTS—HYUNDAI

9. Check belt for proper tension by depressing the belt on its long side with your finger and noting the belt deflection.

The desired reading is 0.20–0.28 in. (5–7mm). If tension is not correct, readjust and check belt deflection.

10. Install the flange, crankshaft and washer to the crankshaft. The flange on the crankshaft sprocket must be installed towards the inner timing belt sprocket. Tighten bolt to 80 to 94 ft. lbs. (110 to 130 Nm).

11. To install the oil pump sprocket, insert a Phillips screwdriver with a shaft 0.31 in. (8mm) in diameter into the plug hole in the left side of the cylinder block to hold the left silent shaft. Tighten the nut to 36 to 43 ft. lbs. (50 to 60 Nm).

12. Using a wrench, hold the camshaft at its hexagon between journal No. 2 and 3 and tighten bolt to 58 to 72 ft. lbs. (80 to 100 Nm). If no hexagon is present between journal No. 2 and 3, hold the sprocket stationary with a wrench while tightening the retainer bolt.

13. Carefully push the auto tensioner rod in until the set hole in the rod is aligned with the hole in the cylinder. Place a wire into the hole to retain the rod.

14. Install the tensioner pulley onto the tensioner arm. Locate the pinhole in the tensioner pulley shaft to the left of the center bolt. Then tighten the center bolt finger-tight.

15. When installing the timing belt, turn the 2 camshaft sprockets so their dowel pins are located on top. Align the timing marks facing each other with the top surface of the cylinder head. When you let go of the exhaust camshaft sprocket, it will rotate 1 tooth in the counterclockwise direction. This should be taken into account when installing the timing belts on the sprocket.

➡Both camshaft sprockets are used for the intake and exhaust camshafts and are provided with 2 timing marks. When the sprocket is mounted on the exhaust camshaft, use the timing mark on the right with the dowel pin hole on top. For the intake camshaft sprocket, use the 1 on the left with the dowel pin hole on top.

16. Align the crankshaft sprocket and oil pump sprocket timing marks.

17. After alignment of the oil pump sprocket timing marks, remove the plug on the cylinder block and insert a Phillips screwdriver with a shaft diameter of 0.31 in. (8mm) through the hole. If the shaft can be inserted 2.4 in. deep, the silent shaft is in the correct position. If the shaft of the tool can only be inserted 0.8 to 1.0 in. (20 to 25mm) deep, turn the oil pump sprocket 1 turn and realign the marks. Reinsert the tool making sure it is inserted 2.4 in. deep. Keep the tool inserted in hole for the remainder of this procedure.

➡The above step assures that the oil pump socket is in correct orientation to the silent shafts. This step must not be skipped or a vibration may develop during engine operation.

18. Install the timing belt as follows:

    a. Install the timing belt around the intake camshaft sprocket and retain it with 2 spring clips or binder clips.

    b. Install the timing belt around the exhaust sprocket, aligning the timing marks with the cylinder head top surface using 2 wrenches. Retain the belt with 2 spring clips.

    c. Install the timing belt around the idler pulley, oil pump sprocket, crankshaft sprocket and the tensioner pulley. Remove the 2 spring clips.

    d. Lift upward on the tensioner pulley in a clockwise direction and tighten the center bolt. Make sure all timing marks are aligned.

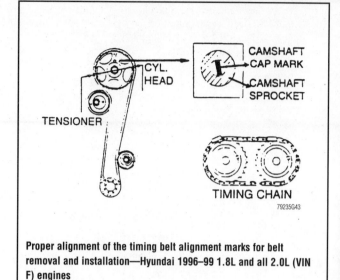

Proper alignment of the timing belt alignment marks for belt removal and installation—Hyundai 1996–99 1.8L and all 2.0L (VIN F) engines

    e. Rotate the crankshaft ¼ turn counterclockwise. Then, turn in clockwise until the timing marks are aligned again.

19. To adjust the timing (outer) belt, turn the crankshaft ¼ turn counterclockwise, then turn it clockwise to move No. 1 cylinder to TDC.

20. Loosen the center bolt. Using tool 09244-28100 or equivalent and a torque wrench, apply a torque of 1.88 to 2.03 ft. lbs. (2.6 to 2.8 Nm). Tighten the center bolt.

21. Screw the special tool into the engine left support bracket until its end makes contact with the tensioner arm. At this point, screw the special tool in some more and remove the set wire attached to the auto tensioner, if the wire was not previously removed. Then remove the special tool.

22. Rotate the crankshaft 2 complete turns clockwise and let it sit for approximately 15 minutes. Then, measure the auto tensioner protrusion (the distance between the tensioner arm and auto tensioner body) to ensure that it is within 0.15 to 0.18 in. (3.8 to 4.5mm). If out of specification, repeat Steps 1 to 4 until the specified value is obtained.

23. If the timing belt tension adjustment is being performed with the engine mounted in the vehicle, and clearance between the tensioner arm and the auto tensioner body cannot be measured, the following alternative method can be used:

    a. Screw in special tool 09244-28000 or equivalent, until its end makes contact with the tensioner arm.

    b. After the special tool makes contact with the arm, screw it in some more to retract the auto tensioner pushrod while counting the number of turns the tool makes until the tensioner arm is brought into contact with the auto tensioner body. Make sure the number of turns the special tool makes conforms with the standard value of 2½ to 3 turns.

    c. Install the rubber plug to the timing belt rear cover.

24. Install the timing belt covers and all related items.
25. Connect the negative battery cable.

2.0L (VIN F) ENGINE

1. Disconnect the negative battery cable.
2. Remove the engine undercover.
3. Using the proper equipment, slightly raise the engine to take the weight off the side engine mount. Remove the engine mount bracket.

HYUNDAI—TIMING BELTS

4. Remove the accessory drive belts, tension pulley brackets, water pump pulley and crankshaft pulley.
5. Remove all attaching screws and remove the upper and lower timing belt covers.
6. Rotate the crankshaft clockwise and align the timing marks so No. 1 piston will be at TDC of the compression stroke.
7. Remove the timing belt tensioner and idler pulley.
8. Mark the timing belt with an arrow showing direction of rotation.
9. Remove the timing belt.

To install:
10. Align the timing marks of the camshaft sprocket and check that the crankshaft timing marks are still in alignment.
11. Install the timing belt tensioner.
12. Install the idler pulley, if equipped. Tighten bolt to 32–41 ft. lbs. (43–55 Nm).
13. Position the timing belt over the camshaft sprocket, then over the crankshaft sprocket.
14. Tension the timing belt and tighten the tensioner pulley bolt to 32–41 ft. lbs. (43–55 Nm). When properly tensioned, the timing belt should deflect 0.16–0.24 in. (4–6mm) when a force of 5 lbs. (2.2kg) is placed on the longest span of the belt.
15. Turn the crankshaft sprocket one turn clockwise and realign the crankshaft sprocket timing mark.
16. Recheck the belt tension and adjust as necessary.
17. Install the timing belt cover and all other applicable components.

3.0L ENGINE

1990–94 Models

1. Disconnect the negative battery cable.
2. To remove the air conditioning compressor belt, loosen the adjustment pulley locknut, turn the screw counterclockwise to reduce the drive belt tension and remove the belt.
3. To remove the serpentine drive belt, insert a ½ in. breaker bar in to the square hole of the tensioner pulley, rotate it counterclockwise to reduce the drive belt tension and remove the belt.
4. Remove the air conditioning compressor and the air compressor bracket, power steering pump and alternator from the mounts and support them to the side. Remove power steering pump/alternator automatic belt tensioner bolt and the tensioner.
5. Raise the vehicle and support safely. Remove the right inner fender splash shield.
6. Remove the crankshaft pulley bolt and the pulley/damper assembly from the crankshaft.
7. Lower the vehicle and place a floor jack under the engine to support it.
8. Separate the front engine mount insulator from the bracket. Raise the engine slightly and remove the mount bracket.
9. Remove the timing belt cover bolts and the upper and lower covers from the engine.
10. Turn the crankshaft until the timing marks on the camshaft sprocket and cylinder head are aligned.
11. Loosen the tensioning bolt, it runs in the slotted portion of the tensioner, and the pivot bolt on the timing belt tensioner.
12. Move the tensioner counterclockwise as far as it will go. Tighten the adjusting bolt.
13. Mark the timing belt with an arrow showing direction of rotation.
14. Remove the timing belt from the camshaft sprocket.
15. Remove the crankshaft pulley. Then, remove the timing belt. Remove the timing belt tensioner. Remove the retainer bolts from the timing sprockets and remove as required.
16. Inspect the belt thoroughly. The back surface must be pliable and rough. If it is hard and glossy, the belt should be replaced. Any cracks in the belt backing or teeth or missing teeth mean the belt must be replaced. The canvas cover should be intact on all the teeth. If rubber is exposed anywhere, the belt should be replaced.
17. Inspect the tensioner for grease leaking from the grease seal and any roughness in rotation. Replace a tensioner for either defect.
18. The sprockets should be inspected and replaced if there is any sign of damaged teeth or cracking anywhere.
19. Do not immerse sprockets in solvent, as solvent that has soaked into the metal may cause deterioration of the timing belt later.
20. Do not clean the tensioner in solvent either, as this may wash the grease out of the bearing.

To install:
21. Align the timing marks of the camshaft sprocket. Check that the crankshaft timing marks are still in alignment, the locating pin on the front of the crankshaft sprocket is aligned with a mark on the front case.
22. Mount the tensioner, spring and spacer with the bottom end of the spring free. Then, install the bolts and tighten the adjusting bolt slightly with the tensioner moved as far as possible away from the water pump. Install the free end of the spring into the locating tang on the front case. Position the belt over the crankshaft sprocket and then over the camshaft sprocket. Slip the back of the belt over the tensioner wheel. Turn the camshaft sprocket in the opposite of its normal direction of rotation until the straight side of the belt is tight and make sure the timing marks align. If not, shift the belt 1 tooth at a time in the appropriate direction until this occurs.
23. Loosen the tensioner mounting bolts so the tensioner works, without the interference of any friction, under spring pressure. Make sure the belt follows the curve of the camshaft pulley so the teeth are engaged all the way around.
24. Correct the path of the belt, if necessary. Tighten the tensioner adjusting bolt to 16–21 ft. lbs. (22–29 Nm). Then, tighten

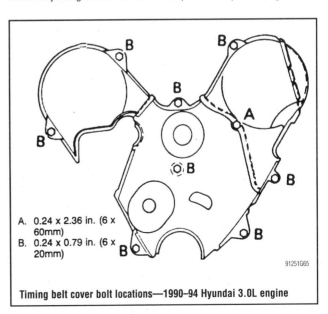

A. 0.24 x 2.36 in. (6 x 60mm)
B. 0.24 x 0.79 in. (6 x 20mm)

Timing belt cover bolt locations—1990–94 Hyundai 3.0L engine

106 TIMING BELTS—HYUNDAI

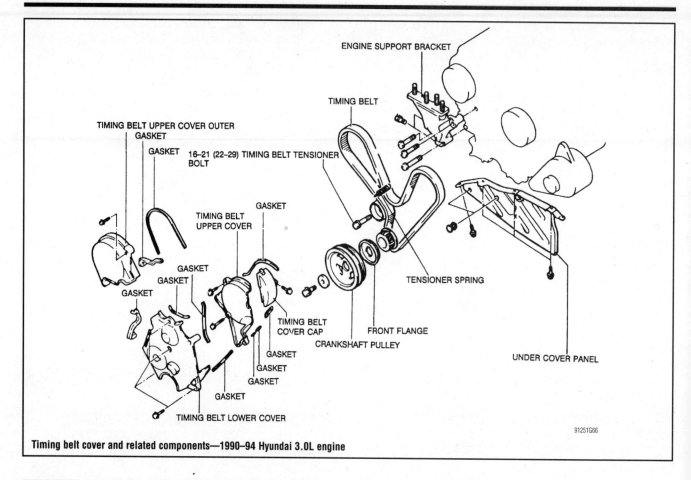

Timing belt cover and related components—1990–94 Hyundai 3.0L engine

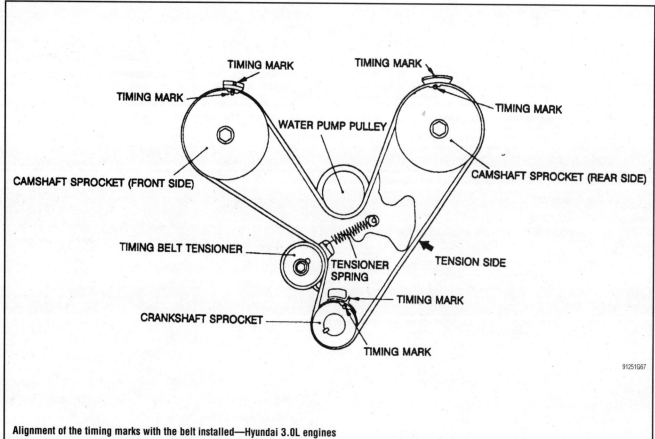

Alignment of the timing marks with the belt installed—Hyundai 3.0L engines

HYUNDAI, INFINITI—TIMING BELTS

the tensioner pivot bolt to the same figure. Bolts must be torqued in that order, or tension won't be correct.

25. Turn the crankshaft 1 turn clockwise until timing marks again align to seat the belt. Then loosen both tensioner attaching bolts and let the tensioner position itself under spring tension as before. Finally, torque the bolts in the proper order exactly as before. Check belt tension by putting a finger on the water pump side of the tensioner wheel and pull the belt toward it. The belt should move toward the pump until the teeth are about 1/4 of the way across the head of the tensioner adjusting bolt. Retension the belt, if necessary.
26. Install the timing belt covers.
27. Install the crankshaft pulley, making sure the pin on the crankshaft sprocket fits through the hole in the rear surface of the pulley. Install the retaining bolt and tighten to 108–116 ft. lbs. (147–157 Nm).
28. Install the engine mount bracket and secure with the mounting hardware.
29. Install the pulley damper assembly to the crankshaft. Tighten the bolt to 110 ft. lbs. (149 Nm). Install the splash shield.
30. Install the power steering pump/alternator automatic belt tensioner.
31. Install the air conditioning compressor bracket, compressor, power steering pump and alternator.
32. Install the accessory drive belt.
33. Connect the negative battery cable and check all disturbed components for proper operation.

1995–99 Models

1. Disconnect the negative battery cable.
2. Remove the engine undercover.
3. Remove the accessory drive belts.
4. Remove the air conditioner compressor tension pulley assembly.
5. Remove the tension pulley bracket.
6. Using the proper equipment, slightly raise the engine to take the weight off the side engine mount.
7. Disconnect the power steering pump pressure switch connector. Remove the power steering pump and wire aside.
8. Remove the engine support bracket.
9. Remove the crankshaft pulley.
10. Remove the timing belt cover cap.
11. Remove the timing belt upper and lower covers.
12. Turn the crankshaft until the timing marks on the camshaft sprocket and cylinder head are aligned.
13. Loosen the timing belt tensioner bolt and turn the tensioner counterclockwise as far as it will go. Tighten the adjusting bolt.
14. Mark the timing belt with an arrow showing direction of rotation.
15. Remove the timing belt.
16. If defective, remove the timing belt tensioner.

To install:

17. If necessary, install the timing belt tensioner.
18. Attach the top of the tensioner spring on the engine coolant pump pin. Ensure the hook on the pin is facing down and the hook on the tensioner is facing away from the engine
19. Rotate the timing belt tensioner to the extreme counterclockwise position. Temporarily lock the tensioner in place.
20. Align the timing marks of the camshaft and crankshaft sprockets.
21. Install the timing belt on the crankshaft sprocket, then onto the rear camshaft sprocket.
22. Route the belt to the coolant pump pulley, the front camshaft sprocket and the timing belt tensioner.
23. Apply force counterclockwise to the rear camshaft sprocket with tension on the tight side of the belt and check that timing marks are aligned.
24. Loosen the tensioner bolt one or two turns and tighten the timing belt to a tension of 57–84 lbs. (260–380 N).
25. Turn the crankshaft two turns clockwise.
26. Readjust the sprocket timing marks and tighten the tensioner bolts.
27. Install the timing covers. Make sure all pieces of packing are positioned in the inner grooves of the covers when installing.
28. Install the crankshaft pulley. Tighten the bolt to 108–116 ft. lbs. (150–160 Nm).
29. Install the engine support bracket.
30. Install the power steering pump and reconnect wire harness at the power steering pump pressure switch.
31. Install the engine mounting bracket and remove the engine support fixture.
32. Install the tension pulleys and drive belts.
33. Install the cruise control actuator.
34. Install the engine undercover.
35. Connect the negative battery cable.

Infiniti

3.0L (VG30) ENGINE

1. Disconnect the negative battery cable.
2. Raise and support the front of the vehicle safely.
3. Remove the engine undercovers.
4. Drain the cooling system.
5. Remove the front right side wheel.
6. Remove the engine side cover.
7. Remove the alternator, power steering and air conditioning compressor drive belts from the engine. When removing the power steering drive belt, loosen the idler pulley from the right side wheel housing.
8. Remove the upper radiator and water inlet hoses; remove the water pump pulley.
9. Remove the idler bracket of the compressor drive belt.
10. Remove the crankshaft pulley with a suitable puller.
11. Remove the upper and lower timing belt covers and gaskets.
12. Rotate the engine with a socket wrench on the crankshaft pulley bolt to align the punch mark on the left hand camshaft pulley with the mark on the upper rear timing belt cover. Align the punchmark on the crankshaft with the notch on the oil pump housing and temporarily install the crankshaft pulley bolt to allow for crankshaft rotation.
13. Use a hex wrench to turn the belt tensioner clockwise and tighten the tensioner locknut just enough to hold the tensioner in position. Then, remove the timing belt.

To install:

14. Before installing the timing belt, confirm that No. 1 cylinder is at TDC of the compression stroke. Install tensioner and tensioner spring. If stud is removed, apply locking sealant to threads before installing.
15. Swing tensioner fully clockwise with hexagon wrench and temporarily tighten locknut.

108 TIMING BELTS—INFINITI

16. Point the arrow on the timing belt toward the front belt cover. Align the white lines on the timing belt with the punch marks on all 3 pulleys.

➡ **There are 133 total timing belt teeth. If timing belt is installed correctly there will be 40 teeth between left hand and right hand camshaft sprocket timing marks. There will be 43 teeth between left hand camshaft sprocket and crankshaft sprocket timing marks.**

17. Loosen tensioner locknut, keeping tensioner steady with a hexagon wrench.
18. Swing tensioner 70–80 degrees clockwise with hexagon wrench and temporarily tighten locknut.
19. Turn crankshaft clockwise 2–3 times, then slowly set No. 1 cylinder at TDC of the compression stroke.
20. Push middle of timing belt between right hand camshaft sprocket and tensioner pulley with a force of 22 lbs.
21. Loosen tensioner locknut, keeping tensioner steady with a hexagon wrench.
22. Insert a 0.138 in. (0.35mm) thick and 0.5 in. (12.7mm) wide feeler gauge between the bottom of tensioner pulley and timing belt. Turn crankshaft clockwise and position gauge completely between tensioner pulley and timing belt. The timing belt will move about 2.5 teeth.
23. Tighten tensioner locknut, keeping tensioner steady with a hexagon wrench.
24. Turn crankshaft clockwise or counterclockwise and remove the gauge.
25. Rotate the engine 3 times, then set No. 1, to TDC, on its compression stroke.
26. Install the upper and lower timing belt covers with new gaskets.
27. Install the crankshaft pulley. Tighten the pulley bolt to 90–98 ft. lbs. (123–132 Nm).
28. Install the compressor drive belt idler bracket.
29. Install the water pump pulley and tighten the nuts to 12–15 ft. lbs. (16–21 Nm). Install the upper radiator and water inlet hoses.
30. Install the drive belts.
31. Install the engine side cover.
32. Mount the front right wheel.
33. Install the engine undercovers.
34. Lower the vehicle.
35. Fill the cooling system and connect the negative battery cable.

3.0L (VG30DE) ENGINE

1993–94 Models

1. Disconnect the negative battery cable.
2. Remove the engine undercover.
3. Drain the cooling system. Remove both cylinder block drain plug to drain coolant from the block.
4. Remove the air ducts.
5. Remove the radiator, drive belts, cooling fan and coupling.
6. Remove the crankshaft pulley bolt. Remove the crankshaft pulley using a puller.
7. Remove the water inlet and outlet. Remove the starter motor.
8. Remove the front timing covers.
9. Install a suitable stopper bolt into the tensioner arm so the auto-tensioner pusher does not spread out.
10. Set the No. 1 cylinder on TDC of the compression stroke.
11. Remove the auto-tensioner and timing belt.

To install:

12. Confirm that the No. 1 cylinder is on TDC of the compression stroke.
13. Align matchmarks on camshaft and crankshaft sprockets with aligning marks on rear belt cover and oil pump housing.
14. Remove all spark plugs.
15. Align white lines on timing belt with matchmarks on camshaft sprocket and crankshaft sprocket. Point arrow on timing belt towards front and install.
16. Adjust the tensioner arm to give 0.16 in. (4mm) clearance with pusher of auto-tensioner using a suitable vise, and then insert stopper bolt into tensioner arm so clearance does not change.
17. Install the auto-tensioner and tighten lower bolts hand tight. Push auto-tensioner toward belt until contact is made. Then push slightly more. Turn the crankshaft 10 degrees clockwise and tighten the tensioner nuts to 12–15 ft. lbs. 16–21 Nm).
18. Turn the crankshaft 120 degrees counterclockwise.
19. Loosen tensioner bolts ½ turn and move tensioner away from timing belt as far as it will move.
20. Turn the crankshaft clockwise and set No. 1 cylinder to TDC on its compression stroke.
21. Push the end of a pusher tool J–38387 with 13 lbs. of force and tighten the auto-tensioner bolts to 12–15 ft. lbs. (16–21 Nm).

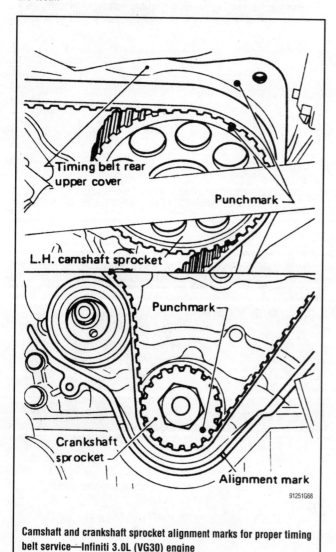

Camshaft and crankshaft sprocket alignment marks for proper timing belt service—Infiniti 3.0L (VG30) engine

22. Turn the crankshaft 120 degrees clockwise, then turn crankshaft 120 degrees counterclockwise. and set No. 1 cylinder to TDC on the compression stroke.
23. Prepare a steel plate measuring 0.12 x 0.39 in. (3 x 10mm). Set the plate on the timing belt and push it using a Pusher tool (J–38387) with 11 lbs. of force at a point midway between the camshaft sprockets. Also use the tool between the camshaft sprockets and the idler/tensioner pulleys. Deflection should be 0.24–0.28 in. (6–7mm). If not, readjust the tensioner.
24. Confirm the auto-tensioner mounting nuts are tightened to 12–15 ft. lbs. (16–21 Nm). Remove the auto-tensioner stopper bolt.
25. After 5 minutes, check the clearance between the tensioner arm and the pusher stays at 0.138–0.205 in. (3.5–5.2mm).
26. Check for proper installation of the timing belt at all pulleys, then install the timing belt covers. Tighten bolts to 24–38 inch lbs. (3–5 Nm).
27. Install the water inlet and outlet.
28. Install the crankshaft pulley and bolt. Tighten the crankshaft pulley bolt to 159–174 ft. lbs. (216–235 Nm).
29. Install the cooling fan and fan coupling, drive belts and radiator.
30. Install the air ducts.
31. Fill the cooling system and install the engine undercover.

1995–99 Models

1. Disconnect the negative battery cable.
2. Remove the engine undercover.
3. Drain the cooling system. Remove both cylinder block drain plug to drain coolant from the block.
4. Remove the air ducts.
5. Remove the radiator, drive belts, cooling fan and coupling.
6. Remove the crankshaft pulley bolt. Remove the crankshaft pulley using a puller.
7. Remove the water inlet and outlet. Remove the starter motor.
8. Remove the front timing covers.
9. Set the No. 1 cylinder on TDC of the compression stroke.
10. The automatic belt tensioner is oil damped and spring operated. Install a 6mm bolt to hold the tensioner back against the spring and release tension on the belt.
11. Remove the auto-tensioner and timing belt.

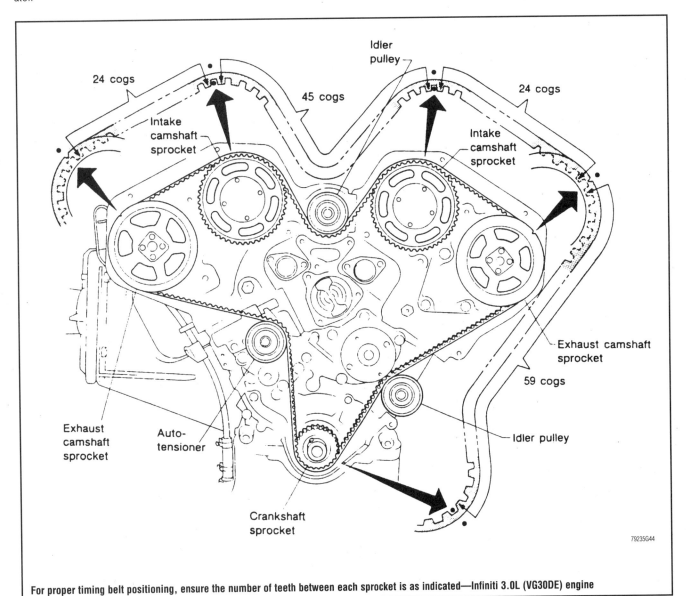

For proper timing belt positioning, ensure the number of teeth between each sprocket is as indicated—Infiniti 3.0L (VG30DE) engine

110 TIMING BELTS—INFINITI

> ※※ **WARNING**
>
> **Do not rotate the crankshaft or camshaft separately because the pistons will strike the valves causing engine damage.**

To install:

12. Confirm that the No. 1 cylinder is at TDC of the compression stroke.
13. Align the marks on the camshaft and crankshaft sprockets with the marks on the rear belt cover and oil pump housing.
14. With the arrows on the timing belt pointing towards the front, align the white lines on the timing belt with the marks on the sprockets and install the belt.
15. To prepare the auto-tensioner for installation, perform the following:
    a. Remove the bolt holding the tensioner in position.
    b. Use a vise to adjust the gap between the tensioner arm and pusher body to 0.160 in. (4mm).
    c. Install the bolt again to hold the arm in this position. Do not try to use the bolt to adjust the gap or the threads will be damaged.
16. Install the auto-tensioner, push it towards the belt to just take up the slack, then tighten the bolts finger-tight.
17. Before adjusting the timing belt tension, the slack must be properly distributed:
    a. Turn the crankshaft 10 degrees clockwise and tighten the tensioner bolts and nut to 12–15 ft. lbs. (16–21 Nm). Do not push the auto-tensioner hard or the belt will be adjusted too tight.
    b. Turn the crankshaft 120 degrees (⅓ turn) counterclockwise.
    c. Loosen the tensioner bolts and nut ½ turn and move the tensioner body away from the timing belt as far as it will move.
    d. Turn the crankshaft clockwise to TDC again.
    e. Push the tensioner against the belt with a force of 13 lbs. (59 N) using a spring scale or similar tool and tighten the bolts again to 12–15 ft. lbs. (16–21 Nm). The pressure specification is important and a special spring scale tool, J-38387, is available to measure the tensioner force.
18. To check the timing belt tension:
    a. Turn the crankshaft 120 degrees (⅓ turn) clockwise, then turn counterclockwise and return the engine to TDC.
    b. Prepare a steel plate that is approximately ⅜ in. (10mm) wide and longer than the width of the belt.
    c. Set the plate on the timing belt between two camshaft sprockets and push against the plate with a force of 11 lbs. (49 N). Note the belt deflection.
    d. Repeat the procedure between the other camshaft sprockets and between the exhaust sprockets and idler/tensioner pulleys. There will be a total of four measurements.
    e. Add the deflection measurements and divide by four. The average deflection must be 0.240–0.280 in. (6–7mm). If belt tension is not correct, start the entire adjustment procedure again.
19. Confirm the auto-tensioner mounting nuts are tightened to 12–15 ft. lbs. (16–21 Nm) and remove the auto-tensioner stopper bolt.
20. After 5 minutes, measure the clearance between the tensioner arm and the pusher. It should be 0.138–0.205 in. (3.5–5.2mm).
21. Be sure all the sprocket timing marks are correctly aligned. Install the timing belt covers and tighten the bolts to 24–38 inch lbs. (3–5 Nm).
22. Install all applicable components.

3.3L (VG33E) ENGINE

1. Remove the engine undercover.
2. Remove the radiator shroud, the fan and the pulleys.
3. Drain the coolant from the radiator and remove the water pump hose.

> ※※ **CAUTION**
>
> **When draining the coolant, keep in mind that cats and dogs are attracted by the ethylene glycol antifreeze, and are quite likely to drink any that is left in an uncovered container or in puddles on the ground. This will prove fatal in sufficient quantity. Always drain the coolant into a sealable container. Coolant should be reused unless it is contaminated or several years old.**

4. Remove the radiator.
5. Remove the power steering, A/C compressor and alternator drive belts.
6. Remove the spark plugs.
7. Remove the distributor protector (dust shield).
8. Remove the A/C compressor drive belt idler pulley and bracket.
9. Remove the fresh air intake tube at the cylinder head cover.
10. Disconnect the radiator hose at the thermostat housing.
11. Remove the crankshaft pulley bolt, then pull off the pulley with a suitable puller.
12. Remove the bolts, then remove the front upper and lower timing belt covers.
13. Set the No. 1 piston at TDC of its compression stroke. Align the punchmark on the left camshaft sprocket with the punchmark on the timing belt upper rear cover. Align the punchmark on the crankshaft sprocket with the notch on the oil pump housing. Temporarily install the crank pulley bolt so the crankshaft can be rotated if necessary.
14. Loosen the timing belt tensioner and return spring, then remove the timing belt.

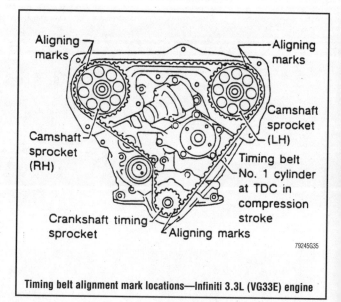

Timing belt alignment mark locations—Infiniti 3.3L (VG33E) engine

INFINITI, ISUZU—TIMING BELTS

To install:

> ※※ **CAUTION**
>
> Before installing the timing belt, confirm that the No. 1 cylinder is set at the TDC of the compression stroke.

15. Remove both cylinder head covers and loosen all rocker arm shaft retaining bolts.

➡ The rocker arm shaft bolts MUST be loosened so that the correct belt tension can be obtained.

16. Install the tensioner and the return spring. Using a hexagon wrench, turn the tensioner clockwise and temporarily tighten the locknut.
17. Be sure that the timing belt is clean and free from oil or water.
18. When installing the timing belt align the white lines on the belt with the punchmarks on the camshaft and crankshaft sprockets. Have the arrow on the timing belt pointing toward the front belt covers.

➡ A good way (although rather tedious!) to check for proper timing belt installation is to count the number of belt teeth between the timing marks. There are 133 teeth on the belt; there should be 40 teeth between the timing marks on the left and right side camshaft sprockets, and 43 teeth between the timing marks on the left side camshaft sprocket and the crankshaft sprocket.

19. While keeping the tensioner steady, loosen the locknut with a hex wrench.
20. Turn the tensioner approximately 70–80 degrees clockwise with the wrench, then tighten the locknut.

> ※※ **WARNING**
>
> If any binding is felt when adjusting the timing belt tension by turning the crankshaft, STOP turning the engine, because the pistons may be hitting the valves.

21. Turn the crankshaft in a clockwise direction several times, then **slowly** set the No. 1 piston to TDC of the compression stroke.
22. Apply 22 lbs. of pressure (push it in!) to the center span of the timing belt between the right side camshaft sprocket and the tensioner pulley, then loosen the tensioner locknut.
23. Using a 0.0138 in. (0.35mm) thick feeler gauge (the actual width of the blade **must** be ½ in. or 13mm!), turn the crankshaft clockwise (**slowly!**). The timing belt should move approximately 2 ½ teeth. Tighten the tensioner locknut, turn the crankshaft slightly and remove the feeler gauge.
24. Slowly rotate the crankshaft clockwise several more times, then set the No. 1 piston to TDC of the compression stroke.
25. Position the two timing covers on the block, then tighten the mounting bolts to 24 ft. lbs. (35 Nm).
26. Press the crankshaft pulley onto the shaft, then tighten the bolt to 90–98 ft. lbs. (123–132 Nm).
27. Connect the radiator hose to the thermostat housing.
28. Reconnect the fresh air intake tube at the cylinder head cover.
29. Install the A/C compressor drive belt idler pulley and bracket.
30. Install the distributor protector (dust shield).
31. Install the spark plugs.
32. Install the power steering, A/C compressor and alternator drive belts.
33. Install the radiator.
34. Reconnect the water pump hose and fill the engine with coolant. Install the fan shroud and pulleys.
35. Install the engine undercover.
36. Start the engine and check for any leaks.

Isuzu

1.6L (SOHC) ENGINE

1. Disconnect the negative battery cable.
2. Use a device to support the engine and remove the right side engine mount.
3. Remove the necessary accessory drive belts.
4. Remove the right side engine mounting bridge bracket and the torque rod.
5. Remove the crank pulley bolt and the crank pulley.
6. Remove the front exhaust pipe, the stud from the transaxle housing, the stiffener and the flywheel dust cover.
7. Remove the upper and lower timing belt covers.
8. Bring the piston in No. 4 cylinder to TDC on the compression stroke. The crankshaft sprocket timing mark should be aligned with the triangular mark on the oil pump housing. The notch on the camshaft sprocket should be aligned with the left upper corner of the cylinder head, with the dowel pin in the up position.
9. Remove the crank pulley bolt and the crank pulley, being careful not to disturb the position of the crankshaft.

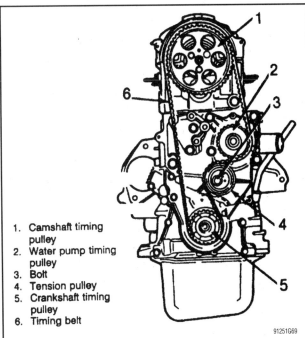

1. Camshaft timing pulley
2. Water pump timing pulley
3. Bolt
4. Tension pulley
5. Crankshaft timing pulley
6. Timing belt

Timing belt, pulley and sprocket identification—Isuzu 1.6L SOHC engines

112 TIMING BELTS—ISUZU

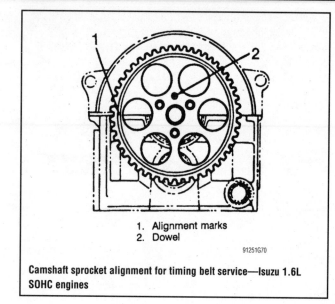

1. Alignment marks
2. Dowel

Camshaft sprocket alignment for timing belt service—Isuzu 1.6L SOHC engines

When installing a new timing belt, position the belt as indicated—Isuzu 1.6L SOHC engines

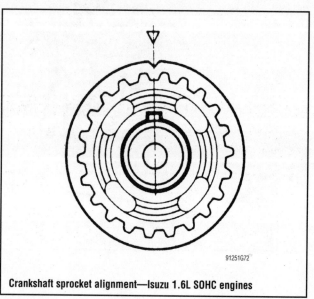

Crankshaft sprocket alignment—Isuzu 1.6L SOHC engines

10. Loosen the bolts retaining the tension pulley. Using a suitable Allen wrench, turn the tension pulley clockwise and relieve the tension on the timing belt.
11. Mark the direction of rotation on the timing belt and remove the timing belt from the vehicle.

To install:

12. If a new belt is used, set the letters marked on the belt in the direction of engine rotation. If the old belt is used, install it in the same direction as before, as indicated by the mark that was made during the removal procedure.
13. Install the belt over the crankshaft sprocket, camshaft sprocket, water pump pulley and tension pulley, in that order.

➥**There must be no slack in the belt after it has been installed. The teeth of the belt and the teeth of the pulley must be in perfect alignment.**

14. Install the crankshaft pulley hub with the taper face to the belt. Tighten the crank pulley bolt to 108 ft. lbs. (150 Nm).
15. Tighten the timing belt cover bolts to 89 inch lbs. (10 Nm).
16. Install the remainder of the components in the reverse order of their removal.

1.6L (DOHC) & 1.8L ENGINES

1. Disconnect the negative battery cable.
2. Use a device to support the engine and remove the right side engine mount.
3. Remove the necessary accessory drive belts.
4. Remove the right side engine mounting bridge bracket and the torque rod.
5. Remove the crank pulley bolt and the crank pulley.
6. Remove the front exhaust pipe, the stud from the transaxle housing, the stiffener and the flywheel dust cover.
7. Remove the upper and lower timing belt covers.
8. Rotate the crankshaft to align the timing marks. The mark on the crankshaft timing sprocket should be aligned with the triangular mark on the oil pump. The keyway should be at the top of the crankshaft, towards the cylinder head. The marks on the camshaft sprockets should be directly across from each other and aligned with the top edge of the cylinder head.

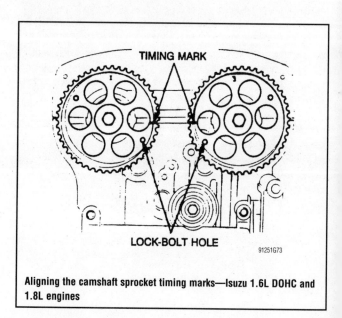

Aligning the camshaft sprocket timing marks—Isuzu 1.6L DOHC and 1.8L engines

ISUZU—TIMING BELTS

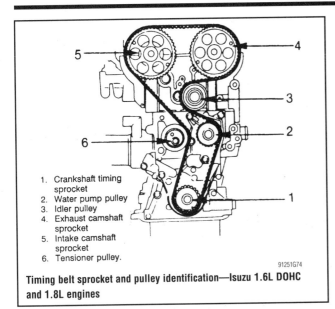

1. Crankshaft timing sprocket
2. Water pump pulley
3. Idler pulley
4. Exhaust camshaft sprocket
5. Intake camshaft sprocket
6. Tensioner pulley.

Timing belt sprocket and pulley identification—Isuzu 1.6L DOHC and 1.8L engines

9. Loosen the tension pulley attaching bolt ½ turn. Insert a hex wrench into the tension pulley hexagonal hole and loosen the timing belt by rotating the tension pulley.
10. Mark the rotational direction of the timing belt and remove it from the vehicle.

To install:

11. Make sure the timing marks are still in alignment.
12. Lock the camshaft sprockets in position by inserting 6mm bolts through the camshaft sprockets and into the cylinder heads.
13. Install the timing belt. A new belt is installed correctly if the lettering can be read while viewing it from the passenger side fender. If the old belt is being used, it must be installed in the same direction as was marked during the removal procedure. The belt must be installed in the following order:
    a. Crankshaft timing sprocket.
    b. Water pump pulley.
    c. Idler pulley.
    d. Exhaust camshaft sprocket.
    e. Intake camshaft sprocket.
    f. Tensioner pulley.

➡**There must be no slack in the belt after it has been installed. The teeth of the belt and the teeth of the sprocket must be in perfect alignment.**

14. Properly tension the belt.
15. Install the remainder of the components in the reverse order of their removal.

2.2L (F22B6) & 2.3L (F23A7) ENGINES

➡**The radio may contain a coded theft protection circuit. Always make note of your code number before disconnecting the battery.**

1. Disconnect the negative and positive battery cables.
2. Remove the valve cover.
3. Remove the upper timing belt cover.
4. Turn the engine to align the timing marks and set cylinder No.1 to TDC for the compression stroke. The white mark on the crankshaft pulley should align with the pointer on the timing belt cover. The words **UP** embossed on the camshaft pulley should be aligned in the upward position. The marks on the edge of the pulley should be aligned with the cylinder head or the back cover upper edge. Once in this position, the engine must NOT be turned or disturbed.
5. Remove the splash shield from below the engine.
6. Remove the wheel well splash shield.
7. Loosen and remove the power steering pump belt. Remove the power steering pump.
8. Loosen the adjusting and mounting bolts for the alternator and remove the drive belt.
9. Support the engine with a floor jack cushioned with a piece of wood.
10. Remove the through-bolt for the side engine mount and remove the mount.
11. Remove the crankshaft pulley bolt and remove the crankshaft pulley. Use a crank pulley holder (part No. 07MAB-PY3010A) and holder handle (part No. 07JAB-001020A), or there equivalents, to hold the crankshaft pulley in place while removing the bolt.
12. Remove the lower timing belt cover.
13. Remove the balancer shaft belt and its drive pulley.
14. Insert a suitable tool into the maintenance hole in the front balancer shaft. Unbolt and remove the balancer driven pulley.

➡**For servicing the balance shafts, front refers to the side of the engine facing the radiator. Rear refers to the side of the engine facing the firewall.**

15. Remove the timing belt.
16. If equipped with a TDC sensor assembly at the crankshaft sprocket, unbolt the assembly and move it to the side before removing the sprocket.
17. Remove the key and the spacers to remove the crankshaft timing sprocket.
18. Unbolt and remove the camshaft timing sprocket.

To install:

19. Install the camshaft timing sprocket so that the **UP** mark is up and the TDC marks are parallel to the cylinder head gasket surface. Install the key and tighten the bolt to 27 ft. lbs. (37 Nm).
20. Install the crankshaft timing sprocket so that the TDC mark aligns with the pointer on the oil pump. Install the spacers with their concave surfaces facing in. Install the key. Install the TDC sensor assembly back into position before installing the timing belt.
21. Install and tension the timing belt.
22. Rotate the crankshaft counterclockwise five to six turns to be sure the belt is properly seated.
23. Set the No. 1 piston at TDC for its compression stroke.

※※ WARNING

If any binding is felt when adjusting the timing belt tension by turning the crankshaft, STOP turning the engine, because the pistons may be hitting the valves.

24. Rotate the crankshaft counterclockwise so that the camshaft pulley moves only three teeth beyond its TDC mark.
25. Tighten the tensioner adjusting nut to 33 ft. lbs. (45 Nm).
26. Tighten the crankshaft pulley bolt to 181 ft. lbs. (245 Nm).
27. Install the balancer shaft belt drive pulley.

114 TIMING BELTS—ISUZU

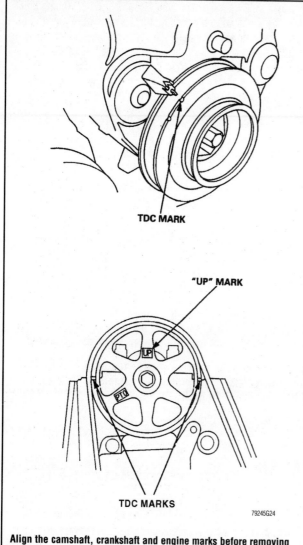

Align the camshaft, crankshaft and engine marks before removing the timing belt and pulleys—Isuzu 2.2L (F22B6) and 2.3L (F23A7) engines

28. Align the groove on the pulley edge to the pointer on the balancer gear case.
29. Check the alignment of the pointer on the balancer pulley to the pointer on the oil pump.
30. Install and tension the balancer shaft belt.
31. Be sure the timing belts have been tensioned correctly and that all TDC and alignment marks are in their proper positions.
32. Install the lower timing cover and the crankshaft pulley. Apply engine oil to the pulley bolt threads and washer surface. Install the pulley bolt and tighten it to 181 ft. lbs. (245 Nm).
33. Install the upper timing cover and the valve cover. Be sure the seals are properly seated.
34. Install the side engine mount. Tighten the through-bolt to 47 ft. lbs. (64 Nm). Tighten the mount nut and bolt to 40 ft. lbs. (55 Nm) each.
35. Remove the floor jack.
36. Install and tension the alternator belt.
37. Install the power steering pump and tension its belt.
38. Install the splash shields.
39. Reconnect the positive and negative battery cables. Enter the radio security code.
40. Check engine operation.

2.6L ENGINE

1. Disconnect the negative battery cable.
2. Loosen and remove the engine accessory drive belts.
3. Remove the cooling fan assembly and the water pump pulley.
4. Drain the fluid from the power steering reservoir.
5. Unbolt and remove the power steering pump. Unbolt the hydraulic line brackets from the upper timing cover and move the pump out of the work area without disconnecting the hydraulic lines.
6. Disconnect and remove the starter motor if a flywheel holder (part No. J-38674 or equivalent) is to be used.
7. Remove the upper timing belt cover.
8. Rotate the crankshaft to set the engine at TDC/compression for the No. 1 cylinder. The arrow mark on the camshaft sprocket will be aligned with the mark on the rear timing cover.
9. Remove the crankshaft pulley.
10. Remove the lower timing belt cover.
11. Verify that the engine is set at TDC/compression for the No. 1 cylinder. The notch on the crankshaft sprocket will be aligned with the pointer on the oil seal retainer.
12. Release and remove the tensioner spring to release the timing belt's tension.
13. Remove the timing belt.
14. Unbolt the tensioner pulley bracket from the engine's front cover.
15. If necessary, unbolt and remove the camshaft sprockets. Use a puller to remove the crankshaft pulley if necessary. Don't lose the crankshaft sprocket key.

To install:
16. If removed, install the camshaft and crankshaft sprockets. Align the camshaft and crankshaft timing marks and be sure to install any keys. Tighten the camshaft sprocket bolt to 43 ft. lbs. (59 Nm).
17. Install the tensioner assembly. Tighten the tensioner mounting bolt to 14 ft. lbs. (19 Nm) and the cap bolt to 9 ft. lbs. (13 Nm).
18. Be sure the crankshaft and the camshaft sprockets are aligned with their timing marks. Install the timing belt onto the sprockets using the following sequence: first around the crankshaft sprocket; second around the oil pump sprocket; third around the camshaft sprocket.
19. Loosen the tensioner mounting bolt. This will allow the tensioner spring to apply pressure to the timing belt.
20. After the spring has pulled the timing belt as far as possible, temporarily tighten the tensioner mounting bolt to 14 ft. lbs. (19 Nm).

➡ Remove the flywheel holder before rotating the crankshaft. Reinstall the holder to tighten the crankshaft pulley bolt.

✶✶ WARNING

If any binding is felt when adjusting the timing belt tension by turning the crankshaft, STOP turning the engine, because the pistons may be hitting the8 valves.

ISUZU—TIMING BELTS 115

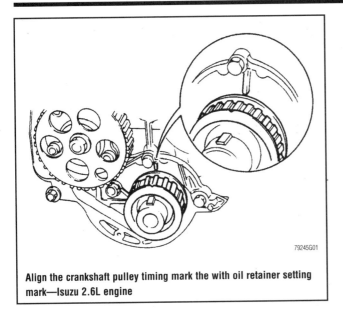

Align the crankshaft pulley timing mark the with oil retainer setting mark—Isuzu 2.6L engine

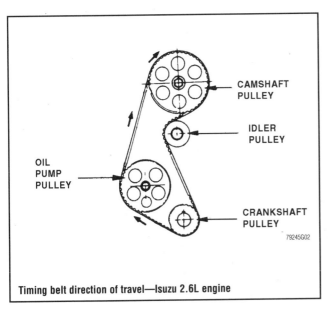

Timing belt direction of travel—Isuzu 2.6L engine

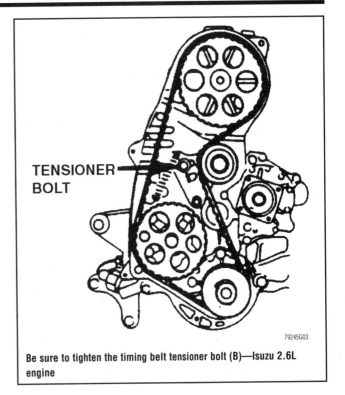

Be sure to tighten the timing belt tensioner bolt (B)—Isuzu 2.6L engine

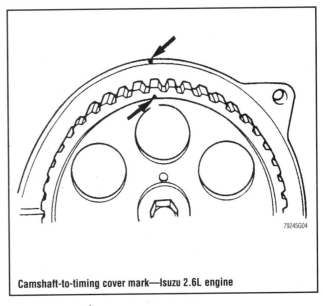

Camshaft-to-timing cover mark—Isuzu 2.6L engine

21. Rotate the crankshaft counterclockwise two complete revolutions to check the rotation of the belt and the alignment of the timing marks. Listen for any rubbing noises which may mean the belt is binding.

22. Loosen the tensioner pulley bolt to allow the spring to adjust the correct tension. Then, retighten the tensioner pulley bolt to 14 ft. lbs. (19 Nm).

23. Install the lower timing cover and the crankshaft pulley.

24. Tighten the crankshaft pulley bolt to 87 ft. lbs. (118 Nm). Tighten the small pulley bolts to 6 ft. lbs. (8 Nm).

25. Install the upper timing cover.

26. Install the starter if it was removed. Tighten the bolts to 30 ft. lbs. (40 Nm).

27. Install the power steering pump. If the hydraulic lines were disconnected, refill and bleed the power steering system.

28. Install the water pump pulley and tighten its nut to 20 ft. lbs. (26 Nm).

29. Install the cooling fan assembly.

30. Install and adjust the accessory drive belts.

31. Connect the negative battery cable.

3.2L & 3.5L ENGINES

1. Disconnect the negative battery cable.
2. Drain the engine coolant into a sealable container.
3. Remove the air cleaner assembly and intake air duct.
4. Disconnect the upper radiator hose from the coolant inlet.
5. Remove the upper fan shroud from the radiator.
6. Remove the four nuts retaining the cooling fan assembly. Remove the cooling fan from the fan pulley.
7. Loosen and remove the drive belts.
8. Remove the upper timing belt covers.
9. Remove the fan pulley assembly.
10. Rotate the crankshaft to align the camshaft timing marks with the pointer dots on the back covers. Verify that the pointer on the crankshaft aligns with the mark on the lower timing cover.

116 TIMING BELTS—ISUZU

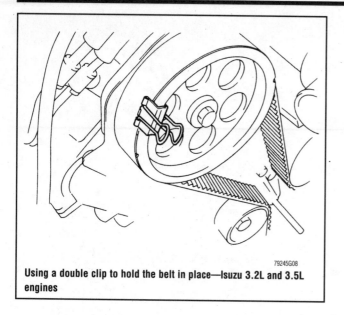

Using a double clip to hold the belt in place—Isuzu 3.2L and 3.5L engines

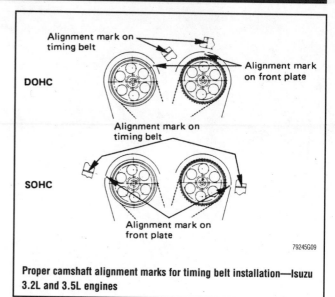

Proper camshaft alignment marks for timing belt installation—Isuzu 3.2L and 3.5L engines

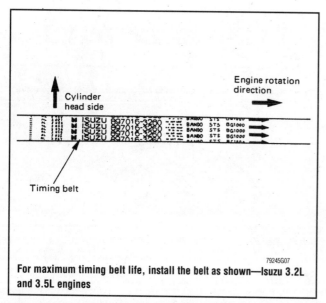

For maximum timing belt life, install the belt as shown—Isuzu 3.2L and 3.5L engines

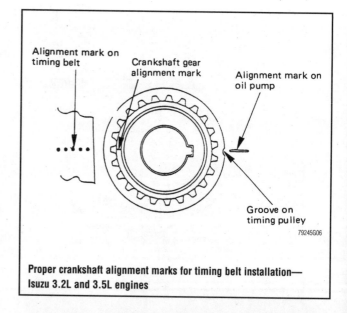

Proper crankshaft alignment marks for timing belt installation—Isuzu 3.2L and 3.5L engines

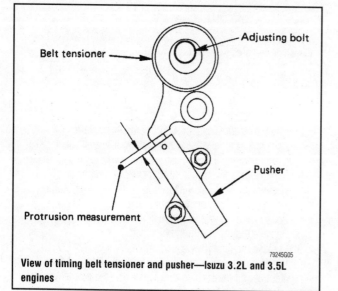

View of timing belt tensioner and pusher—Isuzu 3.2L and 3.5L engines

→ When the timing marks are aligned on 1995 vehicles, no pistons will be at TDC/compression. When the timing marks are aligned on 1995½–99 vehicles, the No. 2 piston is at TDC/compression.

✳✳ WARNING

Align the camshaft and crankshaft sprockets with their alignment marks before removing the timing belt. Failure to align the belt and sprocket marks may result in valve damage.

11. Use tool No. J-8614–01, or a suitable pulley holding tool to remove the crankshaft pulley center bolt. Remove the crankshaft pulley.
12. If present, disconnect the two oil cooler hose bracket bolts on the timing cover. Move the oil cooler hoses and bracket off of the lower timing cover.
13. Remove the lower timing belt cover.
14. Remove the pusher assembly (tensioner) from below the belt tensioner pulley. The pusher rod must always face upward to pre-

ISUZU—TIMING BELTS 117

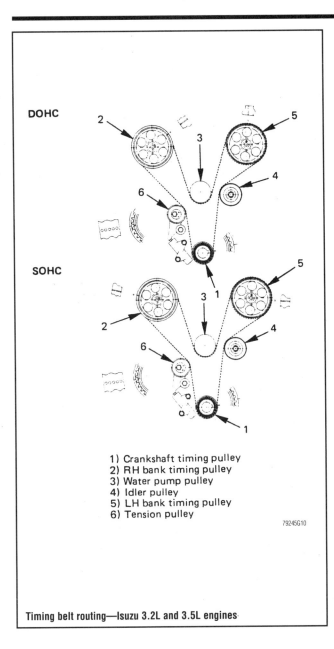

1) Crankshaft timing pulley
2) RH bank timing pulley
3) Water pump pulley
4) Idler pulley
5) LH bank timing pulley
6) Tension pulley

Timing belt routing—Isuzu 3.2L and 3.5L engines

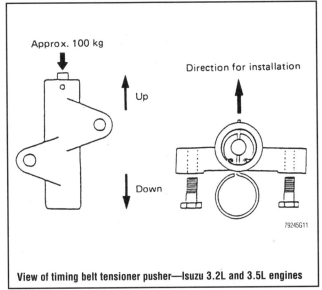

View of timing belt tensioner pusher—Isuzu 3.2L and 3.5L engines

vent oil leakage. Depress the pusher rod, and insert a wire pin into the hole to keep the pusher rod retracted.

15. Remove the timing belt.
16. Inspect the water pump and replace it if there is any doubt about its condition.
17. Repair any oil or coolant leaks before installing a new timing belt. If the timing belt has been contaminated with oil or coolant, or is damaged, it must be replaced.

To install:

18. Verify that the sprocket timing marks are still aligned and that the groove and the keyway on the crankshaft timing sprocket align with the mark on the oil pump. The white pointers on the camshaft timing sprockets should align with the dots on the front plate.
19. Install the timing belt. Use clips to secure the belt onto each sprocket until the installation is complete. Align the dotted marks on the timing belt with the timing mark opposite the groove on the crankshaft sprocket.

→The arrows on the timing belt must follow the belt's direction of rotation. The manufacturer's trademark on the belt's spine should be readable left-to-right when the belt is installed.

20. Align the white line on the timing belt with the alignment mark on the right bank camshaft timing pulley. Secure the belt with a clip.

✶✶ WARNING

If any binding is felt when adjusting the timing belt tension by turning the crankshaft, STOP turning the engine, because the pistons may be hitting the valves.

21. Rotate the crankshaft counterclockwise to remove the slack between the crankshaft sprocket and the right camshaft timing belt sprocket.
22. Install the belt around the water pump pulley.
23. Install the belt on the idler pulley.
24. Align the white alignment mark on the timing belt with the alignment mark on the left bank camshaft timing belt sprocket.
25. Install the crankshaft pulley and tighten the center bolt by hand. Rotate the crankshaft pulley clockwise to give slack between the crankshaft timing belt pulley and the right bank camshaft timing belt pulley.
26. Insert a 1.4mm piece of wire through the hole in the pusher to hold the rod in. Install the pusher assembly while pushing the tension pulley toward the belt.
27. Pull the pin out from the pusher to release the rod.
28. Remove the clamps from the sprockets. Rotate the crankshaft pulley clockwise two turns. Measure the rod protrusion to ensure it is between 0.16–0.24 in. (4–6mm).
29. If the tensioner pulley bracket pivot bolt was removed, tighten it to 31 ft. lbs. (42 Nm).
30. Tighten the pusher bolts to 14 ft. lbs. (19 Nm).
31. Remove the crankshaft pulley. Install the lower and upper timing belt covers and tighten their bolts to 12 ft. lbs. (17 Nm).
32. Fit the oil cooler hose onto the timing cover and tighten its mounting bracket bolts to 16 ft. lbs. (22 Nm).
33. Install the crankshaft pulley and tighten the pulley bolt to 123 ft. lbs. (167 Nm).
34. Install fan pulley assembly and tighten the bolts to 16 ft. lbs. (22 Nm).

118 TIMING BELTS—ISUZU, KIA

35. Install and adjust the accessory drive belts.
36. Install the cooling fan assembly and tighten the bolts to 6 ft. lbs. (8 Nm).
37. Install the upper fan shroud.
38. Install the air cleaner assembly and intake air duct.
39. Connect the upper radiator hose to the coolant inlet.
40. Refill and bleed the cooling system.
41. Connect the negative battery cable.

Kia

SOHC ENGINE

1. Disconnect the negative battery cable.
2. Properly relieve the fuel system pressure.
3. Remove the alternator, power steering and A/C drive belts.
4. Remove the fresh air duct from the top of the radiator.
5. Remove the upper radiator hose.
6. Remove the four attaching nuts to the clutch fan.
7. Remove the five fan shroud bolts. Remove the fan and shroud as an assembly.
8. Remove the six attaching bolts to the crankshaft pulley.
9. Remove the damper pulley.
10. Remove the four upper timing belt cover bolts and remove the cover.
11. Remove the two lower timing belt cover bolts and remove the cover.
12. Align the timing marks. Align the camshaft pulley No. 2 with the matchmark on the front cover and the crankshaft pulley with the matchmark on the oil pump housing.

✱✱ CAUTION

When aligning the timing marks, do not turn the timing gear counterclockwise. Damage to the engine will occur.

13. Loosen the tensioner bolt. Pry the tensioner away from the belt. Tighten tensioner bolt to relieve the pressure against the timing belt.
14. Remove the timing belt.

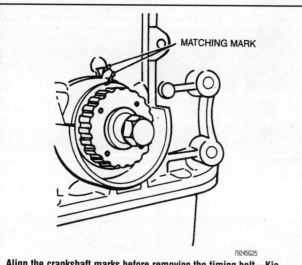
Align the crankshaft marks before removing the timing belt—Kia SOHC and DOHC engines

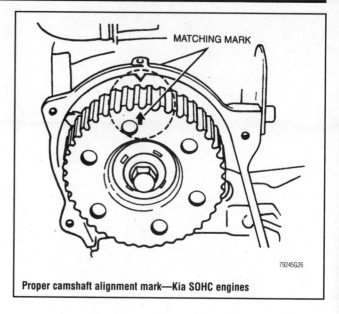
Proper camshaft alignment mark—Kia SOHC engines

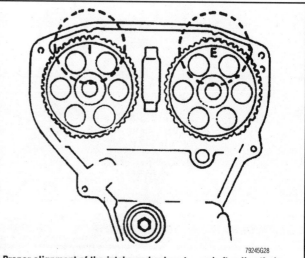
Proper alignment of the intake and exhaust camshaft pulley timing marks—Kia DOHC engines

15. Remove the camshaft pulley attaching bolt. Use a driver placed through one of the holes in the pulley to prevent it from moving when the attaching bolt is removed.

To install:

16. Install the camshaft pulley with the No. 2 pulley aligned with the matchmark on the front cover. Tighten the locking bolt to 42 ft. lbs. (56 Nm).

➡ **Use a driver to stop the pulley from moving when tightening the locking bolt.**

17. Install the crankshaft pulley and belt guide. Tighten the attaching bolt to 132 inch lbs. (15 Nm).
18. Install the timing belt.

✱✱ WARNING

If any binding is felt when adjusting the timing belt tension by turning the crankshaft, STOP turning the engine, because the pistons may be hitting the valves.

KIA, LEXUS—TIMING BELTS

19. Allow the tensioner to press against the timing belt. Tighten tensioner bolt 33 ft. lbs. (45 Nm).
20. Check the timing belt deflection between the crankshaft and camshaft pulleys. The deflection should not exceed 0.43–0.51 in. (11–13 mm) at 22 lbs. (98N).

→ If the deflection exceeds specification, replace the tensioner spring.

21. Install the lower timing belt cover and the two cover bolts.
22. Install the upper timing belt cover. Install the four cover bolts.
23. Install the six attaching bolts to secure the crankshaft damper pulley. Tighten the bolts to 120 ft. lbs. (162 Nm).
24. Install the fan and shroud as an assembly.
25. Install the four attaching nuts to the clutch fan.
26. Install the five fan shroud bolts.
27. Install the upper radiator hose.
28. Install the fresh air duct to the top of the radiator.
29. Install the alternator, power steering and A/C drive belts.
30. Connect the negative battery cable.
31. Start the engine and check for leaks.
32. Road test the vehicle.

DOHC ENGINE

1. Disconnect the negative battery cable.
2. Properly relieve the fuel system pressure.
3. Remove the alternator drive belt.
4. Remove the fresh air duct from the top of the radiator.
5. Remove the upper radiator hose.
6. Remove the four attaching nuts to the clutch fan.
7. Remove the five fan shroud bolts. Remove the fan and shroud as an assembly.
8. Remove the four splash guard mounting bolts and the splash guard.
9. Loosen the lockbolts and loosen the A/C drive belt.
10. Loosen the power steering lock and mounting bolt. Remove the power steering belt.
11. Remove the five upper timing belt cover bolts and remove the cover.
12. Remove the two lower timing belt cover bolts and remove the cover.
13. Align the timing marks.

→When aligning the cam pulleys with the seal plate marks, align the left cam pulley "I" mark and the right cam pulley on the "E" mark.

✼✼ WARNING

When aligning the timing marks, do not turn the timing gear counterclockwise. Damage to the engine will occur.

14. Loosen the tensioner bolt. Pry the tensioner away from the belt. Tighten the tensioner bolt to relieve the pressure against the timing belt.
15. Remove the timing belt.
16. Remove the camshaft pulley attaching bolts. Use a driver placed through one of the holes in the pulley to prevent it from moving when the attaching bolt is removed. Remove and mark the pulleys.
17. Remove the lower timing belt pulley and locking bolt.

To install:
18. Install the camshaft pulleys. Tighten the bolts to 35–48 ft. lbs. (47–65 Nm).
19. Install the lower timing belt pulley and locking bolt. Tighten the bolt to 120 ft. lbs. (162 Nm).
20. If necessary, align the timing marks.

→When aligning the cam pulleys with the seal plate marks, align the left cam pulley "I" mark and the right cam pulley on the "E" mark.

✼✼ WARNING

When aligning the timing marks, do not turn the timing gear counterclockwise. Damage to the engine will occur.

21. Loosen the tensioner bolt. Pry the tensioner away from the belt. Tighten tensioner bolt to relieve the pressure against the timing belt.
22. Install the timing belt.

✼✼ WARNING

If any binding is felt when adjusting the timing belt tension by turning the crankshaft, STOP turning the engine, because the pistons may be hitting the valves.

23. Loosen the tensioner bolt and allow the tensioner to tighten the timing belt. Tighten the tensioner bolt 27–38 ft. lbs. (37–52 Nm).
24. Check the timing belt deflection. If there is more than 0.30–0.33 in. (7.5–8.5 mm) replace the tensioner spring.
25. Install the two lower timing belt cover bolts to the cover.
26. Install the five upper timing belt cover bolts to the cover.
27. Install and adjust the A/C and power steering drive belts.
28. Install the splash guard.
29. Install and tighten the alternator belt.
30. Install the upper radiator hose.
31. Install the fan and shroud as an assembly.
32. Install the four attaching nuts to the clutch fan.
33. Install the five fan shroud bolts.
34. Install the fresh air duct to the top of the radiator.
35. Properly fill the cooling system.
36. Connect the negative battery cable.
37. Start the engine and check for leaks.
38. Road test the vehicle.

Lexus

✼✼ CAUTION

On models with an air bag, wait at least 90 seconds from the time that the ignition switch is turned to the LOCK position and the battery is disconnected before performing any further work.

2.5L (2VZ-FE) & 3.0L (3VZ-FE) ENGINES

1. Disconnect the cable from the negative battery terminal.
2. Remove the power steering pump reservoir and position it out of the way. Remove the right fender apron seal and then remove the alternator and power steering belts. On 2VZ-FE engine, remove the cruise control actuator and vacuum pump.

120 TIMING BELTS—LEXUS

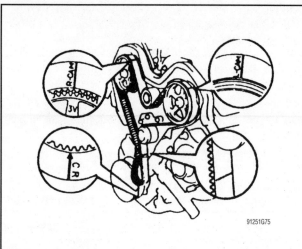

Before removing the timing belt, verify the timing belt alignment marks are visible—Lexus 2.5L (2VZ-FE) and 3.0L (3VZ-FE) engines

Check that the marks on the camshaft sprocket and the No. 3 cover are properly aligned—Lexus 2.5L (2VZ-FE) and 3.0L (3VZ-FE) engines

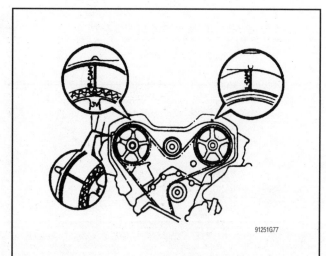

Before removing the timing belt, make sure all marks are aligned—Lexus 2.5L (2VZ-FE) and 3.0L (3VZ-FE) engines

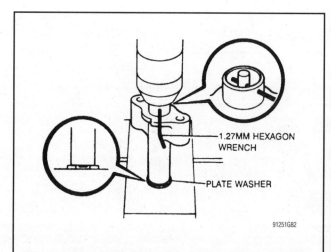

Align the holes on the pushrod and the housing, then insert an Allen wrench to hold the plunger in the depressed position—Lexus 2.5L (2VZ-FE) and 3.0L (3VZ-FE) engines

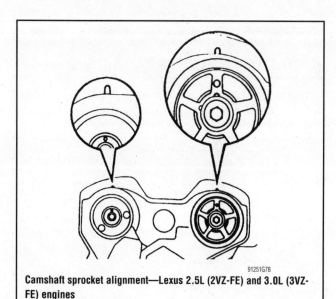

Camshaft sprocket alignment—Lexus 2.5L (2VZ-FE) and 3.0L (3VZ-FE) engines

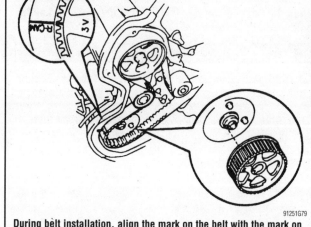

During belt installation, align the mark on the belt with the mark on the right side sprocket—Lexus 2.5L (2VZ-FE) and 3.0L (3VZ-FE) engines

LEXUS—TIMING BELTS

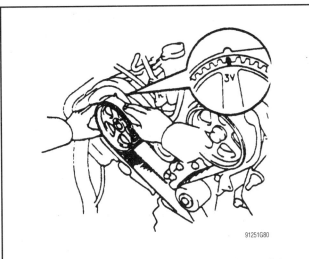

Align the marks on the right sprocket and the No. 3 cover and then slide the belt on—Lexus 2.5L (2VZ-FE) and 3.0L (3VZ-FE) engines

3. On the 3VZ-FE engine, remove the coolant reservoir hose, the washer tank and then the coolant overflow tank.
4. Remove the right side engine mount stays.
5. Position a piece of wood on a floor jack and then slide the jack under the oil pan. Raise the jack slightly until the pressure is off the engine mounts.
6. On the 2VZ-FE engine, remove the right side engine mount insulator. On the 3VZ-FE engine, remove the engine control rod.
7. Remove the spark plugs.
8. Remove the right side engine mounting bracket.
9. Remove the 8 bolts and lift off the upper (No. 2) cover.
10. Paint matchmarks on the timing belt at all points where it meshes with the pulleys and the lower timing cover.
11. Set the No. 1 cylinder to TDC of the compression stroke and check that the timing marks on the camshaft timing pulleys are aligned with those on the No. 3 timing cover. If not, turn the engine 1 complete revolution (360 degrees) and check again.
12. Remove the timing belt tensioner and the dust boot.
13. Turn the right camshaft pulley clockwise slightly to release tension and then remove the timing belt from the pulleys.

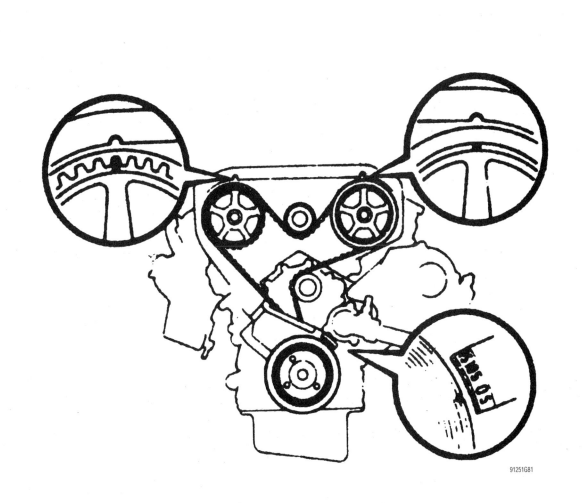

Check that the sprockets are aligned with the timing marks—Lexus 2.5L (2VZ-FE) and 3.0L (3VZ-FE) engines

TIMING BELTS—LEXUS

14. Use a spanner wrench to hold the pulley, loosen the set bolt and then remove the camshaft timing pulleys along with the knock pin. Be sure to keep track of which is which.
15. Remove the No. 2 idler pulley.
16. Remove the crankshaft pulley and then pull off the lower (No. 1) timing belt cover.
17. Remove the timing belt guide.
18. Remove the timing belt.

➡ If the timing belt is to be reused, draw a directional arrow on the timing belt in the direction of engine rotation (clockwise) and place matchmarks on the timing belt and crankshaft gear to match the drilled mark on the pulley.

19. With a 10mm hex wrench, remove the setbolt, plate washer and the No. 1 idler pulley.

To install:
20. Turn the crankshaft until the key groove in the crankshaft timing pulley is facing upward. Slide the timing pulley on so the flange side faces inward.
21. Apply bolt adhesive to the first few threads of the No. 1 idler pulley setbolt, install the plate washer and pulley and then tighten the bolt to 25 ft. lbs. (34 Nm).
22. Install the timing belt on the crankshaft timing, No. 1 idler and water pump pulleys.

➡ If the old timing belt is being reinstalled, make sure the directional arrow is facing in the original direction and that the belt and crankshaft gear matchmarks are properly aligned.

23. Install the lower (No. 1) timing cover and tighten the bolts.
24. Align the crankshaft pulley set key with the key groove on the pulley and slide the pulley on. Tighten the bolt to 181 ft. lbs. (245 Nm).
25. Install the No. 2 idler pulley and tighten the bolt to 29 ft. lbs. (39 Nm). Check that the pulley moves smoothly.
26. Install the left camshaft pulley with the flange side outward. Align the knock pin hole in the camshaft with the knock pin groove on the pulley and then install the pin. Tighten the bolt to 80 ft. lbs. (108 Nm).
27. Set the No. 1 cylinder to TDC again. Turn the right camshaft until the knock pin hole is aligned with the timing mark on the No. 3 belt cover. Turn the left pulley until the marks on the pulley are aligned with the mark on the No. 3 timing cover.
28. Check that the mark on the belt matches with the edge of the lower cover. If not, shift it on the crank pulley until it does. Turn the left pulley clockwise a bit and align the mark on the timing belt with the timing mark on the pulley. Slide the belt over the left pulley. Now move the pulley until the marks on it align with the 1 on the No. 3 cover. There should be tension on the belt between the crankshaft pulley and the left camshaft pulley.
29. Align the installation mark on the timing belt with the mark on the right side camshaft pulley. Hang the belt over the pulley with the flange facing inward. Align the timing marks on the right pulley with the 1 on the No. 3 cover and slide the pulley onto the end of the camshaft. Move the pulley until the camshaft knock pin hole is aligned with the groove in the pulley and then install the knock pin. Tighten the bolt to 55 ft. lbs. (75 Nm).
30. Position a plate washer between the timing belt tensioner and the a block and then press in the pushrod until the holes are aligned between it and the housing. Slide a 1.27mm, except 3VZ-FE engine or 1.5mm for 3VZ-FE Allen wrench through the hole to keep the pushrod set. Install the dust boot and then install the tensioner. Tighten the bolts to 20 ft. lbs. (26 Nm). Don't forget to pull out the Allen wrench.

31. Turn the crankshaft clockwise 2 complete revolutions and check that all marks are still in alignment. If they aren't, remove the timing belt and start over again.
32. Install the right engine mount bracket and tighten it to 30 ft. lbs. (39 Nm).
33. Position a new gasket and then install the upper (No. 2) timing cover.
34. Install the spark plugs.
35. On the 2VZ-FE engine, install the right engine mount insulator. Tighten the bolt to 47 ft. lbs. (64 Nm), the bracket nut to 38 ft. lbs. (52 Nm) and the body nut to 65 ft. lbs. (88 Nm). Install the No. 1 stay and tighten it to 38 ft. lbs. (52 Nm). Install the No. 2 stay and tighten the bolt to 48 ft. lbs. (66 Nm) and the nut to 38 ft. lbs. (52 Nm).
36. On the 3VZ-FE engine, install the control rod and tighten the bolts to 47 ft. lbs. (64 Nm). Install the right stay and tighten it to 23 ft. lbs. (31 Nm).
37. Install and adjust the drive belts.
38. Install the fender apron seal and the wheel.
39. On the 3VZ-FE engine, install the No. 2 stay and tighten the bolt to 55 ft. lbs. (75 Nm), the nut to 46 ft. lbs. (62 Nm). Install the No. 3 stay and tighten it to 54 ft. lbs. (73 Nm).
40. Install the coolant overflow tank and the washer tank.
41. Install the power steering reservoir tank and the cruise control actuator.
42. Connect the battery cable, start the vehicle and check for any leaks.

3.0L (1MZ-FE) ENGINE

1. Disconnect the cable from the negative battery terminal.
2. Remove the power steering pump reservoir and position it out of the way. Remove the right fender apron seal and then remove the alternator and power steering belts.
3. Remove the coolant reservoir hose, the washer tank and then the coolant overflow tank.
4. Remove the right side engine mount stays.
5. Position a piece of wood on a floor jack and then slide the jack under the oil pan. Raise the jack slightly until the pressure is off the engine mounts.
6. Remove the engine control rod.
7. Remove the spark plugs.
8. Remove the right side engine mounting bracket.
9. Remove the 8 bolts and lift off the upper (No. 2) cover.
10. Paint matchmarks on the timing belt at all points where it meshes with the pulleys and the lower timing cover.
11. Set the No. 1 cylinder to TDC of the compression stroke and check that the timing marks on the camshaft timing pulleys are aligned with those on the No. 3 timing cover. If not, turn the engine 1 complete revolution (360 degrees) and check again.
12. Remove the timing belt tensioner and the dust boot.
13. Turn the right camshaft pulley clockwise slightly to release tension and then remove the timing belt from the pulleys.
14. Use a spanner wrench to hold the pulley, loosen the set bolt and then remove the camshaft timing pulleys along with the knock pin. Be sure to keep track of which is which.
15. Remove the No. 2 idler pulley.
16. Remove the crankshaft pulley and then pull off the lower (No. 1) timing belt cover.
17. Remove the upper (No. 3) and lower (No. 1) timing belt covers.
18. Remove the timing belt guide.
19. Remove the timing belt from the engine.

LEXUS—TIMING BELTS

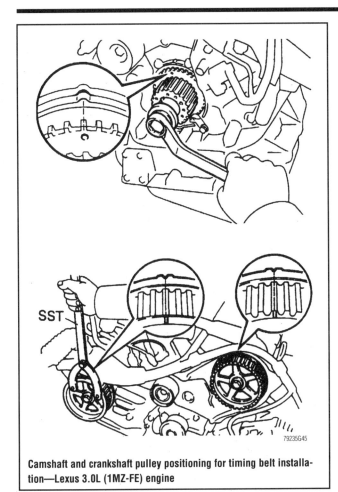

Camshaft and crankshaft pulley positioning for timing belt installation—Lexus 3.0L (1MZ-FE) engine

→ If the timing belt is to be reused, draw a directional arrow on the timing belt in the direction of engine rotation (clockwise) and place matchmarks on the timing belt and crankshaft gear to match the drilled mark on the pulley.

To install:

20. Turn the crankshaft until the key groove in the crankshaft timing pulley is facing upward. Slide the timing pulley on so the flange side faces inward.

21. Apply bolt adhesive to the first few threads of the No. 1 idler pulley setbolt, install the plate washer and pulley and then tighten the bolt to 25 ft. lbs. (34 Nm).

22. Install the timing belt on the crankshaft timing, No. 1 idler and water pump pulleys.

→ If the old timing belt is being reinstalled, make sure the directional arrow is facing in the original direction and that the belt and crankshaft gear matchmarks are properly aligned.

23. Install the lower (No. 1) timing cover and tighten the bolts.

24. Align the crankshaft pulley set key with the key groove on the pulley and slide the pulley on. Tighten the bolt to 181 ft. lbs. (245 Nm).

25. Install the No. 2 idler pulley and tighten the bolt to 29 ft. lbs. (39 Nm). Check that the pulley moves smoothly.

26. Install the left camshaft pulley with the flange side outward. Align the knock pin hole in the camshaft with the knock pin groove on the pulley and then install the pin. Tighten the bolt to 80 ft. lbs. (108 Nm).

27. Set the No. 1 cylinder to TDC again. Turn the right camshaft until the knock pin hole is aligned with the timing mark on the No. 3 belt cover. Turn the left pulley until the marks on the pulley are aligned with the mark on the No. 3 timing cover.

28. Check that the mark on the belt matches with the edge of the lower cover. If not, shift it on the crank pulley until it does. Turn the left pulley clockwise a bit and align the mark on the timing belt with the timing mark on the pulley. Slide the belt over the left pulley. Now move the pulley until the marks on it align with the 1 on the No. 3 cover. There should be tension on the belt between the crankshaft pulley and the left camshaft pulley.

29. Align the installation mark on the timing belt with the mark on the right side camshaft pulley. Hang the belt over the pulley with the flange facing inward. Align the timing marks on the right pulley with the 1 on the No. 3 cover and slide the pulley onto the end of the camshaft. Move the pulley until the camshaft knock pin hole is aligned with the groove in the pulley and then install the knock pin. Tighten the bolt to 55 ft. lbs. (75 Nm).

30. Position a plate washer between the timing belt tensioner and the a block and then press in the pushrod until the holes are aligned between it and the housing. Slide a 1.27mm Allen wrench through the hole to keep the pushrod set. Install the dust boot and then install the tensioner. Tighten the bolts to 20 ft. lbs. (26 Nm). Don't forget to pull out the Allen wrench.

31. Turn the crankshaft clockwise 2 complete revolutions and check that all marks are still in alignment. If they aren't, remove the timing belt and start over again.

32. Install the remaining components. Install and adjust the drive belts.

33. Install the fender apron seal and the wheel.

34. Install the No. 2 stay and tighten the bolt to 55 ft. lbs. (75 Nm), the nut to 46 ft. lbs. (62 Nm). Install the No. 3 stay and tighten it to 54 ft. lbs. (73 Nm).

35. Install the coolant overflow tank and the washer tank.

36. Install the power steering reservoir tank and the cruise control actuator.

37. Connect the battery cable, start the vehicle and check for any leaks.

3.0L (2JZ-GE) ENGINE

1990–94 Models

1. Disconnect the negative battery cable.
2. Drain the engine coolant. Remove the water pump pulley. Remove the radiator.
3. Remove the oil filler cap.
4. Using a 5mm Allen wrench, remove the 9 bolts and lift off the No. 2 and No. 3 timing covers, the top 2.
5. Rotate the crankshaft pulley clockwise so its groove is aligned with the **0** mark in the No. 1 (lower) timing cover. Check that the timing marks on the camshaft timing sprockets are aligned with the marks on the No. 4 (inner) cover, if not, rotate the crankshaft 1 complete revolution (360 degrees).
6. Alternately loosen the 2 tensioner mounting bolts and remove them, the tensioner and the dust boot. Slide the timing belt off of the 2 camshaft sprockets. Its a good idea to matchmark the belt to the pulleys.
7. Making sure the timing belt is securely supported, hold the crankshaft pulley with a spanner wrench and loosen the mounting bolt. Remove the bolt and the pulley.

124 TIMING BELTS—LEXUS

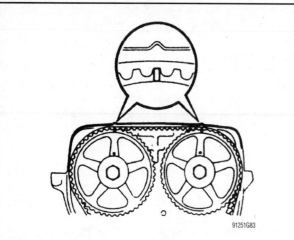

Check that the timing marks on the camshaft sprockets are aligned with the marks on the No. 4 cover—1990–94 Lexus 3.0L (2JZ-GE) engine

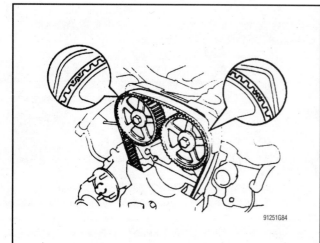

Place matchmarks on the belt and sprockets as shown—1990–94 Lexus 3.0L (2JZ-GE) engine

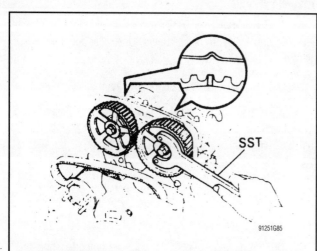

Using the service tool, position the camshaft until the timing marks on the sprockets and No. 4 timing cover are aligned—1990–94 Lexus 3.0L (2JZ-GE) engine

8. Remove the 5 bolts and then lift off the lower No. 1 timing cover.
9. Remove the timing belt guide.
10. Remove the timing belt.

➡ **If the timing belt is to be reused, draw a directional arrow on the timing belt in the direction of engine rotation (clockwise) and place matchmarks on the timing belt and crankshaft gear to match the drilled mark on the pulley.**

11. With a 10mm hex wrench, remove the pivot bolt, plate washer and the idler pulley.

To install:

12. Turn the crankshaft until the key groove in the crankshaft timing pulley is facing upward. Slide the timing pulley on so the flange side faces inward.
13. Apply bolt adhesive to the first few threads of the idler pulley pivot bolt, install the plate washer and pulley and then tighten the bolt to 25 ft. lbs. (34 Nm).
14. Install the timing belt on the crankshaft timing pulley and the idler pulleys.

➡ **If the old timing belt is being reinstalled, make sure the directional arrow is facing in the original direction and that the belt and crankshaft gear matchmarks are properly aligned.**

15. Install the timing belt guide. Install the lower (No. 1) timing cover and tighten the bolts.
16. Align the crankshaft pulley set key with the key groove on the pulley and slide the pulley on. Tighten the bolt to 239 ft. lbs. (324 Nm).
17. Install the camshaft pulleys. Align the knock pin on the camshaft with the key groove on the pulley and then install the pulley. Tighten the bolt to 59 ft. lbs. (79 Nm).
18. Set the No. 1 cylinder to TDC again. Turn the camshaft until the sprocket timing marks are aligned with the timing marks on the No. 4 belt cover.
19. Check that the marks on the belt matches with those on the sprockets and then slide it over the sprockets. If not, shift it on the crank pulley until it does.
20. Position a plate washer between the timing belt tensioner and the a block and then press in the pushrod until the holes are aligned between it and the housing. Slide a 1.5mm Allen wrench through the hole to keep the pushrod set. Install the dust boot and then install the tensioner. Tighten the bolts to 20 ft. lbs. (26 Nm). Don't forget to pull out the Allen wrench.
21. Turn the crankshaft clockwise 2 complete revolutions and check that all marks are still in alignment. If they aren't, remove the timing belt and start over again.
22. Position 2 new gaskets on the lower cover and then install the cover.
23. Align the crankshaft pulley set key with the groove in the pulley and slide it onto the shaft. Secure the pulley with a spanner and tighten the bolt to 239 ft. lbs. (324 Nm).
24. Make sure the crankshaft timing marks and the camshaft sprocket marks are still in alignment and carefully slide the timing belt back over the sprockets. The marks you made previously on the belt and sprockets should still be in alignment.
25. Press the tensioner pushrod into the housing and slide an 1.5mm hex key through the holes to keep it retracted. Install the dust boot and tensioner and tighten the bolts to 20 ft. lbs. (26 Nm). Remove the hex key.

LEXUS—TIMING BELTS

26. Rotate the crankshaft 2 complete revolutions and check that all timing marks are still in alignment. If they aren't, reinstall the belt and try it again.
27. Install the 2 upper covers.
28. Install the radiator and water pump pulley. Refill the engine with coolant. Connect the battery and road test the vehicle.
29. Install and adjust the drive belts.
30. Connect the battery cable, start the vehicle and check for any leaks.

1995–99 Models

1. Disconnect the negative battery cable.
2. Drain the engine coolant. Remove the water pump pulley. Remove the radiator.
3. Remove the oil filler cap.
4. Using a 5mm Allen wrench, remove the 9 bolts and lift off the No. 2 and No. 3 timing covers, the top 2.
5. Rotate the crankshaft pulley clockwise so its groove is aligned with the **0** mark in the No. 1 (lower) timing cover. Check that the timing marks on the camshaft timing sprockets are aligned with the marks on the No. 4 (inner) cover, if not, rotate the crankshaft 1 complete revolution (360 degrees).
6. Alternately loosen the 2 tensioner mounting bolts and remove them, the tensioner and the dust boot. Slide the timing belt off of the 2 camshaft sprockets. Its a good idea to matchmark the belt to the pulleys.
7. Ensuring the timing belt is securely supported, hold the crankshaft pulley with a spanner wrench and loosen the mounting bolt. Remove the bolt and the pulley.
8. Remove the 5 bolts, then lift off the lower No. 1 timing belt cover.
9. Remove the timing belt guide.
10. Remove the timing belt.

➡ **If the timing belt is to be reused, draw a directional arrow on the timing belt in the direction of engine rotation (clockwise) and place matchmarks on the timing belt and crankshaft gear to match the drilled mark on the pulley.**

Set the engine to TDC by aligning the marks before removing the lower timing cover—Lexus 3.0L (2JZ-GE) engines

To install:

11. Install the timing belt on the crankshaft timing pulley and the idler pulleys.

➡ **If the old timing belt is being reinstalled, be sure the directional arrow is facing in the original direction and that the belt and crankshaft gear matchmarks are properly aligned.**

12. Install the timing belt guide. Install the lower (No. 1) timing cover and tighten the bolts.
13. Align the crankshaft pulley set key with the key groove on the pulley and slide the pulley on. Tighten the bolt to 239 ft. lbs. (324 Nm).
14. Set the No. 1 cylinder to TDC again. Turn the camshaft until the sprocket timing marks are aligned with the timing marks on the No. 4 belt cover.
15. Check that the marks on the belt matches with those on the sprockets, then slide it over the sprockets. If not, shift it on the crank pulley until it does.
16. Position a plate washer between the timing belt tensioner and the a block, then press in the pushrod until the holes are aligned between it and the housing. Slide a 1.5mm Allen wrench through the hole to keep the pushrod set. Install the dust boot, then install the tensioner. Tighten the bolts to 20 ft. lbs. (26 Nm). Don't forget to pull out the Allen wrench.
17. Turn the crankshaft clockwise two complete revolutions and check that all marks are still in alignment. If they aren't, remove the timing belt and start over again.
18. Position new gaskets, then install the upper (No. 2 and No. 3) timing covers.

4.0L (1UZ-FE) ENGINE

1990–94 Models

1. Disconnect the negative battery cable and the positive battery cable. Remove the battery.
2. Remove the air duct and dust covers. Remove the engine undercover.
3. Drain the cooling system. Remove the drive belt, fan, fluid coupling and fan pulley.
4. Remove the radiator, air cleaner and throttle body cover. Remove the air intake connector pipe.
5. Remove the air conditioning compressor and power steering pump. Do not disconnect the hoses.
6. Remove the upper high tension cord cover and the right side engine wire cover.
7. Disconnect the PCV hose and remove the left side engine wire cover. Remove the right side No. 3 timing cover.
8. Disconnect and tag the vacuum hoses and remove the left side engine wire cover. Disconnect the spark plug wires.
9. Remove the bolt, cover plate and idler pulley.
10. Disconnect the camshaft position sensor connector and remove the right side No. 2 timing belt cover.
11. Disconnect and remove the ignition coil. Disconnect the hoses and wires from the water bypass pipe. Remove the water bypass pipe.
12. Disconnect the camshaft position sensor connector and remove the left No. 2 timing belt cover.
13. Remove the distributor caps and rotors. Disconnect and remove both distributor housings.

TIMING BELTS—LEXUS

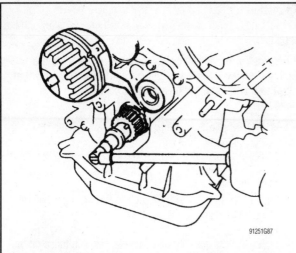

Aligning the timing mark of the crankshaft timing pulley and oil pump body—1990–94 Lexus 4.0L (1UZ-FE) engine

Aligning the timing mark of the camshaft timing pulley and timing belt rear cover—1990–94 Lexus 4.0L (1UZ-FE) engine

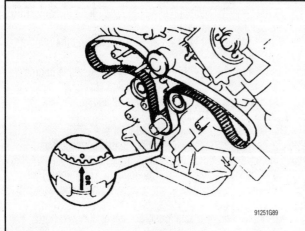

Aligning the installation mark on the timing belt with the drilled mark of the crankshaft timing pulley—1990–94 Lexus 4.0L (1UZ-FE) engine

14. Disconnect and remove the alternator. Remove the drive belt tensioner and the spark plugs.
15. Turn the crankshaft pulley and align it's groove with the timing mark **0** of the No. 1 timing cover. Check that the timing marks of the camshaft timing pulleys and timing belt rear plates are aligned. If not, turn the crankshaft 1 full revolution (360 degrees).
16. Remove the timing belt tensioner. Using the proper tool, loosen the tension between the left side and right side timing pulleys by slightly turning the left side camshaft clockwise.
17. Disconnect the timing belt from the camshaft timing pulleys. Using the proper tool, remove the bolt and the timing pulleys.
18. Remove the bolt and the crankshaft pulley with the proper tool. Remove the fan bracket. Remove the hydraulic pump on the SC400.
19. Remove the mounting bolts and the No. 1 timing belt cover.
20. Remove the 2 upper and lower timing belt covers.
21. Remove the timing belt guide (No. 1 crank position sensor plate).
22. Remove the timing belt.

➡ **If the timing belt is to be reused, draw a directional arrow on the timing belt in the direction of engine rotation (clockwise) and place matchmarks on the timing belt and crankshaft gear to match the drilled mark on the pulley.**

To install:
23. Align the installation mark on the timing belt with the drilled mark of the crankshaft timing pulley. Install the timing belt on the crankshaft timing pulley, No. 1 idler pulley and the No. 2 idler pulley.

➡ **If the old timing belt is being reinstalled, make sure the directional arrow is facing in the original direction and that the belt and crankshaft gear matchmarks are properly aligned.**

24. Install the timing belt guide (No. 1 crank angle sensor plate) with the cup side facing forward. Replace the timing belt cover spacer.
25. Install the No. 1 timing belt cover and tighten the mounting bolts. Install the hydraulic pump on the SC400. Install the fan bracket.
26. Align the pulley set key on the crankshaft with the key groove of the pulley. Install the pulley, using the proper tool to tap in the pulley. Tighten the pulley bolt to 181 ft. lbs. (245 Nm).
27. Align the knock pin on the right side camshaft with the knock pin of the timing pulley. Slide on the timing pulley with the right side mark facing forward. Tighten the bolt to 80 ft. lbs. (108 Nm).
28. Align the knock pin on the left side camshaft with the knock pin of the timing pulley. Slide on the timing pulley with the left side mark facing forward. Tighten the bolt to 80 ft. lbs. (108 Nm).
29. Turn the crankshaft pulley and align it's groove with the **0** timing mark on the No. 1 timing belt cover. Using the proper tool, turn the crankshaft timing pulley and align the timing marks of the camshaft timing pulley and the timing belt rear plate.
30. Install the timing belt to the left side camshaft timing pulley by:

   a. Using the proper tool, slightly turn the left side timing pulley clockwise. Align the installation mark of the timing belt with the timing mark of the camshaft timing pulley and hang the timing belt on the left side camshaft pulley.

   b. Using the proper tool, align the timing marks of the left side camshaft pulley and the timing belt rear plate.

   c. Check that the timing belt has tension between crankshaft timing pulley and the left side camshaft pulley.

LEXUS—TIMING BELTS 127

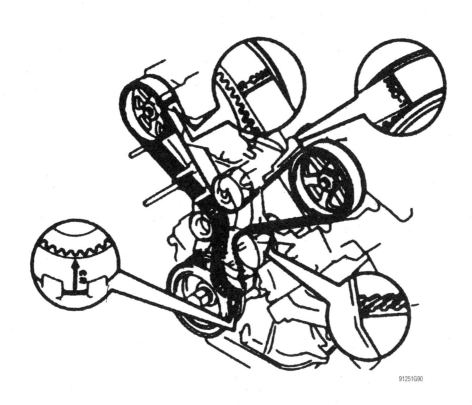

Checking the timing belt installation marks on a used timing belt—1990–94 Lexus 4.0L (1UZ-FE) engine

31. Install the timing belt to the right side camshaft timing pulley by:

    a. Using the proper tool, slightly turn the right side timing pulley clockwise. Align the installation mark of the timing belt with the timing mark of the camshaft timing pulley and hang the timing belt on the right side camshaft pulley.

    b. Using the proper tool, align the timing marks of the right side camshaft pulley and the timing belt rear plate.

    c. Check that the timing belt has tension between crankshaft timing pulley and the right side camshaft pulley.

32. The timing belt tensioner must be set prior to installation. The tensioner can be set by:

    a. Place a plate washer between the tensioner and a block. Using a suitable press, press in the pushrod using 220–2205 lbs. of pressure.

    b. Align the holes of the pushrod and housing, pass the proper tool (1.27mm Allen wrench) through the holes to keep the setting position of the pushrod.

    c. Release the press and install the dust boot to the tensioner.

33. Install the tensioner and tighten the bolts to 20 ft. lbs. (26 Nm). Remove the tool from the tensioner.

34. Turn the crankshaft pulley 2 complete revolutions from TDC to TDC. Always turn the crankshaft clockwise. Check that each pulley aligns with the timing marks.

35. Install the spark plugs and tighten to 13 ft. lbs. (18 Nm). Install the drive belt tensioner and tighten the bolt to 12 ft. lbs. (16 Nm).

36. Install the alternator and engine wire bracket. Tighten the nut and bolt to 26 ft. lbs. (35 Nm) on the LS400 or 27 ft. lbs. (37 Nm) on the SC400. Connect the electrical connections at the alternator.

37. Install both distributor housings and tighten the mounting bolts to 13 ft. lbs. (18 Nm). Replace the distributor rotors and caps.

38. Install the right side No. 2 timing belt cover and tighten the 10mm bolts to 69 inch lbs. (7.8 Nm) and the 12mm bolts to 12 ft. lbs. (16 Nm). Connect the camshaft position sensor connector.

39. Install the left side No. 2 timing belt cover and connect the cam position sensor connector.

40. Install the water bypass pipe and connect the hoses and connectors.

41. Replace the left side ignition coil and connect the coil connector. Install the idler pulley and cover plate. Tighten the bolt to 27 ft. lbs. (37 Nm).

42. Install and secure the ignition wires. Install the right side No. 3 timing belt cover.

43. Install the left side No. 3 timing belt cover and connect the vacuum hose and connectors. Install the right side engine wire cover.

44. Install the left side engine wire cover and connect the vacuum hoses.

45. Install the upper high tension cord covers. Fit the front side claw groove of the upper cover to claw of the lower cover.

46. Install the power steering pump and the air conditioning compressor.

47. Install the throttle body cover and the air cleaner.

48. Install the radiator, fan pulley, fan coupling, fan and drive belt.

49. Install the engine undercover and replace the battery.

50. Install the air ducts and dust covers. Connect the battery cables.

51. Refill the cooling system. Check the ignition timing.

128 TIMING BELTS—LEXUS

1995–99 Models

1. Disconnect the negative battery cable and the positive battery cable. Remove the battery.
2. Remove the air duct and dust covers. Remove the engine undercover.
3. Drain the cooling system. Remove the drive belt, fan, fluid coupling and fan pulley.
4. Remove the radiator, air cleaner and throttle body cover. Remove the air intake connector pipe.
5. Remove the air conditioning compressor and power steering pump. Do not disconnect the hoses.
6. Remove the upper high tension cord cover and the right side engine wire cover.
7. Disconnect the PCV hose and remove the left side engine wire cover. Remove the right side No. 3 timing cover.
8. Disconnect and tag the vacuum hoses and remove the left side engine wire cover. Disconnect the spark plug wires.
9. Remove the bolt, cover plate and idler pulley.
10. Disconnect the camshaft position sensor connector and remove the right side No. 2 timing belt cover.
11. Disconnect and remove the ignition coil. Disconnect the hoses and wires from the water bypass pipe. Remove the water bypass pipe.
12. Disconnect the camshaft position sensor connector and remove the left No. 2 timing belt cover.
13. Remove the distributor caps and rotors. Disconnect and remove both distributor housings.
14. Disconnect and remove the alternator. Remove the drive belt tensioner and the spark plugs.
15. Turn the crankshaft pulley and align it's groove with the timing mark **0** of the No. 1 timing cover. Check that the timing marks of the camshaft timing pulleys and timing belt rear plates are aligned. If not, turn the crankshaft 1 full revolution (360 degrees).
16. Remove the timing belt tensioner. Using the proper tool, loosen the tension between the left side and right side timing pulleys by slightly turning the left side camshaft clockwise.
17. Disconnect the timing belt from the camshaft timing pulleys. Using the proper tool, remove the bolt and the timing pulleys.
18. Remove the bolt and the crankshaft pulley with the proper tool. Remove the fan bracket. Remove the hydraulic pump on the SC400 model.
19. Remove the mounting bolts and the No. 1 timing belt cover.
20. Remove the 2 upper and lower timing belt covers.
21. Remove the timing belt guide (No. 1 crank position sensor plate).
22. Remove the timing belt.

➡ If the timing belt is to be reused, draw a directional arrow on the timing belt in the direction of engine rotation (clockwise) and place matchmarks on the timing belt and crankshaft gear to match the drilled mark on the pulley.

To install:

23. Align the installation mark on the timing belt with the drilled mark of the crankshaft timing pulley. Install the timing belt on the crankshaft timing pulley, No. 1 idler pulley and the No. 2 idler pulley.

➡ If the old timing belt is being reinstalled, be sure the directional arrow is facing in the original direction and that the belt and crankshaft gear matchmarks are properly aligned.

24. Install the timing belt guide (No. 1 crank angle sensor plate) with the cup side facing forward. Replace the timing belt cover spacer.
25. Install the No. 1 timing belt cover and tighten the mounting bolts. Install the hydraulic pump on SC400 models. Install the fan bracket.
26. Align the pulley set key on the crankshaft with the key groove of the pulley. Install the pulley, using the proper tool to tap in the pulley. Tighten the pulley bolt to 181 ft. lbs. (245 Nm).
27. Align the knock pin on the right side camshaft with the knock pin of the timing pulley. Slide on the timing pulley with the right side mark facing forward. Tighten the bolt to 80 ft. lbs. (108 Nm).
28. Align the knock pin on the left side camshaft with the knock pin of the timing pulley. Slide on the timing pulley with the left side mark facing forward. Tighten the bolt to 80 ft. lbs. (108 Nm).
29. Turn the crankshaft pulley and align its groove with the **0** timing mark on the No. 1 timing belt cover. Using the proper tool, turn the crankshaft timing pulley and align the timing marks of the camshaft timing pulley and the timing belt rear plate.
30. Install the timing belt to the left side camshaft timing pulley by:

    a. Using the proper tool, slightly turn the left side timing pulley clockwise. Align the installation mark of the timing belt with the timing mark of the camshaft timing pulley and hang the timing belt on the left side camshaft pulley.

    b. Using the proper tool, align the timing marks of the left side camshaft pulley and the timing belt rear plate.

    c. Check that the timing belt has tension between crankshaft timing pulley and the left side camshaft pulley.

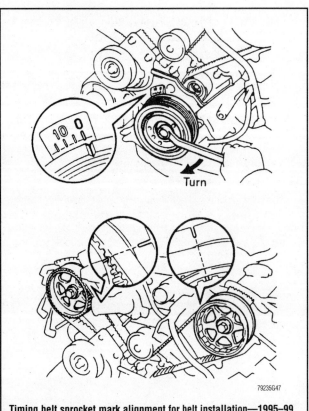

Timing belt sprocket mark alignment for belt installation—1995–99 Lexus 4.0L (1UZ-FE) engine

LEXUS, MAZDA—TIMING BELTS

31. Install the timing belt to the right side camshaft timing pulley by:
   a. Using the proper tool, slightly turn the right side timing pulley clockwise. Align the installation mark of the timing belt with the timing mark of the camshaft timing pulley and hang the timing belt on the right side camshaft pulley.
   b. Using the proper tool, align the timing marks of the right side camshaft pulley and the timing belt rear plate.
   c. Check that the timing belt has tension between the crankshaft timing pulley and the right side camshaft pulley.
32. The timing belt tensioner must be set prior to installation. The tensioner can be set as follows:
   a. Place a plate washer between the tensioner and a block. Using a suitable press, press in the pushrod using 220–2205 lbs. (100–1000kg) of pressure.
   b. Align the holes of the pushrod and housing, pass the proper tool (0.05 in. Allen wrench) through the holes to keep the setting position of the pushrod.
   c. Release the press and install the dust boot on the tensioner.
33. Install the tensioner and tighten the bolts to 20 ft. lbs. (26 Nm). Remove the tool from the tensioner.
34. Turn the crankshaft pulley two complete revolutions from TDC-to-TDC. Always turn the crankshaft clockwise. Check that each pulley aligns with the timing marks.
35. Install the spark plugs and tighten to 13 ft. lbs. (18 Nm). Install the drive belt tensioner and tighten the bolt to 12 ft. lbs. (16 Nm).
36. Install the alternator and engine wire bracket. Tighten the nut and bolt to 26 ft. lbs. (35 Nm) on the LS400 or 27 ft. lbs. (37 Nm) on the SC400. Connect the electrical connections at the alternator.
37. Install both distributor housings and tighten the mounting bolts to 13 ft. lbs. (18 Nm). Replace the distributor rotors and caps.
38. Install the right side No. 2 timing belt cover and tighten the 10mm bolts to 69 inch lbs. (8 Nm) and the 12mm bolts to 12 ft. lbs. (16 Nm). Connect the camshaft position sensor connector.
39. Install the left side No. 2 timing belt cover and connect the camshaft position sensor connector.
40. Install the water bypass pipe and connect the hoses and connectors.
41. Replace the left side ignition coil and connect the coil connector. Install the idler pulley and cover plate. Tighten the bolt to 27 ft. lbs. (37 Nm).
42. Install and secure the ignition wires. Install the right side No. 3 timing belt cover.
43. Install the left side No. 3 timing belt cover and connect the vacuum hose and connectors. Install the right side engine wire cover.
44. Install the left side engine wire cover and connect the vacuum hoses.
45. Install the upper high tension cord covers. Fit the front side claw groove of the upper cover to claw of the lower cover.
46. Install the power steering pump and the air conditioning compressor.
47. Install the throttle body cover and the air cleaner.
48. Install the radiator, fan pulley, fan coupling, fan and drive belt.
49. Install the engine undercover and replace the battery.
50. Install the air ducts and dust covers. Connect the battery cables.
51. Refill the cooling system. Check the ignition timing.

Mazda

1.5L (Z5D), 1.6L (B6) & 1.8L (BP) ENGINES

Except Miata

1. Disconnect the negative battery cable. Remove the engine undercover.
2. Remove the accessory drive belts.
3. Remove the water pump pulley.
4. Remove the crankshaft pulley bolts and remove the crankshaft pulley and baffle plate. Using a suitable tool to hold the crankshaft pulley and remove the pulley lockbolt. Remove the crankshaft pulley boss.
5. Remove the upper and lower timing belt covers.
6. Tag and disconnect the spark plug wires. Remove the spark plugs.

➡ **Spark plugs are removed to make it easier to rotate the engine.**

7. Temporarily reinstall the crankshaft pulley boss and lockbolt.
8. Turn the crankshaft until the timing mark on the crankshaft sprocket aligns with the timing mark on the oil pump and the camshaft sprocket timing marks line up on the camshaft sprockets.
9. Remove the crankshaft pulley lockbolt and pulley boss.
10. Lower the vehicle. Insert a camshaft sprocket holding tool between the camshaft sprockets.

Camshaft and crankshaft sprocket alignment for proper timing belt installation—2.2L engine

130 TIMING BELTS—MAZDA

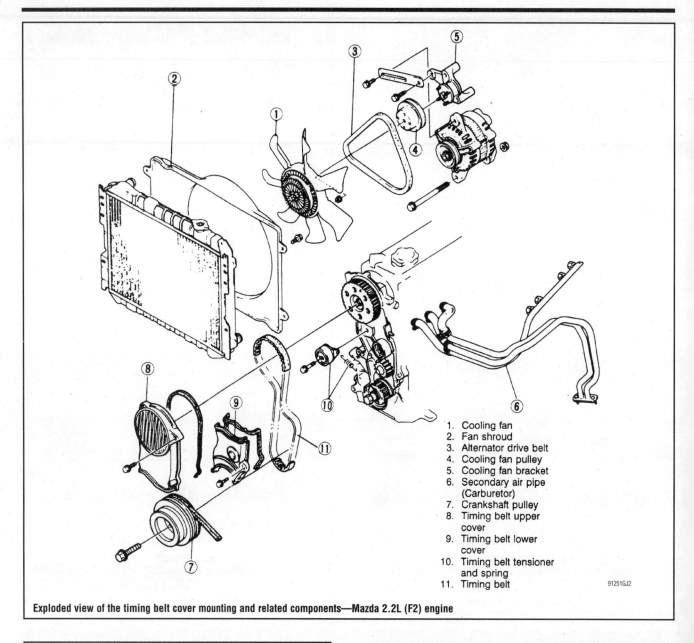

1. Cooling fan
2. Fan shroud
3. Alternator drive belt
4. Cooling fan pulley
5. Cooling fan bracket
6. Secondary air pipe (Carburetor)
7. Crankshaft pulley
8. Timing belt upper cover
9. Timing belt lower cover
10. Timing belt tensioner and spring
11. Timing belt

Exploded view of the timing belt cover mounting and related components—Mazda 2.2L (F2) engine

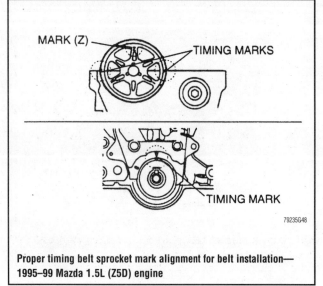

Proper timing belt sprocket mark alignment for belt installation—1995–99 Mazda 1.5L (Z5D) engine

11. Loosen the tensioner pulley lockbolt. Pull the tensioner pulley away from the center of the engine to reduce the tension on the timing belt.

12. If the timing belt is to be reused, mark the direction of rotation on the timing belt. Remove the timing belt.

13. To remove the tensioner, unhook the tensioner spring, and remove the pulley lockbolt and tensioner.

To install:

14. Install the crankshaft sprocket bolt. Install the flywheel locking tool, if equipped with automatic transaxle, or place the shift lever in **4th** gear and apply the parking brake, if equipped with manual transaxle. Tighten the bolt to 108–116 ft. lbs. (147–157 Nm).

15. Be sure the timing marks on the camshaft and crankshaft sprockets are still aligned.

16. If removed, position the tensioner with the spring fully extended, and install the lockbolt tightening the mounting bolt to 28–38 ft. lbs. (38–51 Nm).

17. Install the timing belt. If reusing the original timing belt, be sure it is installed in the same direction of rotation.

MAZDA—TIMING BELTS

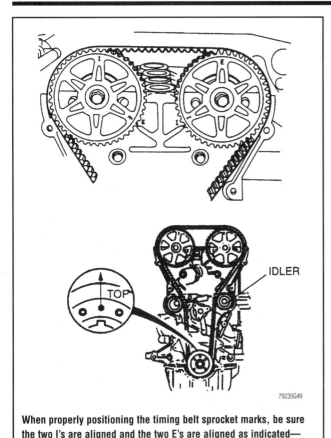

When properly positioning the timing belt sprocket marks, be sure the two I's are aligned and the two E's are aligned as indicated—1995–99 Mazda 1.6L (B6) and Protégé and Miata 1.8L (BP) engines

7. Temporarily reinstall the crankshaft pulley boss and lock bolt.
8. Turn the crankshaft, using the bolt, until the camshaft sprocket and crankshaft sprocket timing marks are aligned. Mark the direction of rotation on the timing belt.
9. Remove the belt tensioner lock bolt, the tensioner wheel and the spring. Remove the timing belt.

➡ **Do not rotate the engine after the timing belt has been removed.**

10. Inspect the belt for wear, peeling, cracking, hardening or signs of oil contamination. Inspect the tensioner for free and smooth rotation. Check the tensioner spring free length; it should not exceed 2.520 in. (64mm). Inspect the sprocket teeth for wear or damage. Replace parts, as necessary.

To install:
11. Make sure the timing marks on the sprockets are properly aligned.
12. Install the timing belt tensioner and spring. Temporarily tighten the bolt with the spring fully extended.
13. Install the timing belt so there is no looseness on the tension side. If reusing the old timing belt, make sure it is reinstalled in the same direction of rotation.
14. Turn the crankshaft 2 turns clockwise and check the timing mark alignment. If the marks are not aligned, repeat Steps 11–14.
15. Loosen the tensioner lock bolt to set the tension, then tighten the bolt to 19 ft. lbs. (25 Nm).
16. Turn the crankshaft 2 turns clockwise and check the alignment of the timing marks. If they are not aligned, repeat Steps 11–16.
17. Apply approximately 22 lbs. pressure to the timing belt on the side opposite the tensioner, at a point midway between the sprockets. The belt should deflect 0.43–0.51 in. (11–13mm). If the tension is not as specified, repeat Steps 14–17 or, if necessary, replace the tensioner spring.
18. Install the spark plugs and connect the spark plug wires.
19. Install the upper and lower timing belt covers. Tighten the bolts to 95 inch lbs. (11 Nm).
20. Install the crankshaft pulley boss and tighten the lock bolt to 123 ft. lbs. (167 Nm), while holding the pulley boss with a suitable tool.

18. Rotate the crankshaft clockwise 1 5/6 turns and align the timing marks. Be sure all marks are still correctly aligned.
19. Loosen the tensioner lockbolt to apply tension to the timing belt. Tighten the tensioner lockbolt to 28–38 ft. lbs. (38–51 Nm). Remove the holding tool from between the camshaft sprockets.
20. Rotate the crankshaft clockwise 2 1/6 turns and be sure all marks are still correctly aligned.
21. Raise and safely support the vehicle. Install the crankshaft pulley lockbolt and boss. Tighten the bolt to 116–122 ft. lbs. (157–166 Nm).
22. Install the timing belt covers.

1.6L (B6E) & 1.8L (BPE) ENGINES

Except Miata

1. Disconnect the negative battery cable. Remove the engine undercover.
2. Remove the accessory drive belts.
3. Remove the water pump pulley.
4. Remove the crankshaft pulley bolts and remove the crankshaft pulley and baffle plate. Using a suitable tool to hold the crankshaft pulley and remove the pulley lock bolt. Remove the crankshaft pulley boss.
5. Remove the upper and lower timing belt covers.
6. Tag and disconnect the spark plug wires. Remove the spark plugs.

➡ **Spark plugs are removed to make it easier to rotate the engine.**

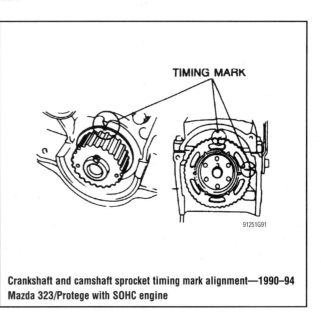

Crankshaft and camshaft sprocket timing mark alignment—1990–94 Mazda 323/Protege with SOHC engine

TIMING BELTS—MAZDA

21. Install the crankshaft pulley and baffle plate.
22. Install the undercover or side cover. Connect the negative battery cable.
23. Start the engine and check for proper operation. Check the ignition timing.

1.6L (B6ZE) ENGINE

Miata

1. Disconnect the negative battery cable. Drain the cooling system.
2. Remove the air intake pipe.
3. Remove the upper radiator hose and disconnect the coolant hoses at the thermostat housing.
4. Remove the accessory drive belts and the water pump pulley.
5. Remove the crankshaft pulley bolts and the crankshaft pulley. On 1992–94 vehicles, hold the pulley boss with a suitable tool and remove the pulley lock bolt. Remove the pulley boss.
6. On 1990–91 vehicles, remove the outer and inner timing belt guide plates.
7. Tag and disconnect the spark plug wires from the spark plugs. Remove the ignition coil and plug wires assembly. Remove the spark plugs.
8. Remove the cylinder head cover. Remove the upper, middle and lower timing belt covers.
9. On 1992–94 vehicles, temporarily reinstall the pulley boss and lock bolt.
10. Turn the crankshaft until the crankshaft and camshaft sprocket timing marks are aligned. On 1992–94 vehicles, the pin on the pulley boss must face upward.
11. On 1992–94 vehicles, remove the pulley boss and lock bolt, being careful not to disturb the crankshaft.
12. Mark the direction of rotation on the timing belt. Loosen the tensioner lock bolt and pry the tensioner outward. Tighten the lock bolt with the tensioner spring fully extended. Remove the timing belt.

➡ **Protect the tensioner with a shop towel before prying on it. Do not rotate the crankshaft after the timing belt has been removed.**

13. Remove the tensioner and spring. If necessary, remove the idler pulley.
14. Inspect the belt for wear, peeling, cracking, hardening or signs of oil contamination. Inspect the tensioner for free and smooth rotation. Check the tensioner spring free length; it should not exceed 2.315 in. (58.8mm). Inspect the sprocket teeth for wear or damage. Replace parts, as necessary.

To install:

15. If removed, install the idler pulley and tighten the bolt to 38 ft. lbs. (52 Nm).
16. Install the tensioner and tensioner spring. Pry the tensioner outward and temporarily tighten the tensioner lock bolt with the tensioner spring fully extended.
17. Make sure the crankshaft sprocket timing mark is aligned with the mark on the oil pump housing and the camshaft sprocket timing marks are aligned with the marks on the seal plate.
18. Install the timing belt so there is no looseness at the idler pulley side or between the camshaft sprockets. If reusing the old belt, make sure it is installed in the same direction of rotation.
19. On 1992–94 vehicles, temporarily install the pulley boss and lock bolt.
20. Turn the crankshaft 2 turns clockwise and align the crankshaft sprocket timing mark. On 1992–94 vehicles, face the pin on the pulley boss upright. Make sure the camshaft sprocket timing marks are aligned. If they are not, repeat Steps 16–20.
21. Turn the crankshaft 1⅚ turns clockwise and align the crankshaft sprocket timing mark with the tension set mark for proper belt tension adjustment. On 1992–94 vehicles, remove the lock bolt and pulley boss.
22. Make sure the crankshaft sprocket timing mark is aligned with the tension set mark. Loosen the tensioner lock bolt and allow the spring to apply tension to the belt. Tighten the tensioner lock bolt to 38 ft. lbs. (52 Nm).
23. On 1992–94 vehicles, install the pulley boss and lock bolt.
24. Turn the crankshaft 2⅙ turns clockwise and make sure the timing marks are correctly aligned.
25. Apply approximately 22 lbs. pressure to the timing belt at a point midway between the camshaft sprockets. The belt should deflect 0.35–0.45 in. (9.0–11.5mm). If the deflection is not correct, repeat Steps 22–25.
26. Install the timing belt covers and tighten the bolts to 95 inch lbs. (11 Nm).
27. Apply silicone sealer to the cylinder head in the area adjacent to the front and rear camshaft caps. Install the cylinder head cover and tighten the bolts to 78 inch lbs. (8.8 Nm).
28. Install the spark plugs. Install the ignition coil and tighten the bolts to 19 ft. lbs. (25 Nm). Connect the spark plug wires.
29. On 1990–91 vehicles, install the inner timing belt guide plate with the dished side facing away from the engine. Install the outer guide plate.

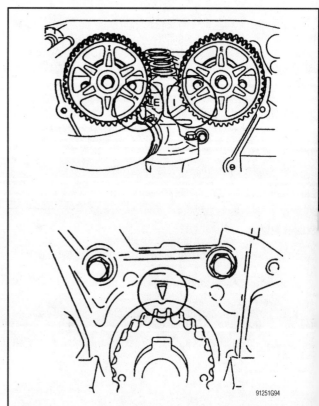

Crankshaft and camshaft sprocket timing mark alignment—Mazda Miata

MAZDA—TIMING BELTS

30. On 1992–94 vehicles, hold the pulley boss with a suitable tool and tighten the lock bolt to 123 ft. lbs. (167 Nm).
31. Install the crankshaft pulley. Tighten the bolts to 13 ft. lbs. (17 Nm).
32. Install the water pump pulley and the accessory drive belts. Adjust the belt tension.
33. Connect the coolant hoses to the thermostat housing and install the upper radiator hose.
34. Install the air intake pipe and connect the negative battery cable. Fill and bleed the cooling system.
35. Start the engine and bring to normal operating temperature. Check for leaks and proper operation. Check the ignition timing.

1.8L (K8) ENGINE

1. Remove the timing belt covers. Temporarily reinstall the crankshaft pulley bolt.
2. On Millenia models, support the engine, and remove the nuts and through-bolt from the right side (number three) engine mount sub bracket. Remove the sub bracket.
3. Turn the crankshaft until the timing mark on the crankshaft sprocket aligns with the timing mark on the oil pump and the camshaft sprocket timing marks align with the marks on the cylinder head. The number one piston should be at TDC of the compression stroke.
4. Remove the two bolts from the automatic tensioner, removing the lower one first. Keep the bolt holes aligned by holding the tensioner to reduce the chance of stripping the threads on the bolts.
5. If the timing belt is to be reused, mark the direction of rotation on the timing belt.
6. Remove the number one idler pulley. Remove the timing belt.

To install:
7. Install the crankshaft sprocket bolt. Install the flywheel locking tool. Tighten the bolt to 116–122 ft. lbs. (157–166 Nm). remove the flywheel locking tool.
8. Position the automatic tensioner in a suitable press. Set a flat washer under the tensioner body to prevent damage to the body plug.
9. Compress the tensioner until the hole in the piston is aligned with the 2nd hole in the tensioner case. Insert a 0.060 in. (1.6mm) diameter wire or pin through the 2nd hole to keep the piston compressed.
10. Be sure the camshaft sprocket timing marks are still aligned. Turn the crankshaft counterclockwise until the timing sprocket is aligned.
11. With the number one idler pulley removed, install the timing belt. If the original belt is being reused, be sure it is installed in the same direction of rotation. The order of installation is: timing belt (crankshaft) sprocket, number two idler pulley, left-hand camshaft sprocket, tensioner pulley and right-hand camshaft sprocket.
12. Install the number one idler pulley while applying pressure on the timing belt. Tighten the bolt to 28–38 ft. lbs. (38–51 Nm).
13. Install the automatic belt tensioner and tighten the bolts to 14–18 ft. lbs. (19–25 Nm). Remove the wire or pin from the tensioner.
14. Turn the crankshaft clockwise, until the crankshaft sprocket timing mark is again at TDC. This should place all of the belt slack in the automatic tensioner portion of the belt.
15. Rotate the crankshaft 2 turns in the normal direction of rotation and align the timing marks. Be sure all marks are still correctly aligned.
16. Inspect timing belt deflection, 0.24–0.31 in. (6–8mm), between the crankshaft sprocket and the tensioner pulley. If it is out of specification, replace the auto-tensioner.
17. On Millenia models, install the right side (number three) engine mount sub bracket. Tighten the nuts to 55–77 ft. lbs.

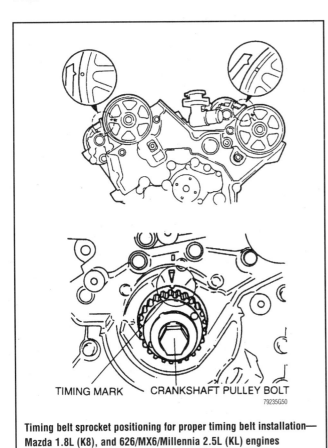

Timing belt sprocket positioning for proper timing belt installation—Mazda 1.8L (K8), and 626/MX6/Millennia 2.5L (KL) engines

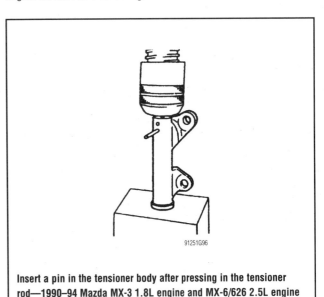

Insert a pin in the tensioner body after pressing in the tensioner rod—1990–94 Mazda MX-3 1.8L engine and MX-6/626 2.5L engine

(75–104 Nm) and the through-bolt to 63–86 ft. lbs. (86–116 Nm). Remove the engine support.

18. Remove the crankshaft damper bolt and install the timing belt covers.

1.8L (K8D) ENGINE

1. Disconnect the negative battery cable. Remove the engine undercover and side cover.
2. Remove the accessory drive belts.
3. Remove the water pump pulley and the accessory drive belt idler pulley bracket.
4. Disconnect the power steering pump pressure switch connector and disconnect the power steering hose from the engine.
5. Remove the power steering pump reservoir bolts and secure the reservoir aside.
6. Keep the power steering pump pulley from turning, by installing a socket on the end of a breaker bar through one of the pulley holes and engaging a pump mounting bolt. Remove the pulley nut and the pulley.
7. Remove the power steering pump mounting bolts and remove the power steering pump. Secure the pump aside, leaving the hoses connected.
8. Hold the crankshaft pulley with a suitable tool and remove the pulley bolt. Remove the crankshaft pulley, being careful not to damage the crank angle sensor rotor on the rear of the pulley.
9. Disconnect the crank angle sensor connector and remove the clip from the engine oil dipstick tube. Remove the dipstick and tube. Plug the hole after removal to prevent the entry of dirt or foreign material.
10. Remove the knock sensor harness bracket and wiring harness from the timing belt cover.
11. Support the engine with engine support tool 49 G017 5A0 or equivalent.
12. Remove the right side engine mount.
13. Remove the right and left timing belt covers.
14. Install the crankshaft pulley bolt and turn the crankshaft until the No. 1 piston is at TDC on the compression stroke. Mark the direction of rotation on the timing belt.
15. Loosen the automatic tensioner bolts and remove the lower bolt. Hold the tensioner so the bolt threads are not damaged during removal.
16. Hold the upper idler pulley to reduce the belt resistance and remove the pulley bolts and pulley. Remove the timing belt. Remove the automatic tensioner and the remaining idler pulley.

➥**Do not rotate the engine after the timing belt has been removed.**

17. Inspect the belt for wear, peeling, cracking, hardening or signs of oil contamination. Inspect the tensioner pulley for free and smooth rotation. Check the automatic tensioner for oil leakage. Check the tensioner rod projection (free length); it should be 0.55–0.63 in. (14–16mm). Inspect the sprocket teeth for wear or damage. Replace parts, as necessary.

To install:

18. Position the automatic tensioner in a suitable press. Place a flat washer under the tensioner body to prevent damage to the body plug.
19. Slowly press in the tensioner rod, but do not exceed 2200 lbs. force. Insert a pin into the tensioner body to hold the rod in place.
20. Install the tensioner and loosely tighten the upper bolts so the tensioner can move.

➥**This is done to reduce the timing belt resistance when the upper idler pulley is installed.**

21. If removed, install the lower idler pulley and tighten the bolt to 38 ft. lbs. (52 Nm).
22. Make sure the crankshaft and camshaft sprocket timing marks are aligned.
23. Install the timing belt over the crankshaft sprocket, lower idler pulley, left camshaft sprocket, tensioner pulley and right camshaft sprocket, in that order. Make sure the belt has no looseness at the tension side. If reusing the old timing belt, make sure it is installed in the same direction of rotation.
24. Install the upper idler pulley while applying pressure on the timing belt. Be careful not to damage the pulley bolt threads when installing. Tighten the upper idler pulley bolt to 34 ft. lbs. (46 Nm).
25. Push the bottom of the automatic tensioner away from the belt and tighten the mounting bolts to 19 ft. lbs. (25 Nm). Remove the pin from the tensioner, applying tension to the belt.
26. Turn the crankshaft twice in the normal direction of rotation and make sure the timing marks are aligned. If the timing marks are not aligned, repeat Steps 18–26.
27. Apply approximately 22 lbs. (98 N) pressure to the timing belt at a point midway between the automatic tensioner and the crankshaft sprocket. The belt should deflect 0.24–0.31 in. (6–8mm). If the deflection is not as specified, replace the automatic tensioner.
28. Install new gaskets and the right and left timing belt covers. Tighten the bolts to 95 inch lbs. (11 Nm).
29. Install the crank angle sensor and harness brackets and tighten the bolts to 95 inch lbs. (11 Nm).
30. Install the right side engine mount. Tighten the mount-to-engine nuts to 76 ft. lbs. (103 Nm) and the mount through bolt to 69 ft. lbs. (93 Nm).
31. Remove the engine support tool.
32. Apply clean engine oil to a new O-ring and install on the dipstick tube. Remove the plug and install the dipstick tube and dipstick. Tighten the tube bracket bolt to 95 inch lbs. (11 Nm).
33. Install the crank angle sensor harness and clip to the dipstick tube. Connect the electrical connector.
34. Remove the crankshaft pulley bolt and install the crankshaft pulley. Reinstall the bolt and hold the pulley with a suitable tool. Tighten the bolt to 123 ft. lbs. (167 Nm).
35. Install the power steering pump. Tighten the mounting bolts to 34 ft. lbs. (46 Nm) except the bolt to the right of the idler pulley. Tighten that bolt to 19 ft. lbs. (25 Nm).
36. Install the power steering pump pulley and loosely tighten the nut. Hold the pulley with the socket and breaker bar and tighten the nut to 69 ft. lbs. (93 Nm).
37. Connect the power steering hose to the engine and connect the pressure switch connector.
38. Install the power steering fluid reservoir and engine ground. Tighten to 87 inch lbs. (9.8 Nm).
39. Install the water pump pulley and loosely tighten the bolts. Install the accessory drive belts and adjust the belt tension.

MAZDA—TIMING BELTS

40. Tighten the water pump pulley bolts to 95 inch lbs. (11 Nm).
41. Install the engine undercover and side cover. Connect the negative battery cable.
42. Start the engine and check for proper operation. Check the ignition timing.

1.8L (BPD) ENGINE

1. Disconnect the negative battery cable. Remove the engine undercover.
2. Remove the accessory drive belts.
3. Remove the crankshaft pulley bolts and remove the crankshaft pulley.
4. Remove the outer timing belt guide plate. Remove the inner timing belt guide plate if so equipped.
5. Tag and disconnect the spark plug wires. Remove the spark plugs.

➡ **Spark plugs are removed to make it easier to rotate the engine.**

6. Remove the engine oil dipstick.
7. Remove the upper, middle and lower timing belt covers.
8. Turn the crankshaft until the timing marks on the crankshaft and camshaft sprockets are aligned. On 1992–94 vehicles, the pin on the pulley boss must face upward.
9. On 1992–94 vehicles, hold the crankshaft pulley boss with a suitable tool and remove the pulley lock bolt, being careful not to rotate the crankshaft. Remove the crankshaft pulley boss.
10. Mark the direction of rotation on the timing belt. Loosen the tensioner lock bolt and pry the tensioner outward. Tighten the lock bolt with the tensioner spring fully extended. Remove the timing belt.

➡ **Protect the tensioner with a shop towel before prying on it. Do not rotate the crankshaft after the timing belt has been removed.**

11. Remove the tensioner and spring. If necessary, remove the idler pulley.
12. Inspect the belt for wear, peeling, cracking, hardening or signs of oil contamination. Inspect the tensioner for free and smooth rotation. Check the tensioner spring free length; it should not exceed 2.315 in. (58.8mm). Inspect the sprocket teeth for wear or damage. Replace parts, as necessary.

To install:

13. If removed, install the idler pulley and tighten the bolt to 38 ft. lbs. (52 Nm).
14. Install the tensioner and tensioner spring. Pry the tensioner outward and temporarily tighten the tensioner lock bolt with the tensioner spring fully extended.
15. Make sure the crankshaft sprocket timing mark is aligned with the mark on the oil pump housing and the camshaft sprocket timing marks are aligned with the marks on the seal plate.
16. Install the timing belt so there is no looseness at the idler pulley side or between the camshaft sprockets. If reusing the old belt, make sure it is installed in the same direction of rotation.
17. On 1992–94 vehicles, temporarily install the pulley boss and lock bolt.
18. Turn the crankshaft 2 turns clockwise and align the crankshaft sprocket timing mark. On 1992–94 vehicles, face the pin on the pulley boss upright. Make sure the camshaft sprocket timing marks are aligned. If they are not, repeat Steps 15–19.

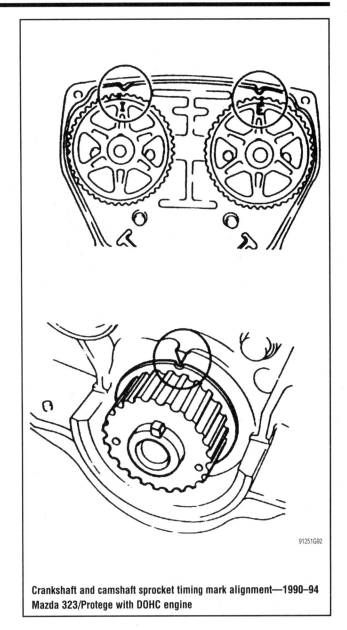

Crankshaft and camshaft sprocket timing mark alignment—1990–94 Mazda 323/Protege with DOHC engine

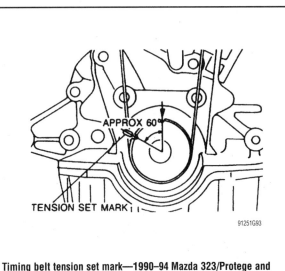

Timing belt tension set mark—1990–94 Mazda 323/Protege and Miata with DOHC engine

136 TIMING BELTS—MAZDA

19. Turn the crankshaft 1⅚ turns clockwise and align the crankshaft sprocket timing mark with the tension set mark for proper belt tension adjustment. On 1992–94 vehicles, remove the lock bolt and pulley boss.

20. Make sure the crankshaft sprocket timing mark is aligned with the tension set mark. Loosen the tensioner lock bolt and allow the spring to apply tension to the belt. Tighten the tensioner lock bolt to 38 ft. lbs. (52 Nm).

21. On 1992–94 vehicles, install the pulley boss and lock bolt.

22. Turn the crankshaft 2⅙ turns clockwise and make sure the timing marks are correctly aligned.

23. Apply approximately 22 lbs. pressure to the timing belt at a point midway between the camshaft sprockets. The belt should deflect 0.35–0.45 in. (9.0–11.5mm). If the deflection is not correct, repeat Steps 21–24.

24. On 1992–94 vehicles, hold the pulley boss with a suitable tool and tighten the lock bolt to 123 ft. lbs. (167 Nm).

25. Install the timing belt covers and tighten the bolts to 95 inch lbs. (11 Nm). Install the engine oil dipstick.

26. Install the spark plugs and connect the spark plug wires.

27. Install the timing belt inner guide plate, if equipped. Make sure the dished side of the plate faces away from the timing belt. Install the outer guide plate, if equipped.

28. Install the crankshaft pulley and tighten the bolts to 13 ft. lbs. (17 Nm).

29. Install the water pump pulley and the accessory drive belts. Adjust the belt tension.

30. Install the engine side or undercover, as necessary. Connect the negative battery cable.

31. Start the engine and check for proper operation. Check the ignition timing.

2.0L (FS) ENGINE

1. Disconnect the negative battery cable.
2. Raise and safely support the vehicle. Remove the right front wheel and tire assembly.
3. Remove the engine undercover.
4. Remove the accessory drive belts and the water pump pulley. Remove the power steering pump pulley shield.
5. Remove the power steering pump and position aside, leaving the hoses connected.
6. Hold the crankshaft pulley using a suitable tool and remove the bolt. Remove the crankshaft pulley and the guide plate.
7. Disconnect the spark plug wires and remove the spark plugs.
8. Loosen the cylinder head bolt cover bolts, in 2–3 steps, in the reverse order of the tightening sequence. Remove the cylinder head cover.
9. Remove the engine oil dipstick and dipstick tube.
10. Remove the upper and lower timing belt covers.
11. Support the engine using engine support tool 49 G017 5A0. Remove the right side engine mount.
12. Turn the crankshaft, in the normal direction of rotation, until the crankshaft and camshaft sprocket timing marks are aligned. Mark the direction of rotation on the belt.
13. Turn the belt tensioner clockwise and disconnect the tensioner spring from the hook pin. Remove the timing belt.

➡ **Do not rotate the engine after the timing belt has been removed.**

14. Inspect the belt for wear, peeling, cracking, hardening or signs of oil contamination. Inspect the tensioner pulley for free and smooth rotation and for oil leaks. Check the spring bracket and grommet for looseness or damage. Measure the tensioner spring free length; it should not exceed 1.441 in. (36.6mm). Check the sprockets for worn teeth or other damage.

To install:

15. Make sure the crankshaft and camshaft timing marks are aligned.

➡ **It may be easier to check the camshaft sprocket alignment by looking behind the sprockets. The camshafts are properly aligned when the grooves on the rear of the sprockets are even with the cylinder head surface.**

16. Install the timing belt so there is no looseness at the idler side or between the camshaft sprockets. If reusing the old timing belt, make sure it is installed in the same direction of rotation.

17. Turn the crankshaft clockwise 2 turns and make sure the timing marks are correctly aligned. If the marks are not aligned, repeat Steps 15–17.

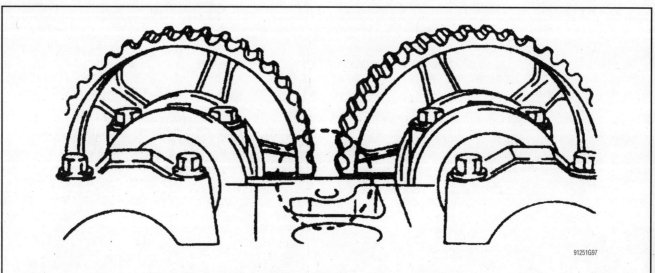

Camshaft sprocket alignment as seen from the rear of the sprockets—1990–94 Mazda 626 and MX-6 2.0L engine

MAZDA—TIMING BELTS

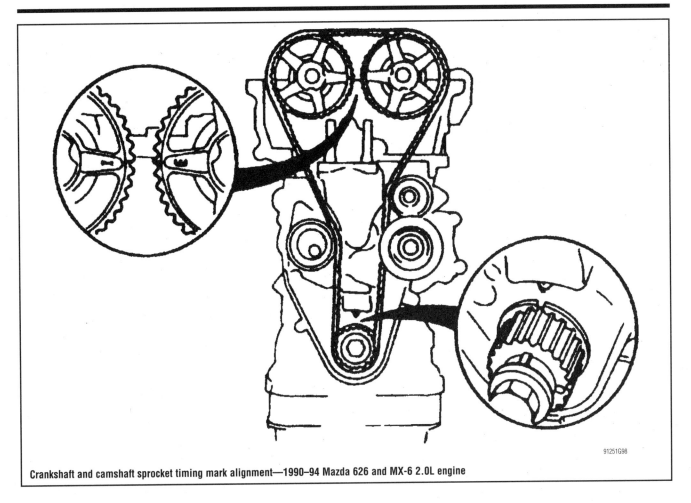

Crankshaft and camshaft sprocket timing mark alignment—1990–94 Mazda 626 and MX-6 2.0L engine

18. Turn the tensioner clockwise and connect the tensioner spring to the hook pin. Make sure tension is applied to the timing belt.
19. Turn the crankshaft clockwise 2 turns and make sure the timing marks are correctly aligned. If the marks are not aligned, repeat Steps 15–19.
20. Install the right side engine mount. Tighten the mount-to-engine nuts to 76 ft. lbs. (102 Nm) and the mount through bolt to 86 ft. lbs. (116 Nm). Install the ground harness and tighten the nut to 65 ft. lbs. (89 Nm).
21. Remove the engine support tool.
22. Install the timing belt covers and tighten the bolts to 95 inch lbs. (10.7 Nm). Install the dipstick tube and dipstick.
23. Apply silicone sealant to the contact surfaces of the cylinder head cover. Also apply sealant to the cylinder head surface in the area adjacent to the front camshaft caps.
24. Install the cylinder head cover and tighten the bolts in 2–3 steps to 69 inch lbs. (7.8 Nm), in the proper sequence.
25. Install the spark plugs and connect the spark plug wires.
26. Install the guide plate and the crankshaft pulley. Hold the pulley with a suitable tool and tighten the lock bolt to 122 ft. lbs. (166 Nm).
27. Install the power steering pump and tighten the bolts to 33 ft. lbs. (46 Nm).
28. Install the power steering pulley shield and the water pump pulley.
29. Install the accessory drive belts and adjust the belt tension.
30. Install the engine undercover and the right front wheel and tire assembly. Lower the vehicle.
31. Connect the negative battery cable. Start the engine and check for proper operation. Check the ignition timing.

2.2L (F2) ENGINE

1. Disconnect the negative battery cable. Tag and disconnect the spark plug wires and remove the spark plugs.
2. Remove the engine side cover from the fenderwell.
3. Remove the accessory drive belts.
4. Remove the retaining bolts and remove the crankshaft pulley.
5. Remove the upper and lower timing belt covers. Remove the baffle plate from in front of the crankshaft sprocket.
6. Turn the crankshaft clockwise until the `arrow'1 mark on the camshaft is aligned with the mark on top of the front housing. Unbolt and remove the tensioner and the tensioner spring.
7. Remove the timing belt. If the timing belt is to be reused, mark the direction of rotation.

➡**Do not rotate the engine after the timing belt has been removed.**

8. Inspect the belt for wear, peeling, cracking, hardening or signs of oil contamination. Inspect the tensioner pulley for free and smooth rotation. Measure the tensioner spring free length; it should not exceed 2.480 in. (63mm). Check the sprockets for worn teeth or other damage.

To install:

9. Make sure the crankshaft and camshaft timing marks are aligned.

138 TIMING BELTS—MAZDA

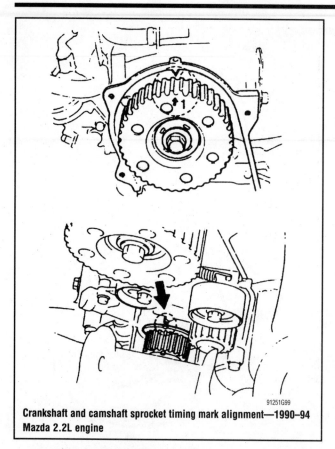

Crankshaft and camshaft sprocket timing mark alignment—1990–94 Mazda 2.2L engine

2.3L (KJ) ENGINE

1. Remove the timing belt covers. Temporarily reinstall the crankshaft pulley bolt.
2. Remove the power steering auto-tensioner and pulley.
3. Turn the crankshaft until the timing mark on the crankshaft sprocket aligns with the timing mark on the oil pump and the camshaft sprocket timing marks align with the marks on the cylinder head. The number one piston should be at TDC of the compression stroke.
4. Remove the two bolts from the automatic tensioner, removing the lower one first. Keep the bolt holes aligned by holding the tensioner to reduce the chance of stripping the threads on the bolts.
5. If the timing belt is to be reused, mark the direction of rotation on the timing belt.
6. Remove the timing belt.

To install:

7. Install the crankshaft sprocket bolt. Install the flywheel locking tool. Tighten the bolt to 116–122 ft. lbs. (157–166 Nm). remove the flywheel locking tool.
8. Position the automatic tensioner in a press. Set a flat washer under the tensioner body to prevent damage to the body plug.
9. Compress the tensioner until the hole in the piston is aligned with the 2nd hole in the tensioner case. Insert a 0.063 in. (1.6mm) diameter wire or pin through the 2nd hole to keep the piston compressed.

10. Install the timing belt tensioner and spring. Move the tensioner until the spring is fully extended and temporarily tighten the tensioner bolt to hold it in place.
11. Install the timing belt. Make sure there is no slack at the side of the water pump and idler pulleys. If reusing the old belt, it should be installed in the original direction of rotation.
12. Turn the crankshaft 2 turns clockwise and make sure the timing marks are aligned. If the marks are not aligned, repeat Steps 9–12.
13. Loosen the tensioner lock bolt to apply tension to the belt. Tighten the tensioner bolt to 38 ft. lbs. (52 Nm).
14. Turn the crankshaft 2 turns clockwise and make sure the timing marks are aligned. If the marks are not aligned, repeat Steps 9–14.
15. Apply approximately 22 lbs. (98 N) pressure to the timing belt at a point midway between the idler pulley and camshaft sprocket. A new belt should deflect 0.31–0.35 in. (8–9mm). A used belt should deflect 0.35–0.39 in. (9–10mm). If the deflection is not as specified, repeat Steps 12–15 or, if necessary, replace the tensioner spring.
16. Install the baffle plate with the dished side facing away from the engine.
17. Install the timing belt covers and tighten the bolts to 87 inch lbs. (10 Nm).
18. Install the crankshaft pulley and tighten the bolts to 13 ft. lbs. (17 Nm).
19. Install the accessory drive belts and adjust the belt tension.
20. Install the engine side cover in the fenderwell.
21. Install the spark plugs and connect the spark plug wires.
22. Connect the negative battery cable. Start the engine and check for proper operation. Check the ignition timing.

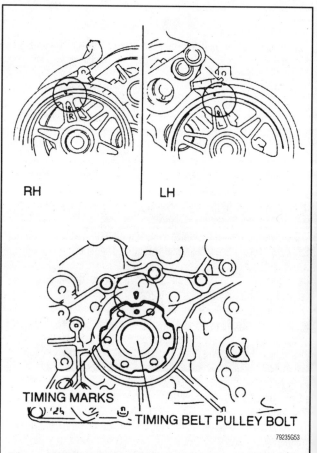

Proper crankshaft and camshaft timing belt sprocket alignment mark positioning—Mazda 2.3L (KJ) engines

MAZDA—TIMING BELTS

10. Be sure the camshaft sprocket timing marks are still aligned. Turn the crankshaft clockwise until the timing sprocket is aligned.
11. Install the timing belt. If the original belt is being reused, be sure it is installed in the same direction of rotation. The order of installation is: timing belt (crankshaft) sprocket, number two idler pulley, LEFT-HAND camshaft sprocket, both number one idler pulleys, right-hand camshaft sprocket and the tensioner pulley.
12. Install the automatic belt tensioner and tighten the bolts to 14–18 ft. lbs. (19–25 Nm). Remove the wire or pin from the tensioner.
13. Turn the crankshaft clockwise, until the crankshaft sprocket timing mark is again at TDC. This should place all of the belt slack in the automatic tensioner portion of the belt.
14. Rotate the crankshaft two turns in the normal direction of rotation and align the timing marks. Be sure all marks are still correctly aligned.
15. Inspect timing belt deflection, 0.24–0.31 in. (6–8mm), between the crankshaft sprocket and the tensioner pulley. If it is out of specification, replace the auto-tensioner.
16. Install the power steering auto-tensioner and tighten the bolts to 14–18 ft. lbs. (19–25 Nm). Install the pulley, and tighten the bolt to 29–34 ft. lbs. (40–47 Nm).
17. Remove the crankshaft damper bolt and install the timing belt covers.

2.3L (VIN A) & 2.5L (VIN C) ENGINES

1. Rotate the engine so that No. 1 cylinder is at TDC on the compression stroke. Check that the timing marks are aligned on the camshaft and crankshaft pulleys. An access plug is provided in the cam belt cover so that the camshaft timing can be checked without removal of the cover or any other parts. Set the crankshaft to TDC by aligning the timing mark on the crank pulley with the TDC mark on the belt cover. Look through the access hole in the belt cover to be sure that the timing mark on the cam drive sprocket is lined up with the pointer on the inner belt cover.

➡**Always turn the engine in the normal direction of rotation. Backward rotation may cause the timing belt to jump time, due to the arrangement of the belt tensioner.**

2. Drain cooling system. Remove the upper radiator hose as necessary. Remove the fan blade and water pump pulley bolts.

✳✳ CAUTION

When draining the coolant, keep in mind that cats and dogs are attracted by ethylene glycol antifreeze, and are quite likely to drink any that is left in an uncovered container or in puddles on the ground. This will prove fatal in sufficient quantity. Always drain the coolant into a sealable container. Coolant should be reused unless it is contaminated or several years old.

3. Loosen the alternator retaining bolts and remove the drive belt from the pulleys. Remove the water pump pulley.
4. Remove the power steering pump and set it aside.
5. Remove the four timing belt outer cover retaining bolts and remove the cover. Remove the crankshaft pulley and belt guide.
6. Loosen the belt tensioner pulley assembly, then position a camshaft belt adjuster tool (T74P-6254-A or equivalent) on the tension spring rollpin and retract the belt tensioner away from the timing belt. Tighten the adjustment bolt to lock the tensioner in the retracted position.

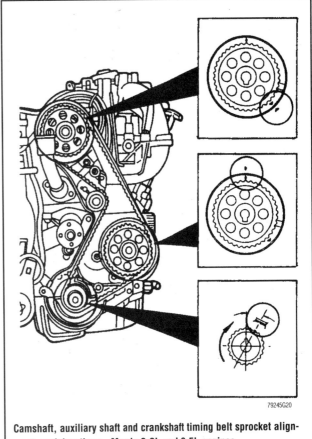

Camshaft, auxiliary shaft and crankshaft timing belt sprocket alignment mark locations—Mazda 2.3L and 2.5L engines

7. If the belt is to be reused, mark the direction of rotation on the belt for installation reference.
8. Remove the timing belt.

To install:

9. Install the new belt over the crankshaft sprocket and then counterclockwise over the auxiliary and camshaft sprockets, making sure the lugs on the belt properly engage the sprocket teeth on the pulleys. Be careful not to rotate the pulleys when installing the belt.
10. Release the timing belt tensioner pulley, allowing the tensioner to take up the belt slack. If the spring does not have enough tension to move the roller against the belt (belt hangs loose), it might be necessary to manually push the roller against the belt and tighten the bolt.

➡**The spring cannot be used to set belt tension; a wrench must be used on the tensioner assembly.**

✳✳ WARNING

If any binding is felt when adjusting the timing belt tension by turning the crankshaft, STOP turning the engine, because the pistons may be hitting the valves.

11. Rotate the crankshaft two complete turns by hand (in the normal direction of rotation) to remove slack from the belt, then tighten the tensioner adjustment to 26–33 ft. lbs. (35–45 Nm) and pivot bolts to 30–40 ft. lbs. (40–55 Nm). Be sure the belt is seated properly on the pulleys and that the timing marks are still in alignment when No. 1 cylinder is again at TDC/compression.

TIMING BELTS—MAZDA

12. Install the crankshaft pulley and belt guide.
13. Install the timing belt cover.
14. Install the water pump pulley and fan blades. Install the upper radiator hose if necessary. Refill the cooling system.
15. Install the accessory drive belts.
16. Start the engine and check the ignition timing. Adjust the timing, if necessary.

2.5L (KL) ENGINE

1. Disconnect the negative battery cable. Remove the engine undercover and side cover.
2. Remove the accessory drive belts.
3. Remove the water pump pulley and the accessory drive belt idler pulley bracket.
4. Remove the power steering pump reservoir bolts and secure the reservoir aside.
5. Keep the power steering pump pulley from turning, by installing a socket on the end of a breaker bar through one of the pulley holes and engaging a pump mounting bolt. Remove the pulley nut and the pulley.
6. Remove the power steering pump mounting bolts and remove the power steering pump. Secure the pump aside, leaving the hoses connected.
7. Hold the crankshaft pulley with a suitable tool and remove the pulley bolt. Remove the crankshaft pulley, being careful not to damage the crank position sensor rotor on the rear of the pulley.
8. Disconnect the crank position sensor connector and remove the clip from the engine oil dipstick tube. Remove the dipstick and tube. Plug the hole after removal to prevent the entry of dirt or foreign material.
9. Remove the crank position sensor harness bracket and wiring harness bracket from the timing belt cover.
10. Support the engine with engine support tool 49 G017 5A0 or equivalent.
11. Remove the right side engine mount.
12. Remove the right and left timing belt covers.
13. Install the crankshaft pulley bolt and turn the crankshaft until the No. 1 piston is at TDC on the compression stroke. Mark the direction of rotation on the timing belt.
14. Loosen the automatic tensioner bolts and remove the lower bolt. Hold the tensioner so the bolt threads are not damaged during removal.
15. Hold the upper idler pulley to reduce the belt resistance and remove the pulley bolts and pulley. Remove the timing belt. Remove the automatic tensioner and the remaining idler pulley.

➡ **Do not rotate the engine after the timing belt has been removed.**

16. Inspect the belt for wear, peeling, cracking, hardening or signs of oil contamination. Inspect the tensioner pulley for free and smooth rotation. Check the automatic tensioner for oil leakage. Check the tensioner rod projection (free length); it should be 0.55–0.63 in. (14–16mm). Inspect the sprocket teeth for wear or damage. Replace parts, as necessary.

To install:

17. Position the automatic tensioner in a suitable press. Place a flat washer under the tensioner body to prevent damage to the body plug.
18. Slowly press in the tensioner rod, but do not exceed 2200 lbs. force. Insert a pin into the tensioner body to hold the rod in place.
19. Install the tensioner and loosely tighten the upper bolt so the tensioner can move.

➡ **This is done to reduce the timing belt resistance when the upper idler pulley is installed.**

20. If removed, install the lower idler pulley and tighten the bolt to 38 ft. lbs. (52 Nm).
21. Make sure the crankshaft and camshaft sprocket timing marks are aligned.
22. Install the timing belt over the crankshaft sprocket, lower idler pulley, left camshaft sprocket, tensioner pulley and right camshaft sprocket, in that order. Make sure the belt has no looseness at the tension side. If reusing the old timing belt, make sure it is installed in the same direction of rotation.
23. Install the upper idler pulley while applying pressure on the timing belt. Be careful not to damage the pulley bolt threads when installing. Tighten the upper idler pulley bolt to 34 ft. lbs. (46 Nm).
24. Push the bottom of the automatic tensioner away from the belt and tighten the mounting bolts to 19 ft. lbs. (25 Nm). Remove the pin from the tensioner, applying tension to the belt.
25. Turn the crankshaft twice in the normal direction of rotation and make sure the timing marks are aligned. If the timing marks are not aligned, repeat Steps 17–25.
26. Apply approximately 22 lbs. (98 N) pressure to the timing belt at a point midway between the automatic tensioner and the crankshaft sprocket. The belt should deflect 0.24–0.31 in. (6–8mm). If the deflection is not as specified, replace the automatic tensioner.
27. Install the right and left timing belt covers. Tighten the bolts to 95 inch lbs. (11 Nm).
28. Install the crank position sensor and harness brackets and tighten the bolts to 95 inch lbs. (11 Nm).
29. Install the right side engine mount. Tighten the mount-to-engine nuts to 76 ft. lbs. (103 Nm) and the mount through bolt to 86 ft. lbs. (116 Nm).
30. Remove the engine support tool.
31. Apply clean engine oil to a new O-ring and install on the dipstick tube. Remove the plug and install the dipstick tube and dipstick. Tighten the tube bracket bolt to 95 inch lbs. (11 Nm).
32. Install the crank angle sensor harness and clip to the dipstick tube. Connect the electrical connector.
33. Remove the crankshaft pulley bolt and install the crankshaft pulley. Reinstall the bolt and hold the pulley with a suitable tool. Tighten the bolt to 122 ft. lbs. (166 Nm).
34. Install the power steering pump. Tighten the mounting bolts to 34 ft. lbs. (46 Nm) except the bolt to the right of the idler pulley. Tighten that bolt to 19 ft. lbs. (25 Nm).
35. Install the power steering pump pulley and loosely tighten the nut. Hold the pulley with the socket and breaker bar and tighten the nut to 69 ft. lbs. (93 Nm).
36. Install the power steering fluid reservoir and engine ground. Tighten to 87 inch lbs. (9.8 Nm).
37. Install the water pump pulley and loosely tighten the bolts. Install the accessory drive belts and adjust the belt tension.
38. Tighten the water pump pulley bolts to 95 inch lbs. (11 Nm).
39. Install the engine undercover and side cover. Connect the negative battery cable.

Mazda—Timing Belts

40. Start the engine and check for proper operation. Check the ignition timing.

3.0L (JE) ENGINE

SOHC Engine

1. Disconnect the negative battery cable. Drain the cooling system.
2. Tag and disconnect the spark plug wires. Remove the spark plugs.
3. Remove the fresh air duct. Remove the cooling fan and the fan shroud.
4. Remove the accessory drive belts and the A/C compressor idler pulley.
5. Remove the crankshaft pulley and baffle plate.
6. Remove the coolant bypass hose and the upper radiator hose.
7. Remove the timing belt covers and gaskets.
8. Turn the crankshaft, in the normal direction of rotation, and align the crankshaft and camshaft sprocket timing marks. Mark the direction of rotation on the timing belt.
9. Remove the upper idler pulley and remove the timing belt. Remove the automatic tensioner.

➡ **Do not rotate the engine after the timing belt has been removed.**

10. Inspect the belt for wear, peeling, cracking, hardening or signs of oil contamination. Inspect the tensioner pulley for free and smooth rotation. Check the automatic tensioner for oil leakage. Check the tensioner rod projection (free length); it should be 0.47–0.55 in. (12–14mm). Inspect the sprocket teeth for wear or damage. Replace parts, as necessary.

To install:

11. Position the automatic tensioner in a suitable press. Place a flat washer under the tensioner body to prevent damage to the body plug.
12. Slowly press in the tensioner rod, but do not exceed 2200 lbs. force. Insert a pin into the tensioner body to hold the rod in place.
13. Install the tensioner and tighten the bolts to 19 ft. lbs. (25 Nm).

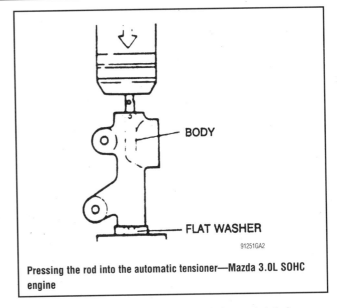

Pressing the rod into the automatic tensioner—Mazda 3.0L SOHC engine

14. Make sure the crankshaft and camshaft sprocket timing marks are aligned.
15. Install the timing belt over the crankshaft sprocket, lower idler pulley, left camshaft sprocket, right camshaft sprocket and tensioner pulley, in that order. If reusing the old timing belt, make sure it is installed in the same direction of rotation.
16. Install the upper idler pulley and tighten the bolt to 38 ft. lbs. (52 Nm).
17. Turn the crankshaft 2 turns, in the normal direction of rotation, and align the timing marks. If the timing marks are not aligned, repeat Steps 14–17.
18. Remove the pin from the automatic tensioner. Turn the crankshaft 2 turns, in the normal direction of rotation, and make sure the timing marks are aligned.
19. Apply approximately 22 lbs. (98 N) pressure to the timing belt at a point midway between the right camshaft sprocket and tensioner pulley. The belt should deflect 0.20–0.28 in. (5–7mm). If the deflection is not as specified, replace the timing belt or the automatic tensioner.
20. Install the timing belt covers with new gaskets. Tighten the bolts to 95 inch lbs. (11 Nm).
21. Install the upper radiator hose and coolant bypass hose.
22. Install the baffle plate and crankshaft pulley. Tighten the bolts to 130 inch lbs. (15 Nm).
23. Install the A/C compressor idler pulley and tighten the bolts to 19 ft. lbs. (25 Nm). Install the accessory drive belts and adjust the tension.
24. Install the fan shroud and cooling fan. Install the fresh air duct.
25. Install the spark plugs and connect the spark plug wires.
26. Connect the negative battery cable. Fill and bleed the cooling system.
27. Start the engine and bring to normal operating temperature. Check for leaks and for proper operation. Check the ignition timing.

DOHC Engine

1. Disconnect the negative battery cable and drain the cooling system.
2. Remove the fresh air duct.
3. Remove the cooling fan and fan shroud.

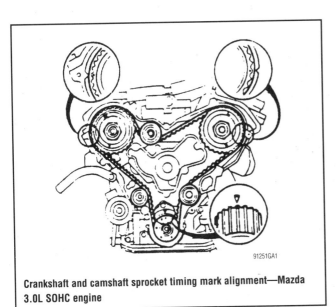

Crankshaft and camshaft sprocket timing mark alignment—Mazda 3.0L SOHC engine

4. Remove the air intake pipe from the throttle body and air cleaner.
5. Disconnect the spark plug wires and remove the spark plugs.
6. Remove the idler pulleys and the accessory drive belts.
7. Remove the coolant bypass hose and the upper radiator hose.
8. Disconnect the electrical connector and remove the distributor.
9. Hold the crankshaft pulley with a suitable tool and remove the bolt. Remove the crankshaft pulley. On 1992–94 vehicles, be careful not to damage the sensor rotor.
10. Remove the timing belt covers. Rotate the crankshaft until the camshaft and crankshaft sprocket timing marks are aligned.
11. Remove the upper idler pulley and the automatic tensioner and pulley. Mark the direction of rotation on the timing belt and remove the timing belt.

➡ Do not rotate the engine after the timing belt has been removed.

12. Inspect the belt for wear, peeling, cracking, hardening or signs of oil contamination. Inspect the tensioner pulley for free and smooth rotation. Check the automatic tensioner for oil leakage. Check the tensioner rod projection (free length); it should be 0.47–0.55 in. (12–14mm). Inspect the sprocket teeth for wear or damage. Replace parts, as necessary.

To install:
13. Position the automatic tensioner in a suitable press. Place a flat washer under the tensioner body to prevent damage to the body plug.
14. Slowly press in the tensioner rod, but do not exceed 2200 lbs. force. Insert a pin into the tensioner body to hold the rod in place.
15. Install the tensioner and tighten the bolts to 19 ft. lbs. (25 Nm).
16. Make sure the crankshaft and camshaft sprocket timing marks are aligned.
17. Install the timing belt over the crankshaft sprocket, lower idler pulley, left exhaust camshaft sprocket, left intake camshaft sprocket, tensioner pulley, right exhaust camshaft sprocket and right intake camshaft sprocket, in that order.
18. Push the belt down and install the upper idler pulley. Tighten the bolt to 38 ft. lbs. (52 Nm). Make sure the timing marks are still aligned after installing the upper idler pulley.
19. Turn the crankshaft 2 revolutions in the direction of normal rotation and realign the timing marks. If the timing marks do not align, repeat Steps 16–20.
20. Remove the pin from the automatic tensioner. Again rotate the crankshaft 2 turns and make sure the timing marks are aligned.
21. Apply approximately 22 lbs. (98 N) pressure to the timing belt at a point midway between the right exhaust camshaft sprocket and

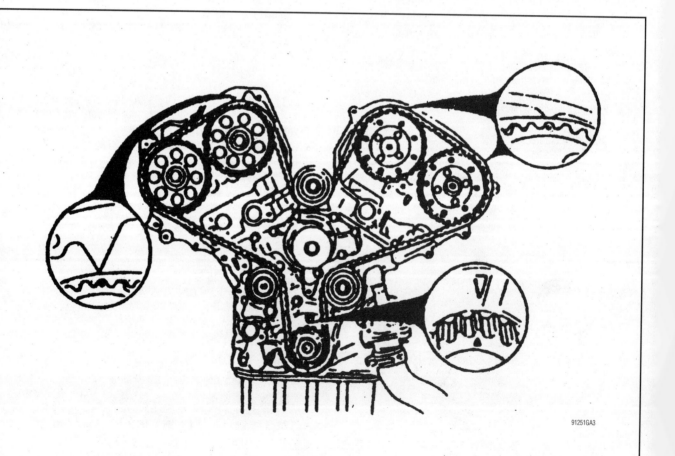

Crankshaft and camshaft sprocket timing mark alignment—Mazda 3.0L DOHC engine

MAZDA—TIMING BELTS 143

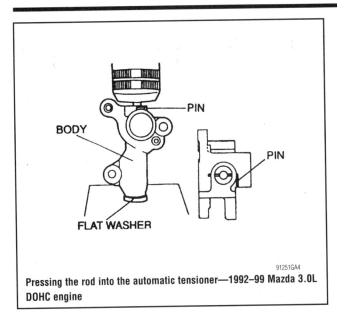

Pressing the rod into the automatic tensioner—1992-99 Mazda 3.0L DOHC engine

the tensioner pulley. The belt should deflect 0.20–0.28 in. (5–7mm). If the deflection is not as specified, replace the automatic tensioner.

22. Install the timing belt cover and tighten the bolts to 95 inch lbs. (11 Nm).
23. Install the crankshaft pulley. Hold the pulley with a suitable tool and tighten the lock bolt to 123 ft. lbs. (167 Nm). On 1992–99 vehicles, be careful not to damage the sensor rotor.
24. Install the distributor and connect the electrical connector.
25. Install the upper radiator hose and coolant bypass hose.
26. Install the idler pulleys and the accessory drive belts. Adjust the belt tension.
27. Install the spark plugs and connect the spark plug wires.
28. Install the air intake pipe to the throttle body and air cleaner.
29. Install the cooling fan and radiator shroud. Install the fresh air duct.
30. Connect the negative battery cable. Fill and bleed the cooling system.
31. Start the engine and bring to normal operating temperature. Check for leaks and for proper operation. Check the ignition timing.

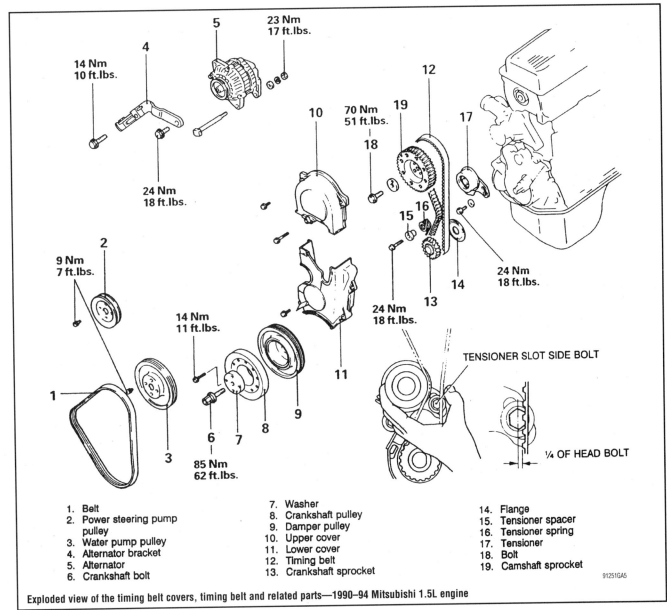

1. Belt
2. Power steering pump pulley
3. Water pump pulley
4. Alternator bracket
5. Alternator
6. Crankshaft bolt
7. Washer
8. Crankshaft pulley
9. Damper pulley
10. Upper cover
11. Lower cover
12. Timing belt
13. Crankshaft sprocket
14. Flange
15. Tensioner spacer
16. Tensioner spring
17. Tensioner
18. Bolt
19. Camshaft sprocket

Exploded view of the timing belt covers, timing belt and related parts—1990-94 Mitsubishi 1.5L engine

TIMING BELTS—MITSUBISHI

Mitsubishi

1.5L ENGINE

1990—94 Models

1. Disconnect the negative battery cable. Remove the engine under cover.
2. Raise and safely support the weight of the engine using the appropriate equipment. Remove the front engine mount bracket and accessory drive belts.
3. If necessary, remove the coolant reservoir tank.
4. Using the proper equipment, slightly raise the engine to take the weight off the side engine mount. Remove the engine mount bracket.
5. Remove the drive belts, tension pulley brackets, water pump pulley and crankshaft pulley.
6. Remove all attaching screws and remove the upper and lower timing belt covers.
7. Make a mark on the back of the timing belt indicating the direction of rotation so it may be reassembled in the same direction if it is to be reused. Loosen the timing belt tensioner and remove the timing belt.

➡ If coolant or engine oil comes in contact with the timing belt, they will drastically shorten its life. Also, do not allow engine oil or coolant to contact the timing belt sprockets or tensioner assembly.

8. Remove the tensioner spacer, tensioner spring and tensioner assembly.
9. Inspect the timing belt for cracks on back surface, sides, bottom and check for separated canvas. Check the tensioner pulley for smooth rotation.

To install:

10. Position the tensioner, tensioner spring and tensioner spacer on engine block.
11. Align the timing marks on the camshaft sprocket and crankshaft sprocket. This will position No. 1 piston on TDC on the compression stroke.
12. Position the timing belt on the crankshaft sprocket and keeping the tension side of the belt tight, set it on the camshaft sprocket.
13. Apply counterclockwise force to the camshaft sprocket to give tension to the belt and make sure all timing marks are aligned.
14. Loosen the pivot side tensioner bolt and the slot side bolt. Allow the spring to take up the slack.
15. Tighten the slot side tensioner bolt and then the pivot side bolt. If the pivot side bolt is tightened first, the tensioner could turn with bolt, causing over tension.
16. Turn the crankshaft clockwise. Loosen the pivot side tensioner bolt and then the slot side bolt to allow the spring to take up any remaining slack. Tighten the slot bolt and then the pivot side bolt to 14–20 ft. lbs. (20–27 Nm).
17. Check the belt tension by holding the tensioner and timing belt together by hand and give the belt a slight thumb pressure at a point level with tensioner center. Make sure the belt cog crest comes as deep as about ¼ of the width of the slot side tensioner bolt head. Do not manually overtighten the belt or it will howl.
18. Install the timing belt covers and all related items.
19. Connect the negative battery cable.

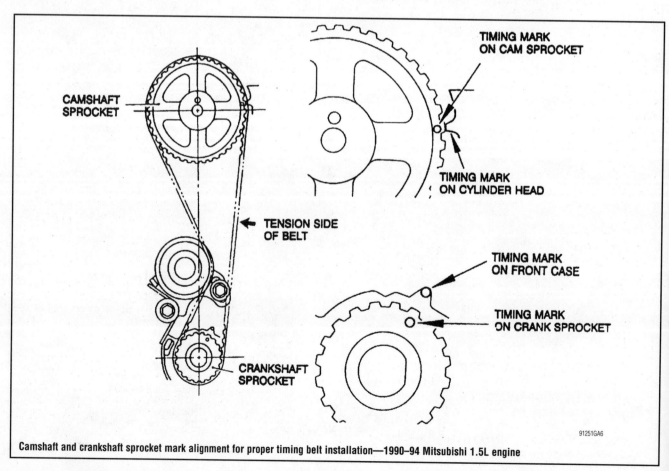

Camshaft and crankshaft sprocket mark alignment for proper timing belt installation—1990–94 Mitsubishi 1.5L engine

MITSUBISHI—TIMING BELTS

1995–99 Models

1. Disconnect the negative battery cable. Remove the engine undercover.
2. Rotate crankshaft clockwise and position engine at TDC compression stroke.
3. Raise and safely support the weight of the engine using the appropriate equipment. Remove the A/C clamp, front engine mount bracket and accessory drive belts.
4. Remove the crankshaft pulley.
5. Remove timing belt upper and lower covers.
6. Make a mark on the back of the timing belt indicating the direction of rotation so it may be reassembled in the same direction if it is to be reused. Loosen the timing belt tensioner and move the tensioner to provide slack to the timing belt. Tighten the tensioner in this position.
7. Remove the timing belt.

⁂⁂ WARNING

Coolant and engine oil will damage the rubber in the timing belt, drastically reducing its life. Do not allow engine oil or coolant to contact the timing belt, the sprockets or tensioner assembly.

8. If defective, remove the tensioner spacer, tensioner spring and tensioner assembly.

To install:

9. Position the tensioner, tensioner spring and tensioner spacer on engine block.
10. Align the timing marks on the camshaft sprocket and crankshaft sprocket. This will position No. 1 piston on TDC on the compression stroke.
11. Position the timing belt on the crankshaft sprocket and keeping the tension side of the belt tight, set it on the camshaft sprocket, then the tensioner.
12. Apply slight counterclockwise force to the camshaft sprocket to give tension to the belt and be sure all timing marks are aligned.
13. Loosen the pivot side tensioner bolt and the slot side bolt. Allow the spring to remove the slack.
14. Tighten the slot side tensioner bolt, then the pivot side bolt. If the pivot side bolt is tightened first, the tensioner could turn with bolt, causing over tension.
15. Turn the crankshaft clockwise. Loosen the pivot side tensioner bolt, then the slot side bolt to allow the spring to take up any remaining slack. Tighten the slot bolt, then the pivot side bolt to 17 ft. lbs. (24 Nm).
16. Install the timing belt covers and tighten the cover bolts to 84–96 inch lbs. (10–11 Nm). Install all other applicable components.

1.6L & 2.0L (NON-TURBO) DOHC ENGINES

➡ **The 1.6L engine is not equipped with silent shafts. Disregard all instructions pertaining to silent shafts if working on that engine.**

1. Disconnect the negative battery cable.
2. Remove the engine undercover.
3. If necessary, remove the coolant reservoir.
4. Using the proper equipment, slightly raise the engine to take the weight off the side engine mount. Remove the engine mount bracket.
5. Remove the drive belts, tension pulley brackets, water pump pulley and crankshaft pulley.
6. Remove all attaching screws and remove the upper and lower timing belt covers.
7. Rotate the crankshaft clockwise and align the timing marks so No. 1 piston will be at TDC of the compression stroke. At this time the timing marks on the camshaft sprocket and the upper surface of the cylinder head should coincide, and the dowel pin of the camshaft sprocket should be at the upper side.

➡ **Always rotate the crankshaft in a clockwise direction. Make a mark on the back of the timing belt indicating the direction of rotation so it may be reassembled in the same direction if it is to be reused.**

8. Remove the auto tensioner and remove the outermost timing belt.
9. Remove the timing belt tensioner pulley, tensioner arm, idler pulley, oil pump sprocket, special washer, flange and spacer.
10. Remove the silent shaft (inner) belt tensioner and remove the belt.

To install:

11. Align the timing marks on the crankshaft sprocket and the silent shaft sprocket. Fit the inner timing belt over the crankshaft and silent shaft sprocket. Ensure that there is no slack in the belt.
12. While holding the inner timing belt tensioner with your fingers, adjust the timing belt tension by applying a force towards the center of the belt, until the tension side of the belt is taut. Tighten the tensioner bolt.

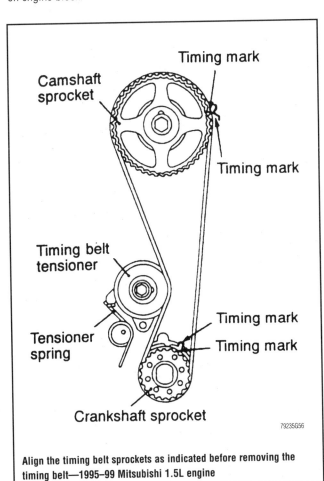

Align the timing belt sprockets as indicated before removing the timing belt—1995–99 Mitsubishi 1.5L engine

TIMING BELTS—MITSUBISHI

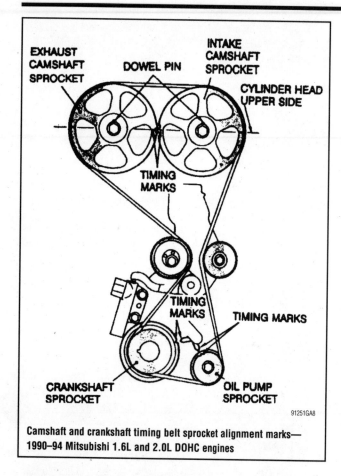

Camshaft and crankshaft timing belt sprocket alignment marks—1990–94 Mitsubishi 1.6L and 2.0L DOHC engines

➡ **When tightening the bolt of the tensioner, ensure that the tensioner pulley shaft does not rotate with the bolt. Allowing it to rotate with the bolt can cause excessive tension on the belt.**

13. Check belt for proper tension by depressing the belt on its' long side with your finger and noting the belt deflection. The desired reading is 0.20–0.28 in. (5–7mm). If tension is not correct, readjust and check belt deflection.
14. Install the flange, crankshaft and washer to the crankshaft. The flange on the crankshaft sprocket must be installed towards the inner timing belt sprocket. Tighten bolt to 80–94 ft. lbs. (110–130 Nm).
15. To install the oil pump sprocket, insert a Phillips screwdriver with a shaft 0.31 in. (8mm) in diameter into the plug hole in the left side of the cylinder block to hold the left silent shaft. Tighten the nut to 36–43 ft. lbs. (50–60 Nm).
16. Using a wrench, hold the camshaft at its' hexagon between journal No. 2 and 3 and tighten bolt to 58–72 ft. lbs. (80–100 Nm). If no hexagon is present between journal No. 2 and 3, hold the sprocket stationary with a spanner wrench while tightening the retainer bolt.
17. Carefully push the auto tensioner rod in until the set hole in the rod aligned up with the hole in the cylinder. Place a wire into the hole to retain the rod.
18. Install the tensioner pulley onto the tensioner arm. Locate the pinhole in the tensioner pulley shaft to the left of the center bolt. Then, tighten the center bolt finger-tight.
19. When installing the timing belt, turn the 2 camshaft sprockets so their dowel pins are located on top. Align the timing marks facing each other with the top surface of the cylinder head. When you let go of the exhaust camshaft sprocket, it will rotate 1 tooth in the counterclockwise direction. This should be taken into account when installing the timing belts on the sprocket.

➡ **Both camshaft sprockets are used for the intake and exhaust camshafts and are provided with 2 timing marks. When the sprocket is mounted on the exhaust camshaft, use the timing mark on the right with the dowel pin hole on top. For the intake camshaft sprocket, use the 1 on the left with the dowel pin hole on top.**

20. Align the crankshaft sprocket and oil pump sprocket timing marks.
21. After alignment of the oil pump sprocket timing marks, remove the plug on the cylinder block and insert a Phillips screw driver with a shaft diameter of 0.31 in. (8mm) through the hole. If the shaft can be inserted 2.4 in. deep, the silent shaft is in the correct position. If the shaft of the tool can only be inserted 0.8–1.0 in. (20–25mm) deep, turn the oil pump sprocket 1 turn and realign the marks. Reinsert the tool making sure it is inserted 2.4 in. deep. Keep the tool inserted in hole for the remainder of this procedure.

➡ **The above step assures that the oil pump socket is in correct orientation to the silent shafts. This step must not be skipped or a vibration may develop during engine operation.**

22. Install the timing belt as follows:
    a. Install the timing belt around the intake camshaft sprocket and retain it with 2 spring clips or binder clips.
    b. Install the timing belt around the exhaust sprocket, aligning the timing marks with the cylinder head top surface using 2 wrenches. Retain the belt with 2 spring clips.
    c. Install the timing belt around the idler pulley, oil pump sprocket, crankshaft sprocket and the tensioner pulley. Remove the 2 spring clips.
    d. Lift upward on the tensioner pulley in a clockwise direction and tighten the center bolt. Make sure all timing marks are aligned.
    e. Rotate the crankshaft ¼ turn counterclockwise. Then, turn in clockwise until the timing marks are aligned again.
23. To adjust the timing (outer) belt, turn the crankshaft ¼ turn counterclockwise, then turn it clockwise to move No. 1 cylinder to TDC.
24. Loosen the center bolt. Using tool MD998738 or equivalent and a torque wrench, apply a torque of 1.88–2.03 ft. lbs. (2.6–2.8 Nm). Tighten the center bolt.
25. Screw the special tool into the engine left support bracket until its end makes contact with the tensioner arm. At this point, screw the special tool in some more and remove the set wire attached to the auto tensioner, if the wire was not previously removed. Then remove the special tool.
26. Rotate the crankshaft 2 complete turns clockwise and let it sit for approximately 15 minutes. Then, measure the auto tensioner protrusion (the distance between the tensioner arm and auto tensioner body) to ensure that it is within 0.15–0.18 in. (3.8–4.5mm). If out of specification, repeat Step 1–4 until the specified value is obtained.
27. If the timing belt tension adjustment is being performed with the engine mounted in the vehicle, and clearance between the tensioner arm and the auto tensioner body cannot be measured, the following alternative method can be used:

MITSUBISHI—TIMING BELTS

a. Screw in special tool MD998738 or equivalent, until its end makes contact with the tensioner arm.

b. After the special tool makes contact with the arm, screw it in some more to retract the auto tensioner pushrod while counting the number of turns the tool makes until the tensioner arm is brought into contact with the auto tensioner body. Make sure the number of turns the special tool makes conforms with the standard value of 2½–3 turns.

c. Install the rubber plug to the timing belt rear cover.

28. Install the timing belt covers and all related items.
29. Connect the negative battery cable.

1.8L, 2.0L & 2.4L SOHC ENGINES

1. Position the engine so the No. 1 piston is at TDC of the compression stroke.
2. Disconnect the negative battery cable. On 2.4L engine, remove the coolant reservoir and the power steering and air conditioner hose clamp bolt.
3. Remove the engine undercover.
4. Using the proper equipment, slightly raise the engine to take the weight off the side engine mount. Remove the engine mount bracket.
5. Remove the drive belts, tension pulley brackets, water pump pulley and crankshaft pulley.
6. Remove all attaching screws and remove the upper and lower timing belt covers.
7. Remove the timing belt covers.
8. For 1995–97 models, loosen the timing belt tensioner and remove the timing belt.
9. For 1998–99 models, place 8mm Allen wrench into the belt tensioner, then using the long end of a 3mm(⅛)in. Allen wrench, rotate the tensioner counterclockwise until it slides into the locking hole.
10. Remove the outer crankshaft sprocket and flange.
11. Remove the silent shaft (inner) belt tensioner and remove the belt.

To install:

12. Align the timing marks of the silent shaft sprockets and the crankshaft sprocket with the timing marks on the front case. Wrap the timing belt around the sprockets so there is no slack in the upper span of the belt and the timing marks are still aligned.
13. Install the tensioner pulley and move the pulley by hand so the long side of the belt deflects about ¼ in.
14. Hold the pulley tightly so the pulley cannot rotate when the bolt is tightened. Tighten the bolt to 15 ft. lbs. (20 Nm) and recheck the deflection amount.
15. Install the timing belt tensioner fully toward the water pump and tighten the bolts. Place the upper end of the spring against the water pump body.
16. Align the timing marks of the camshaft, crankshaft and oil pump sprockets with their corresponding marks on the front case or rear cover.

➡ There is a possibility to align all timing marks and have the oil pump sprocket and silent shaft out of time, causing an engine vibration during operation. If the following step is not followed exactly, there is a 50 percent chance that the silent shaft alignment will be 180 degrees off.

17. Before installing the timing belt, ensure that the left side (rear) silent shaft (oil pump sprocket) is in the correct position as follows:

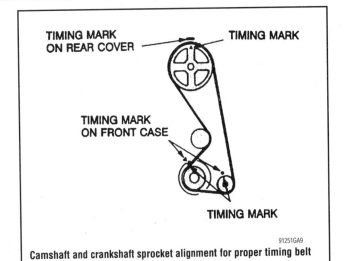

Camshaft and crankshaft sprocket alignment for proper timing belt replacement—1990–94 Mitsubishi 1.8L, 2.0L and 2.4L SOHC engines

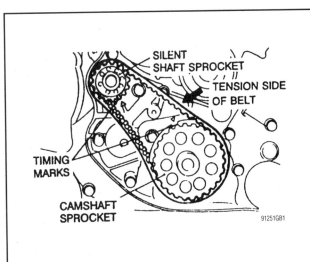

Silent shaft sprocket alignment marks for belt replacement—1990–94 Mitsubishi 1.8L, 2.0L and 2.4L SOHC engines

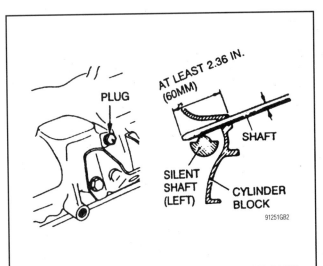

Checking the rear silent shaft for proper positioning—1990–94 Mitsubishi 1.8L, 2.0L and 2.4L SOHC engines

148 TIMING BELTS—MITSUBISHI

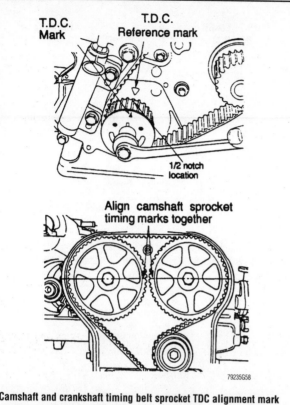

Camshaft and crankshaft timing belt sprocket TDC alignment mark positioning for timing belt removal and installation—1995–97 Mitsubishi 2.0L non-turbo engine

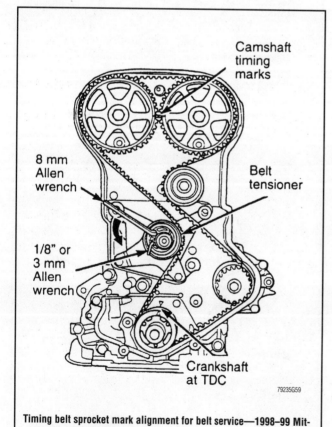

Timing belt sprocket mark alignment for belt service—1998–99 Mitsubishi 2.0L non-turbo engines

a. Remove the plug from the rear side of the block and insert a tool with shaft diameter of 0.31 in. (8mm) into the hole.

b. With the timing marks still aligned, the shaft of the tool must be able to go in at least 2⅓ in. If the tool can only go in about 1 in., the shaft is not in the correct orientation and will cause a vibration during engine operation. Remove the tool from the hole and turn the oil pump sprocket 1 complete revolution. Realign the timing marks and insert the tool. The shaft of the tool must go in at least 2⅓ in.

c. Recheck and realign the timing marks.

d. Leave the tool in place to hold the silent shaft while continuing.

18. Install the belt to the crankshaft sprocket, oil pump sprocket, then camshaft sprocket, in that order. While doing so, make sure there is no slack between the sprocket except where the tensioner is installed.

19. Recheck the timing marks' alignment. If all are aligned, loosen the tensioner mounting bolt and allow the tensioner to apply tension to the belt.

20. Remove the tool that is holding the silent shaft and rotate the crankshaft a distance equal to 2 teeth on the camshaft sprocket. This will allow the tensioner to automatically apply the proper tension on the belt. Do not manually overtighten the belt or it will howl.

21. Tighten the lower mounting bolt first, then the upper spacer bolt.

22. To verify correct belt tension, check that the deflection at the longest span of the belt is about ½ in.

23. The installation of the timing belt covers and all related items, is the reverse of the removal procedure. Make sure all pieces of packing are positioned in the inner grooves of the covers when installing.

24. Connect the negative battery cable.

2.0L (TURBO) DOHC ENGINE

1. Disconnect the negative battery cable.
2. Remove the engine undercover.
3. Remove the engine mount bracket.
4. Remove the drive belts.
5. Remove the belt tensioner pulley.
6. Remove the water pump pulleys.
7. Remove the crankshaft pulley.
8. Remove the stud bolt from the engine support bracket and remove the timing belt covers.
9. Rotate the crankshaft clockwise to line up the camshaft timing marks. Always turn the crankshaft in the normal direction of rotation only.
10. Loosen the tension pulley center bolt.

➡ **If the timing belt is to be reused, mark the direction of rotation on the flat side of the belt with an arrow.**

11. Move the tension pulley towards the water pump and remove the timing belt.

12. Remove the crankshaft sprocket center bolt using special tool MB990767 to hold the crankshaft sprocket while removing the center bolt. Then, use MB998778 or equivalent puller to remove the sprocket.

13. Mark the direction of rotation on the timing belt B with a arrow.

14. Loosen the center bolt on the tensioner and remove the belt.

MITSUBISHI—TIMING BELTS

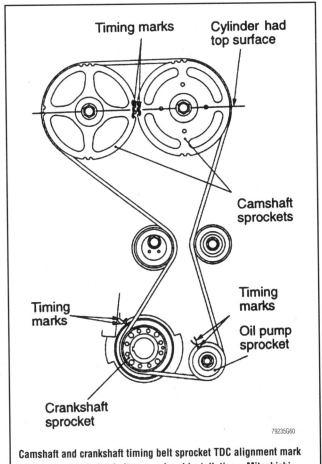

Camshaft and crankshaft timing belt sprocket TDC alignment mark positioning for timing belt removal and installation—Mitsubishi 2.0L turbo engine

※※ **WARNING**

Do not rotate the camshafts or the crankshaft while the timing belt is removed.

To install:

15. Place the crankshaft sprocket on the crankshaft. Use tool MB990767 or equivalent to hold the crankshaft sprocket while tightening the center bolt. Tighten the center bolt to 80–94 ft. lbs. (108–127 Nm).
16. Align the timing marks on the crankshaft sprocket B and the balance shaft.
17. Install timing belt B on the sprockets. Position the center of the tensioner pulley to the left and above the center of the mounting bolt.
18. Push the pulley clockwise toward the crankshaft to apply tension to the belt and tighten the mounting bolt to 14 ft. lbs. (19 Nm). Do not let the pulley turn when tightening the bolt because it will cause excessive tension on the belt. The belt should deflect 0.20–0.28 in. (5–7mm) when finger pressure is applied between the pulleys.
19. Install the crankshaft sensing blade and the crankshaft sprocket. Apply engine oil to the mounting bolt and tighten the bolt to 80–94 ft. lbs. (108–127 Nm).

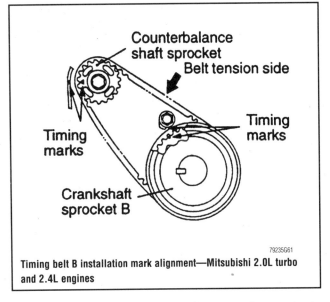

Timing belt B installation mark alignment—Mitsubishi 2.0L turbo and 2.4L engines

20. Use a press or vise to compress the auto-tensioner pushrod. Insert a set pin when the holes are lined up.

※※ **WARNING**

Do not compress the pushrod too quickly, damage to the pushrod can occur.

21. Install the auto-tensioner on the engine.
22. Align the timing marks on the camshaft sprocket, crankshaft sprocket and the oil pump sprocket.
23. After aligning the mark on the oil pump sprocket, remove the cylinder block plug and insert a prytool in the hole to check the position of the counterbalance shaft. The prytool should go in at least 2.36 in. (60mm) or more, if not, rotate the oil pump sprocket once and realign the timing mark so the prytool goes in. Do not remove the prytool until the timing belt is installed.
24. Install the timing belt on the intake camshaft and secure it with a clip.
25. Install the timing belt on the exhaust camshaft. Align the timing marks with the cylinder head top surface using two wrenches. Secure the belt with another clip.
26. Install the belt around the idler pulley, oil pump sprocket, crankshaft sprocket and the tensioner pulley.
27. Turn the tensioner pulley so the pinholes are at the bottom. Press the pulley lightly against the timing belt.
28. Screw the special tool into the left engine support bracket until it contacts the tensioner arm, then screw the tool in a little more and remove the pushrod pin from the auto-tensioner. Remove the special tool and tighten the center bolt to 35 ft. lbs. (48 Nm).
29. Turn the crankshaft ¼ turn counterclockwise, then clockwise until the timing marks are aligned.
30. Loosen the center bolt. Install Mitsubishi Special Tool MD998767, or equivalent, on the tensioner pulley. Turn the tensioner pulley counterclockwise with a torque of 2.6 ft. lbs. (3.5 Nm) and tighten the center bolt to 35 ft. lbs. (48 Nm). Do not let the tensioner pulley turn when tightening the bolt.
31. Turn the crankshaft clockwise two revolutions and align the timing marks. After 15 minutes, measure the protrusion of the

pushrod on the auto-tensioner. The standard measurement is 0.150–0.177 in (3.8–4.5mm). If the protrusion is out of specification, loosen the tensioner pulley, apply the proper torque to the belt and retighten the center bolt.

32. Install the timing belt covers and all applicable components.

2.4L (VIN W) ENGINE

1. Disconnect the negative battery cable.
2. Remove the splash shield under the engine.
3. Safely support the weight of the engine and remove the engine mount and bracket assembly.
4. Remove the drive belts and the timing belt covers.
5. Position the engine so that the No. 1 piston is at Top Dead Center (TDC).

➡ If the timing belts are going to be reused, mark the direction of rotation on the belt. This will ensure the belt is reinstalled in same direction, extending belt life.

6. To loosen the timing (outer) belt tensioner, install Mitsubishi Special Tool MD998738 or equivalent, to the slot and screw inward to move the tensioner toward the water pump. Once the tension has been relieved, remove the outer timing belt.
7. If tensioner replacement is required, align the pin hole in the tensioner rod to the hole in the tensioner cylinder. Insert a 0.055 in. (1.4mm) wire in the hole and remove the special tool from the slot.

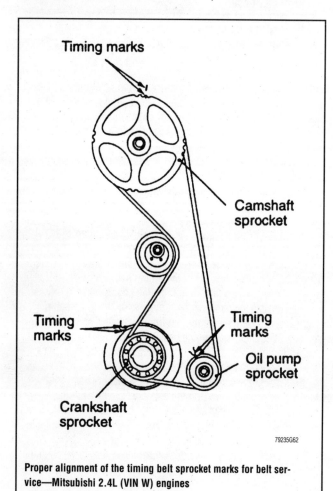

Proper alignment of the timing belt sprocket marks for belt service—Mitsubishi 2.4L (VIN W) engines

With the cylinder tension relieved, remove the auto-tensioner cylinder assembly two mounting bolts.

8. Remove the outer crankshaft sprocket and flange.
9. Loosen the silent shaft (inner) belt tensioner and remove the belt.

To install:

✳✳ WARNING

Do not spray or immerse the sprockets or tensioners in cleaning solvent. The sprocket may absorb the solvent and transfer it to the belt. The tensioners are internally lubricated and the solvent will dilute or dissolve the lubricant.

10. Align the timing marks of the silent shaft sprockets and the crankshaft sprocket with the timing marks on the front case. Route the timing belt around the sprockets so there is no slack in the upper span of the belt and the timing marks are still aligned.
11. Install the tensioner pulley and move the pulley by hand so the long side of the belt deflects approximately ¼ in. (6mm).
12. Hold the pulley tightly so the pulley cannot rotate when the bolt is tightened. Tighten the bolt to 14 ft. lbs. (19 Nm) and recheck the deflection.
13. Align the timing marks of the camshaft, crankshaft and oil pump sprockets with their corresponding marks on the front case or rear cover.

➡ There is a possibility to align all timing marks and have the oil pump sprocket and silent shaft out of time, causing an engine vibration during operation. If the following step is not followed exactly, there is a 50 percent chance that the silent shaft alignment will be 180 degrees (½ turn) off.

14. Before installing the timing belt, ensure that the left side (rear) silent shaft (oil pump sprocket) is in the correct position as follows:

   a. Remove the plug from the rear side of the block and insert a tool with shaft diameter of 0.31 in. (8mm) into the hole.
   b. With the timing marks still aligned, the shaft of the tool must be able to go in at least 2 ½ in. (63.5mm). If the tool can only go in approximately 1 in. (25mm), the shaft is not in the correct orientation and will cause a vibration during engine operation. Remove the tool from the hole and turn the oil pump sprocket 1 complete revolution. Realign the timing marks and insert the tool. The shaft of the tool must go in at least 2 ¼ in. (63.5mm).
   c. Recheck and realign the timing marks.
   d. Leave the tool in place to hold the silent shaft while continuing.

15. If the camshaft belt tensioner was removed, use a vise to carefully push the auto-tensioner rod in until the set hole in the rod is aligned with the hole in the cylinder. Place a wire into the hole to retain the rod. Mount the tensioner to the engine block and tighten the mounting bolt to 17 ft. lbs. (23 Nm).
16. Install the belt to the crankshaft sprocket, oil pump sprocket, then camshaft sprocket, in that order. While doing so, be sure there is no slack between the sprocket except where the tensioner is installed.
17. To adjust the timing (outer) belt perform the following steps:

a. Turn the crankshaft ¼ turn counterclockwise, then turn it clockwise to move No. 1 cylinder to TDC.

b. Loosen the center bolt. Using tool MD998752 (or equivalent) and a torque wrench, apply a torque of 2.6 ft. lbs. (3.6 Nm) to the tensioner. Tighten the center bolt.

c. Screw the special tool into the engine left support bracket until its end makes contact with the tensioner arm. At this point, screw the special tool in some more and remove the set wire attached to the auto-tensioner, if the wire was not previously removed. Then, remove the special tool.

d. Rotate the crankshaft two complete turns clockwise and let it sit for approximately 15 minutes. Then, measure the auto-tensioner protrusion (the distance between the tensioner arm and auto-tensioner body) to ensure that it is within 0.15–0.18 in. (3.8–4.5mm). If out of specification, repeat Substeps a through d until the specified value is obtained.

➡ Do not manually overtighten the belt or it will howl.

18. Install the upper and lower timing belt covers.

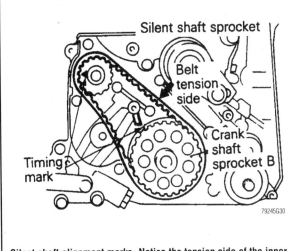

Silent shaft alignment marks. Notice the tension side of the inner (silent shaft) belt—Mitsubishi 2.4L (VIN G) engine

2.4L (VIN G) ENGINE

1. Be sure that the engine's No. 1 piston is at TDC in the compression stroke.

✷✷ CAUTION

Wait at least 90 seconds after the negative battery cable is disconnected to prevent possible deployment of the air bag.

2. Disconnect the negative battery cable.
3. Remove the spark plug wires from the tree on the upper cover.
4. Drain the cooling system.
5. Remove the shroud, fan and accessory drive belts.
6. Remove the radiator as required.
7. Remove the power steering pump, alternator, air conditioning compressor, tension pulley and accompanying brackets, as required.
8. Remove the upper front timing belt cover.
9. Remove the water pump pulley and the crankshaft pulley(s).
10. Remove the lower timing belt cover mounting screws and remove the cover.
11. If the belt(s) are to be reused, mark the direction of rotation on the belt.
12. Remove the timing (outer) belt tensioner and remove the belt. Unbolt the tensioner from the block and remove.
13. Remove the outer crankshaft sprocket and flange.
14. Remove the silent shaft (inner) belt tensioner and remove the inner belt. Unbolt the tensioner from the block and remove it.
15. To remove the camshaft sprockets, use SST MB990767–01 and MIT308239, or their equivalents.

To install:

16. Install the camshaft sprockets and tighten the center bolt to 65 ft. lbs. (90 Nm).
17. Align the timing mark of the silent shaft belt sprockets on the crankshaft and silent shaft with the marks on the front case. Wrap the silent shaft belt around the sprockets so there is no slack in the upper span of the belt and the timing marks are still in line.

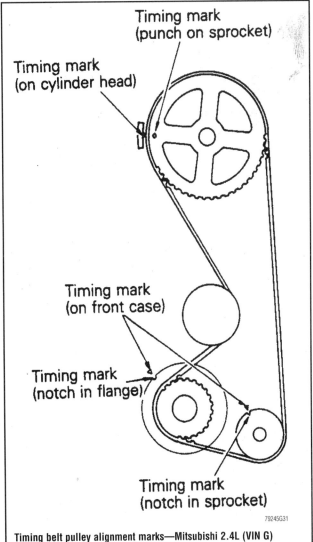

Timing belt pulley alignment marks—Mitsubishi 2.4L (VIN G) engine

18. Install the tensioner initially so the actual center of the pulley is above and to the left of the installation bolt.
19. Move the pulley up by hand so the center span of the long side of the belt deflects about ¼ in. (6mm).
20. Hold the pulley tightly so it does not rotate when the bolt is tightened. Tighten the bolt to 15 ft. lbs. (20 Nm). If the pulley has moved, the belt will be too tight.
21. Install the timing belt tensioner fully toward the water pump and temporarily tighten the bolts. Place the upper end of the spring against the water pump body. Align the timing marks of the cam, crankshaft and oil pump sprockets with the corresponding marks on the front case or head.

➡ **If the following steps are not followed exactly, there is a chance that the silent shaft alignment will be 180 degrees off. This will cause a noticeable vibration in the engine and the entire procedure will have to be repeated.**

22. Before installing the timing belt, ensure that the left side silent shaft is in the correct position:

➡ **It is possible to align the timing marks on the camshaft sprocket, crankshaft sprocket and the oil pump sprocket with the left balance shaft out of alignment.**

23. With the timing mark on the oil pump pulley aligned with the mark on the front case, check the alignment of the left balance shaft to assure correct shaft timing.
    a. Remove the plug located on the left side of the block in the area of the starter.
    b. Insert a tool having a shaft diameter of 0.3 in. (8mm) into the hole.
    c. With the timing marks still aligned, the tool must be able to go in at least 2⅓ in. If it can only go in about 1 in. turn the oil pump sprocket one complete revolution.
    d. Recheck the position of the balance shaft with the timing marks realigned. Leave the tool in place to hold the silent shaft while continuing.
24. Install the belt to the crankshaft sprocket, oil pump sprocket and the camshaft sprocket, in that order. While doing so, be sure there is no slack between the sprockets except where the tensioner will take it up when released.
25. Recheck the timing marks' alignment.
26. If all are aligned, loosen the tensioner mounting bolt and allow the tensioner to apply tension to the belt.
27. Remove the tool that is holding the silent shaft in place and turn the crankshaft clockwise a distance equal to two teeth of the camshaft sprocket. This will allow the tensioner to automatically tension the belt the proper amount.

✲✲ WARNING

Do not manually apply pressure to the tensioner. This will overtighten the belt and will cause a howling noise.

28. First tighten the lower mounting bolt and then tighten the upper spacer bolt.

✲✲ WARNING

If any binding is felt when adjusting the timing belt tension by turning the crankshaft, STOP turning the engine, because the pistons may be hitting the valves.

29. To verify that belt tension is correct, check that the deflection of the longest span (between the camshaft and oil pump sprockets) is ½ in. (13mm).
30. Install the lower timing belt cover. Be sure the packing is properly positioned in the inner grooves of the covers when installing.
31. Install the water pump pulley and the crankshaft pulley(s).
32. Install the upper front timing belt cover.
33. Install the power steering pump, alternator, air conditioning compressor, tension pulley and accompanying brackets, as required.
34. Install the radiator, shroud, fan and accessory drive belts.
35. Install the spark plug wires to the tree on the upper cover.
36. Refill the cooling system.
37. Connect the negative battery cable. Start the engine and check for leaks.

3.0L (6G72) SOHC ENGINE

1990–94 Models

1. Disconnect the negative battery cable.
2. Remove the engine undercover.
3. Remove the cruise control actuator.
4. Remove the accessory drive belts.
5. Remove the air conditioner compressor tension pulley assembly.
6. Remove the tension pulley bracket.
7. Using the proper equipment, slightly raise the engine to take the weight off the side engine mount. Remove the engine mounting bracket.
8. Disconnect the power steering pump pressure switch connector. Remove the power steering pump and wire aside.
9. Remove the engine support bracket.
10. Remove the crankshaft pulley.
11. Remove the timing belt cover cap.
12. Remove the timing belt upper and lower covers.
13. If the same timing belt will be reused, mark the direction of the timing belt's rotation for installation in the same direction. Make sure the engine is positioned so the No. 1 cylinder is at the TDC of its compression stroke and the sprockets' timing marks are aligned with the engine's timing mark indicators.
14. Loosen the timing belt tensioner bolt and remove the belt. If the tensioner is not being removed, position it as far away from the center of the engine as possible and tighten the bolt.
15. If the tensioner is being removed, paint the outside of the spring to ensure that it is not installed backwards. Unbolt the tensioner and remove it along with the spring.

To install:
16. Install the tensioner, if removed, and hook the upper end of the spring to the water pump pin and the lower end to the tensioner

MITSUBISHI—TIMING BELTS

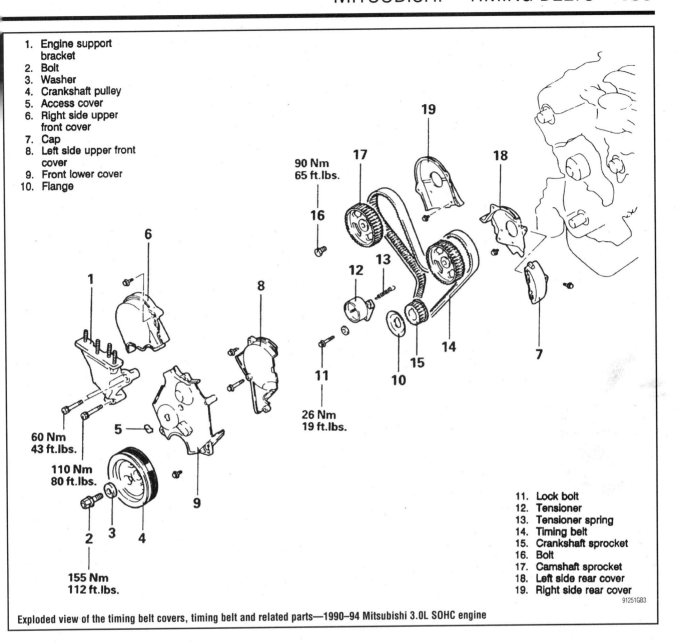

Exploded view of the timing belt covers, timing belt and related parts—1990-94 Mitsubishi 3.0L SOHC engine

in exactly the same position as originally installed. If not already done, position both camshafts so the marks align with those on the rear. Rotate the crankshaft so the timing mark aligns with the mark on the oil pump.

17. Install the timing belt on the crankshaft sprocket and while keeping the belt tight on the tension side, install the belt on the front camshaft sprocket.
18. Install the belt on the water pump pulley, then the rear camshaft sprocket and the tensioner.
19. Rotate the front camshaft counterclockwise to tension the belt between the front camshaft and the crankshaft. If the timing marks became misaligned, repeat the procedure.
20. Install the crankshaft sprocket flange.
21. Loosen the tensioner bolt and allow the spring to apply tension to the belt.
22. Turn the crankshaft 2 full turns in the clockwise direction until the timing marks align again. Now that the belt is properly tensioned,

torque the tensioner lock bolt to 21 ft. lbs. (29 Nm). Measure the belt tension between the rear camshaft sprocket and the crankshaft with belt tension gauge. The specification is 46–68 lbs. (210–310 N).

23. Install the timing covers. Make sure all pieces of packing are positioned in the inner grooves of the covers when installing.
24. Install the crankshaft pulley. Tighten the bolt to 108–116 ft. lbs. (150–160 Nm).
25. Install the engine support bracket.
26. Install the power steering pump and reconnect wire harness at the power steering pump pressure switch.
27. Install the engine mounting bracket and remove the engine support fixture.
28. Install the tension pulleys and drive belts.
29. Install the cruise control actuator.
30. Install the engine undercover.
31. Connect the negative battery cable and road test the vehicle.

154 TIMING BELTS—MITSUBISHI

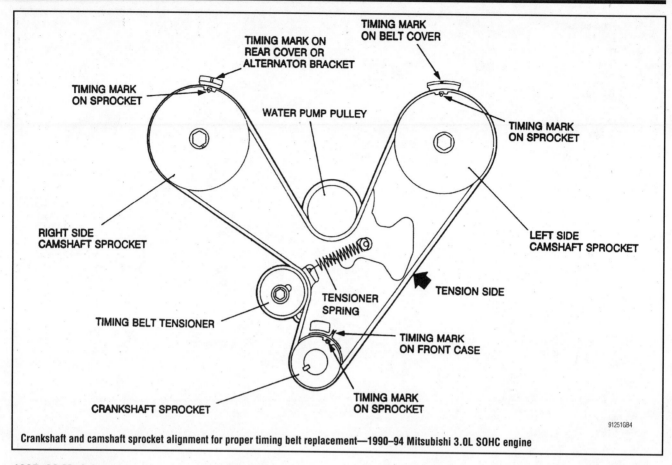

Crankshaft and camshaft sprocket alignment for proper timing belt replacement—1990–94 Mitsubishi 3.0L SOHC engine

1995–99 Models

1. Position the engine so the No. 1 cylinder is at TDC of its compression stroke.
2. Disconnect the negative battery cable.

⁕⁕ CAUTION

Wait at least 90 seconds after the negative battery cable is disconnected to prevent possible deployment of the air bag.

3. Remove the engine undercover.
4. Remove the front undercover panel.
5. Remove the cruise control pump and the link assembly.
6. Remove the alternator.
7. Raise and suspend the engine so that force is not applied to the engine mount.
8. Remove the timing covers from the engine.
9. If the same timing belt will be reused, mark the direction of the timing belt's rotation for installation in the same direction. Make sure the engine is positioned so the No. 1 cylinder is at the TDC of its compression stroke and the timing marks are aligned with the engine's timing mark indicators on the valve covers or head.
10. Loosen the center bolt of tensioner pulley and unbolt auto-tensioner assembly. The auto-tensioner assembly must be reset to correctly adjust belt tension. Remove the timing belt.
11. Using a wrench, hold the camshaft at its hexagon and remove the camshaft sprocket bolt.
12. Remove and position the auto-tensioner into a vise with soft jaws. The plug at the rear of tensioner protrudes, be sure to use a washer as a spacer to protect the plug from contacting vise jaws.
13. Slowly push the rod into the tensioner until the set hole in rod is aligned with set hole in the auto-tensioner.
14. Insert a 0.055 in. (1.4mm) wire into the aligned set holes. Unclamp the tensioner from the vise and install it on the engine. Tighten tensioner to 17 ft. lbs. (24 Nm).
15. Clean and inspect both auto tensioner mounting bolts. Coat the threads of the old bolts with thread sealer. If new bolts are installed, inspect the heads of the new bolts. If there is white paint on the bolt head, no sealer is required. If there is no paint on the head of the bolt, apply a coat of thread sealer to the bolt. Install both bolts and tighten to 17 ft. lbs. (24 Nm).

To install:

16. Install the tensioner, if removed, and hook the upper end of the spring to the water pump pin and the lower end to the tensioner in exactly the same position as originally installed.
17. Ensure both camshafts are still positioned so the timing marks align with those on the rear timing covers. Rotate the crankshaft so the timing mark aligns with the mark on the front cover.
18. Install the timing belt on the crankshaft sprocket and while keeping the belt tight on the tension side, install the belt on the front (left) camshaft sprocket.
19. Install the belt on the water pump pulley, then the rear (right) camshaft sprocket and the tensioner.
20. Loosen the bolt that secures the adjustment of the tensioner and lightly press the tensioner against the timing belt.

MITSUBISHI—TIMING BELTS

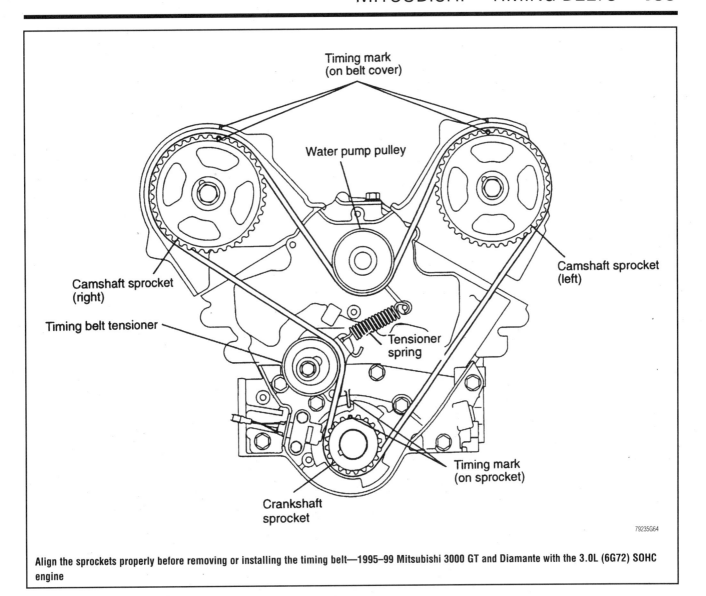

Align the sprockets properly before removing or installing the timing belt—1995–99 Mitsubishi 3000 GT and Diamante with the 3.0L (6G72) SOHC engine

21. Check that the timing marks are in alignment.
22. Rotate the crankshaft 2 full turns in the clockwise direction only, then realign the timing marks.
23. Tighten the bolt that secures the tensioner to 19 ft. lbs. (26 Nm).
24. Install the lower and the upper timing belt covers, along with all other applicable components.

3.0L (6G72) DOHC ENGINE

1990–94 Models

1. Position the engine so the No. 1 cylinder is at TDC of its compression stroke. Disconnect the negative battery cable.
2. Remove the engine undercover.
3. Remove the cruise control actuator.
4. Remove the alternator. Remove the air hose and pipe.
5. Remove the belt tensioner assembly and the power steering belt.
6. Remove the crankshaft pulley.
7. Disconnect the brake fluid level sensor.
8. Remove the timing belt upper cover.
9. Using the proper equipment, slightly raise the engine to take the weight off the side engine mount. Remove the engine mount bracket.
10. Remove the alternator/air conditioner idler pulley.
11. Remove the engine support bracket. The mounting bolts are different lengths; mark them for proper installation.
12. Remove the timing belt lower cover. Timing bolt cover mounting bolts are different in length, note their position during removal.
13. If the same timing belt will be reused, mark the direction of the timing belt's rotation for installation in the same direction. Make sure the engine is positioned so the No. 1 cylinder is at the TDC of its compression stroke and the sprockets' timing marks are aligned with the engine's timing mark indicators on the valve covers or head.
14. Loosen the timing belt tensioner bolt and remove the belt.
15. Remove the tensioner assembly.

To install:

16. If the auto tensioner rod is fully extended, reset it as follows:
    a. Clamp the tensioner in a soft-jaw vice in level position.
    b. Slowly push the rod in with the vice until the set hole in the rod is aligned with the hole in the cylinder.
    c. Insert a stiff wire into the set holes to retain the position.
    d. Remove the assembly from the vice.
17. Leave the retaining wire in the tension and install to the engine.

TIMING BELTS—MITSUBISHI

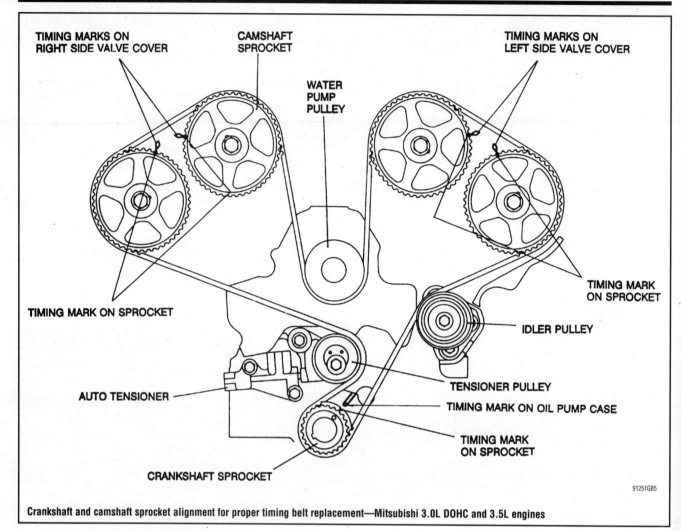

Crankshaft and camshaft sprocket alignment for proper timing belt replacement—Mitsubishi 3.0L DOHC and 3.5L engines

➡ On 1991 DOHC 3.0L engines, clean and inspect both auto tensioner mounting bolts. Coat the threads of the old bolts with thread sealer. If new bolts are installed, inspect the heads of the new bolts. If there is white paint on the bolt head, no sealer is required. If there is no paint on the head of the bolt, apply a coat of thread sealer to the bolt. Install both bolts and tighten to 17 ft. lbs. (24 Nm).

18. If the timing marks of the camshaft sprockets and crankshaft sprocket are not aligned at this point, proceed as follows:

➡ **Keep fingers out from between the camshaft sprockets. The sprockets may move unexpectedly because of valve spring pressure and could pinch fingers.**

   a. Align the mark on the crankshaft sprocket with the mark on the front case. Then move the sprocket 2 teeth clockwise to lower the piston so the valve can't touch the piston when the camshafts are being moved.

   b. Turn each camshaft sprocket 1 at a time to align the timing marks with the mark on the valve cover or head. If the intake and exhaust valves of the same cylinder are opened simultaneously, they could interfere with each other. Therefore, if any resistance is felt, turn the other camshaft to move the valve.

   c. Align the timing mark of the crankshaft sprocket, then continue 1 tooth farther in the counterclockwise direction to facilitate belt installation.

19. Using 4 spring loaded paper clips to hold the belt on the cam sprockets, install the belt to the sprockets in the following order:
- **1st**—exhaust camshaft sprocket for the front head
- **2nd**—intake camshaft sprocket for the front head
- **3rd**—water pump pulley
- **4th**—intake camshaft sprocket for the rear head
- **5th**—exhaust camshaft sprocket for the rear head
- **6th**—idler pulley
- **7th**—crankshaft sprocket
- **8th**—tensioner pulley

20. Turn the tensioner pulley so its pin holes are located above the center bolt. Then press the tensioner pulley against the timing belt and simultaneously tighten the center bolt.

21. Make certain that all timing marks are still aligned. If so, remove the 4 clips.

22. Turn the crankshaft ¼ turn counterclockwise, then turn it clockwise until all timing marks are aligned.

23. Loosen the center bolt on the tensioner pulley. Using tool MD998767 or equivalent and a torque wrench, apply a torque of 7.2 ft. lbs. (10 Nm). Tighten the tensioner bolt; make sure the tensioner doesn't rotate with the bolt.

24. Remove the set wire attached to the auto tensioner, if the wire was not previously removed.

25. Rotate the crankshaft 2 complete turns clockwise and let it sit for approximately 5 minutes. Then, make sure the set pin can easily be inserted and removed from the hole in the tensioner.

MITSUBISHI—TIMING BELTS

26. Measure the auto tensioner protrusion (the distance between the tensioner arm and auto tensioner body) to ensure that it is within 0.15–0.18 in. (3.8–4.5mm). If out of specification, repeat Steps 1–4 until the specified value is obtained.

27. Make sure all pieces of packing are positioned in the inner grooves of the lower cover, position cover on engine and install mounting bolts in their original location.

28. Install the engine support bracket and secure using mounting bolts in their original location. Lubricate the reaming area of the reamer bolt and tighten slowly.

29. Install the idler pulley.

30. Install the engine mount bracket. Remove the engine support fixture.

31. Make sure all pieces of packing are positioned in the inner grooves of the upper cover and install.

32. Connect the brake fluid level sensor.

33. Install the crankshaft pulley. Tighten the bolt to 130–137 ft. lbs. (180–190 Nm).

34. Install the belt tensioner assembly and the power steering belt.

35. Install the air hose and pipe.

36. Install the alternator.

37. Install the cruise control actuator.

38. Install the engine undercover.

39. Connect the negative battery cable.

1995–99 Models

1. Position the engine so the No. 1 cylinder is at TDC of its compression stroke. Disconnect the negative battery cable.

2. Remove the engine undercover.

3. Remove the cruise control actuator.

4. Remove the alternator. Remove the air hose and pipe.

5. Remove the belt tensioner assembly and the power steering belt.

6. Remove the crankshaft pulley.

7. Disconnect the brake fluid level sensor.

8. Remove the timing belt upper cover.

9. Using the proper equipment, slightly raise the engine to take the weight off the side engine mount. Remove the engine mount bracket.

10. Remove the alternator/air conditioner idler pulley.

11. Remove the engine support bracket. The mounting bolts are different lengths; mark them for proper installation.

12. Remove the timing belt lower cover. Timing bolt cover mounting bolts are different in length, note their position during removal.

✷✷ CAUTION

Be sure to disconnect the negative battery cable. Wait at least 90 seconds after the negative battery cable is disconnected to prevent possible deployment of the air bag.

13. If the same timing belt will be reused, mark the direction of the timing belt's rotation for installation in the same direction. Be sure the engine is positioned so the No. 1 cylinder is at the TDC of its compression stroke and the timing marks are aligned with the engine's timing mark indicators on the rear timing covers.

✷✷ WARNING

Turning the camshaft sprocket when the timing belt is removed could cause the valves to contact with the pistons, resulting in severe engine damage.

14. Remove the bolts that secure the auto-tensioner to the engine block and remove the tensioner.

To install:

➡ The auto-tensioner assembly must be reset to correctly adjust belt tension.

15. Loosen the center bolt of tensioner pulley to provide timing belt slack. Remove the timing belt assembly.

16. Position the auto-tensioner into a vise with soft jaws. The plug at the rear of tensioner protrudes, be sure to use a washer as a spacer to protect the plug from contacting vise jaws.

17. Slowly push the rod into the tensioner until the set hole in rod is aligned with set hole in the auto-tensioner.

18. Insert a 0.055 in. (1.4mm) wire into the aligned set holes. Unclamp the tensioner from the vise and install it on the engine. Tighten tensioner mounting bolts to 17 ft. lbs. (24 Nm).

✷✷ WARNING

DO NOT rotate or turn the camshafts when removing the sprockets or severe engine damage will result from internal component interference.

19. Align the mark on the crankshaft sprocket with the mark on the front case. Then, move the crankshaft sprocket 1 tooth counterclockwise.

20. Align the timing marks of the camshafts with the marks on the rear covers.

21. Using large paper clips to secure the timing belt to the sprockets, install the timing belt in the following order. Be sure camshafts-to-cylinder heads and crankshaft-to-front cover timing marks are aligned. Install the timing belt around the pulleys in the following order:
   a. Exhaust camshaft sprocket (front bank).
   b. Intake camshaft sprocket (front bank).
   c. Water pump pulley.
   d. Intake camshaft sprocket (rear bank).
   e. Exhaust camshaft sprocket (rear bank).
   f. Tensioner pulley.
   g. Crankshaft pulley.
   h. Idler pulley.

➡ Since the camshaft sprockets turn easily, secure them with box wrenches when installing the timing belt.

22. Align all timing mark on the crankshaft and raise tensioner pulley against belt to remove slack, snug tensioner bolt.

23. Check the alignment of all the timing marks and remove the clips that secure the timing belt to the camshaft sprockets.

24. Rotate the engine ¼ turn counterclockwise, then rotate the engine clockwise to align the timing marks. Check that all the timing marks are in alignment.

25. Loosen the center bolt on the tensioner pulley. Using tool MD998752 or equivalent and a torque wrench, apply 84 inch lbs.

158 TIMING BELTS—MITSUBISHI, NISSAN

(10 Nm) to the tool on the tensioner. Tighten the tensioner bolt to 35 ft. lbs. (49 Nm) and be sure the tensioner does not rotate with the bolt.

26. Rotate the crankshaft two complete turns clockwise and let it sit for approximately five minutes. Then, check that the set pin can easily be inserted and removed from the hole in the auto-tensioner.

27. Remove the set wire attached to the auto-tensioner.

28. Measure the auto-tensioner protrusion (the distance between the tensioner arm and auto-tensioner body) to ensure that it is within 0.15–0.18 in. (3.8–4.5mm). If out of specification, repeat adjustment procedure until the specified value is obtained.

29. Check again that the timing marks on all sprockets are in proper alignment.

30. Install the timing belt covers and all other applicable components.

3.5L ENGINE

1. Disconnect the negative battery cable.
2. Drain the cooling system. Remove the drive belts.

✱✱ CAUTION

Never open, service or drain the radiator or cooling system when hot; serious burns can occur from the steam and hot coolant. Also, when draining engine coolant, keep in mind that cats and dogs are attracted to ethylene glycol antifreeze and could drink any that is left in an uncovered container or in puddles on the ground. This will prove fatal in sufficient quantities. Always drain coolant into a sealable container. Coolant should be reused unless it is contaminated or is several years old.

3. Remove the upper radiator shroud.
4. Remove the fan and fan pulley.
5. Without disconnecting the lines, remove the power steering pump from its bracket and position it to the side. Remove the pump brackets.
6. Remove the belt tensioner pulley bracket.
7. Without releasing the refrigerant, remove the air conditioning compressor from its bracket and position it to the side. Remove the bracket.
8. Remove the cooling fan bracket.
9. On some vehicles it may be necessary to remove the pulley from the crankshaft to access the lower cover bolts.
10. Remove the timing belt cover bolts and the upper and lower covers from the engine.
11. Remove the crankshaft position sensor connector.
12. Using SST MB990767–01 and MD998754, or their equivalents, remove the crankshaft pulley from the crankshaft.
13. Use a shop rag to clean the timing marks to assist in properly aligning the timing marks.
14. Loosen the center bolt on the tension pulley and remove the timing belt.

➡ If the same timing belt will be reused, mark the direction of timing belt's rotation, for installation in the same direction. Be sure engine is positioned so No. 1 cylinder is at the TDC of it's compression stroke and the sprockets timing marks are aligned with the engine's timing mark indicators.

15. Remove the auto-tensioner, the tension pulley and the tension arm assembly.
16. Remove the sprockets by holding the hexagonal portion of the camshaft with a wrench while removing the sprocket bolt.

To install:

17. Install the crankshaft pulley and turn the crankshaft sprocket timing mark forward (clockwise) three teeth to move the piston slightly past No. 1 cylinder top dead center.
18. If removed, install the camshaft sprockets and tighten the bolts to 64 ft. lbs. (88 Nm).
19. Align the timing mark of the left bank side camshaft sprocket.
20. Align the timing mark of the right bank side camshaft sprocket, and hold the sprocket with a wrench so that it doesn't turn.
21. Set the timing belt onto the water pump pulley.
22. Check that the camshaft sprocket timing mark of the left bank side is aligned and clamp the timing belt with double clips.
23. Set the timing belt onto the idler pulley.

✱✱ WARNING

If any binding is felt when adjusting the timing belt tension by turning the crankshaft, STOP turning the engine, because the pistons may be hitting the valves.

24. Turn the crankshaft one turn counterclockwise and set the timing belt onto the crankshaft sprocket.
25. Set the timing belt on the tension pulley.
26. Place the tension pulley pin hole so that it is towards the top. Press the tension pulley onto the timing belt, and then provisionally tighten the fixing bolt. Tighten the bolt to 35 ft. lbs. (48 Nm).
27. Slowly turn the crankshaft two full turns in the clockwise direction until the timing marks align. Remove the four double clips.
28. Install the crankshaft position sensor connector.
29. Install the upper and lower covers on the engine and secure them with the retaining screws. Be sure the packing is properly positioned in the inner grooves of the covers when installing.
30. Install the crankshaft pulley if it was removed. Tighten the bolt to 110 ft. lbs. (150 Nm).
31. Install the air conditioning bracket and compressor on the engine. Install the belt tensioner.
32. Install the power steering pump into position. Install the fan pulley and fan.
33. Install the fan shroud on the radiator.
34. Refill the cooling system.
35. Connect the negative battery cable. Start the engine and check for fluid leaks.

Nissan

3.0L (VG30DE & VG30DETT) ENGINES

1990–94 Models

1. Disconnect the negative battery cable.
2. Remove the engine undercover.
3. Drain the cooling system.
4. Remove the radiator.
5. Remove the drive belts.

NISSAN—TIMING BELTS

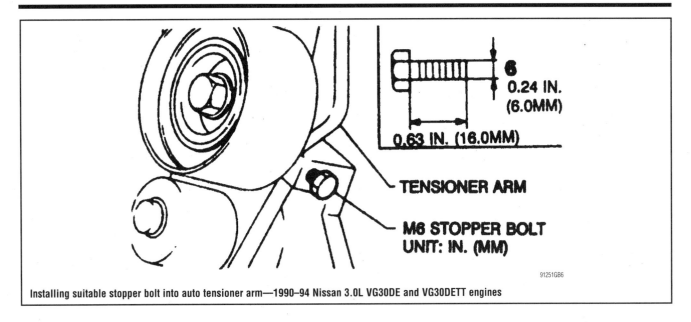

Installing suitable stopper bolt into auto tensioner arm—1990–94 Nissan 3.0L VG30DE and VG30DETT engines

6. Remove the cooling fan and cooling fan coupling.
7. Remove the crankshaft pulley bolt.
8. Remove the starter and lock the flywheel ring gear using a suitable locking device. This is done to prevent the crankshaft gear from turning during REMOVAL AND INSTALLATION.
9. Remove the crankshaft pulley using a suitable puller, then remove the locking device.
10. Remove the water inlet and outlet housings.
11. Remove the timing belt covers and gaskets.
12. Install a suitable 6mm stopper bolt in the tenioner arm of the auto tensioner so the length of the pusher does not change.
13. Set the No. 1 piston at TDC of the compression stroke.
14. Remove the auto-tensioner and the timing belt.

To install:
15. Check the auto-tensioner for oil leaks in the pusher rod and diaphragm. If oil is evident, replace the auto-tensioner assembly.
16. Verify that the No. 1 piston is at TDC of the compression stroke.

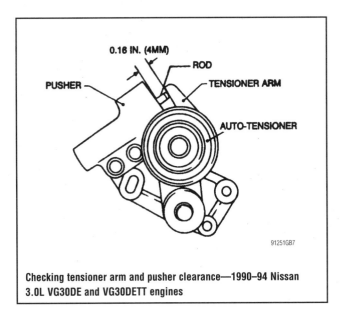

Checking tensioner arm and pusher clearance—1990–94 Nissan 3.0L VG30DE and VG30DETT engines

17. Align the timing marks on the camshaft and crankshaft sprockets with the timing marks on the rear timing belt cover and the oil pump housing.
18. Remove all the spark plugs.
19. With a feeler gauge, check the clearance between the tensioner arm and the pusher of the auto-tensioner. The clearance should be 0.16 in. (4mm) with a slight drag on the feeler gauge. If the clearance is not as specified, mount the tensioner in a vise and adjust the clearance. When the clearance is set, insert the stopper bolt into the tensioner arm to retain the adjustment.

➡ **When adjusting the clearance, do not push the tensioner arm with the stopper bolt fitted, because damage to the threaded portion of the bolt will result.**

20. Mount the auto-tensioner and tighten nuts and bolts by hand.
21. Install the timing belt. Ensure the timing sprockets are free of oil and water. Do not bend or twist the timing belt. Align the white lines on the belt with the timing marks on the camshaft and crankshaft sprockets. Point the arrow on the belt towards the front.
22. Push the auto-tensioner slightly towards the timing belt to prevent the belt from slipping. At the same time, turn the crankshaft 10 degrees clockwise and torque the tensioner fasteners to 12–15 ft. lbs. (16–21 Nm).

➡ **Do not push the tensioner too hard because it will create excessive tension on the belt.**

23. Turn the crankshaft 120 degrees counterclockwise.
24. Turn the crankshaft clockwise and set the No. 1 piston at TDC of the compression stroke.
25. Back off on the auto-tensioner fasteners ½ turn.
26. Using push-pull gauge No. EG1486000 (J–38387) or equivalent, apply approximately 15.2–18.3 lbs. (67.7–81.4 N) of force to the tensioner.
27. Turn the crankshaft 120 degrees clockwise.
28. Turn the crankshaft counterclockwise and set the No. 1 piston at TDC of the compression stroke.

160 TIMING BELTS—NISSAN

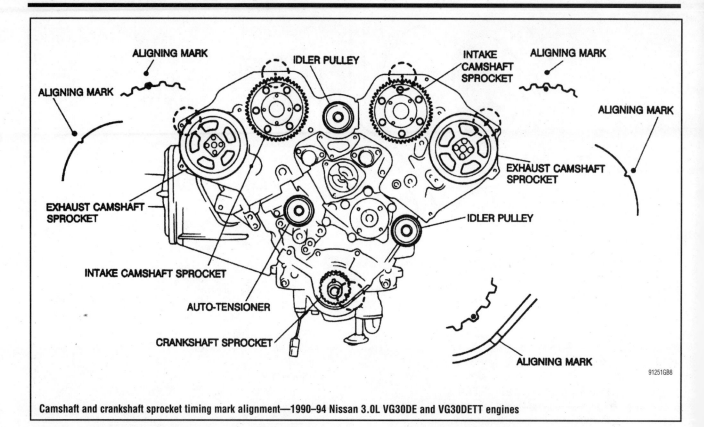

Camshaft and crankshaft sprocket timing mark alignment—1990–94 Nissan 3.0L VG30DE and VG30DETT engines

29. Fabricate a 0.35 in. (9mm) wide x 0.10 in. (2mm) deep steel plate. The length of the plate should be slightly longer than the width of the belt.

30. Set the steel plate at positions **A**, **B**, **C** and **D** of the timing belt mid-way between the pulleys as shown. Using the push-pull gauge or equivalent, apply approximately 11 lbs. (49 N) of force to the tensioner and check (and record) the belt deflection at each position with the steel plate in place. The timing belt deflection at each position should be 0.217–0.256 in. (5.5–6.5mm). Another means of determining the belt deflection is to add all deflection readings and divide them by 4. This average deflection should be 0.217–0.256 in. (5.5–6.5mm).

31. If the belt deflection is not as specified, repeat Steps 22–30 until the belt deflection is correct.

32. Once the belt is properly tensioned, tighten the auto-tensioner fasteners to 12–15 ft. lbs. (16–21 Nm).

33. Remove the stopper bolt from the tensioner and wait 5 minutes. After 5 minutes, check the clearance between the tensioner arm and the pusher of the auto-tensioner. The clearance should remain at 0.138–0.205 in. (3.5–5.2mm).

34. Make sure the belt is installed and aligned properly on each pulley and timing sprocket. There must be no slippage or misalignment.

35. Install the timing belt covers with new gaskets. Tighten the covers bolts to 2–4 ft. lbs. (3–5 Nm).

36. Install the water inlet and outlet housings with new gaskets.

37. Install the crankshaft pulley. Tighten the pulley bolt to 159–174 ft. lbs. (21–235 Nm).

38. Remove the flywheel locking device and install the starter.

39. Install the cooling fan and cooling fan coupling.

40. Install and tension the drive belts.

41. Install the radiator.
42. Fill the cooling system to the proper level.
43. Connect the negative battery cable.

1995–99 Models

1. Remove the spark plugs and position the engine so that No. 1 piston is at TDC of the compression stroke.
2. Disconnect the negative battery cable.
3. Remove the engine undercover.
4. Drain the cooling system.
5. Remove the radiator.
6. Remove the drive belts.
7. Remove the cooling fan and cooling fan clutch assembly from the front of the water pump.
8. Remove the water pump pulley.
9. Remove the starter and lock the flywheel ring gear using a suitable locking device.
10. Remove the crankshaft pulley bolt.

➡ **The engine is locked to prevent the crankshaft gear from turning during removal and installation. Remove the lock during engine timing portion of this procedure.**

11. Remove the crankshaft pulley using a suitable puller.
12. Remove the water inlet and outlet housings.
13. Remove the timing belt covers and gaskets.

✱✱ WARNING

After the timing belt has been removed, DO NOT rotate the camshafts or the crankshaft. Severe internal engine damage will result from piston and valve contact.

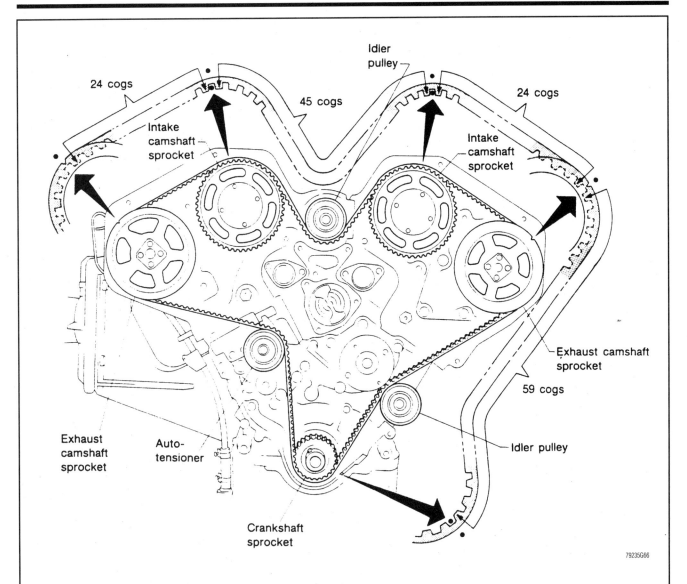

To ensure proper installation of the timing belt, be sure the proper number of belt teeth are between each sprocket, as indicated—1995–99 Nissan 3.0L (VG30DE) engines

14. Install a suitable 6mm stopper bolt in the tensioner arm of the auto-tensioner so the length of the pusher does not change.
15. Remove the automatic tensioner and the timing belt.
16. Check the automatic tensioner for oil leaks in the pusher rod and diaphragm. If oil is evident, replace the automatic tensioner assembly.
17. Inspect the timing gear teeth for wear.

To install:
18. Verify that the No. 1 piston is at TDC of the compression stroke.
19. Align the timing marks on the camshaft and crankshaft sprockets with the timing marks on the rear timing belt cover and the oil pump housing.
20. With a feeler gauge, check the clearance between the tensioner arm and the pusher of the automatic tensioner. The clearance should be 0.16 in. (4mm) with a slight drag on the feeler gauge. If the clearance is not as specified, mount the tensioner in a vise and adjust the clearance. When the clearance is set, insert the stopper bolt into the tensioner arm to retain the adjustment.

➡ **When adjusting the clearance, do not push the tensioner arm with the stopper bolt fitted, because damage to the threaded portion of the bolt will result.**

21. Mount the automatic tensioner and tighten the nuts and bolts by hand.
22. Install the timing belt. Ensure the timing sprockets are free of oil and water. Do not bend or twist the timing belt. Align the white lines on the belt with the timing marks on the camshaft and crankshaft sprockets. Point the arrow on the belt towards the front.
23. Push the automatic tensioner slightly towards the timing belt to prevent the belt from slipping. At the same time, turn the crankshaft 10 degrees clockwise and tighten the tensioner fasteners to 12–15 ft. lbs. (16–21 Nm).

➡ **Do not push the tensioner too hard because it will create excessive tension on the belt.**

24. Turn the crankshaft 120 degrees counterclockwise.
25. Back off on the automatic tensioner fasteners ½ turn.

162 TIMING BELTS—NISSAN

26. Turn the crankshaft clockwise and set the No. 1 piston at TDC of the compression stroke.
27. Using push-pull gauge No. EG1486000 (J-38387) or equivalent, apply approximately 15.2–18.3 lbs. (67.7–81.4 N) of force to the tensioner.
28. Tighten the tensioner mounting bolts to 12–15 ft. lbs. (17–21 Nm).
29. Turn the crankshaft 120 degrees clockwise.
30. Turn the crankshaft 120 degrees counterclockwise and set the No. 1 piston at TDC of the compression stroke.

➡ **If the timing belt deflection exceeds the specification, change the applied pushing force.**

31. Fabricate a 0.35 in. (9mm) wide x 0.10 in. (2mm) steel plate. The length of the plate should be slightly longer than the width of the belt.
32. Set the steel plate at positions mid-way between the camshaft sprockets on each head, between the left exhaust camshaft sprocket and the idler pulley, and between the right exhaust camshaft sprocket and the tensioner.
33. Using the push-pull gauge or equivalent, apply approximately 11 lbs. (49 N) of force to the tensioner.
34. Check and record the belt deflection at each position with the steel plate in place. The timing belt deflection at each position should be 0.24–0.28 in. (6–7mm). Another means of determining the belt deflection is to add all deflection readings and divide them by 4. This average deflection should be 0.24–0.28 in. (6–7mm).
35. If the belt deflection is not as specified, repeat the timing belt adjusting procedure until the belt deflection is correct.
36. Once the belt is properly tensioned, tighten the automatic tensioner fasteners to 12–15 ft. lbs. (16–21 Nm).
37. Remove the stopper bolt from the tensioner and wait 5 minutes. After 5 minutes, check the clearance between the tensioner arm and the pusher of the automatic tensioner. The clearance should remain at 0.138–0.205 in. (3.5–5.2mm).
38. Be sure the belt is installed and aligned properly on each pulley and the timing sprocket. There must be no slippage or misalignment.
39. Install the timing belt covers with new gaskets. Tighten the covers bolts to 2–4 ft. lbs. (3–5 Nm).
40. Install the remaining components in the reverse order of removal.

3.0L (VG30E) & 3.3L (VG33E) ENGINES

Except Quest

1990–94 MODELS—EXCEPT MAXIMA

1. Disconnect the negative battery cable.
2. Drain the cooling system.
3. Remove the engine undercovers.
4. Remove the radiator shroud and fan.
5. Remove the power steering, alternator and air conditioning compressor drive belts.

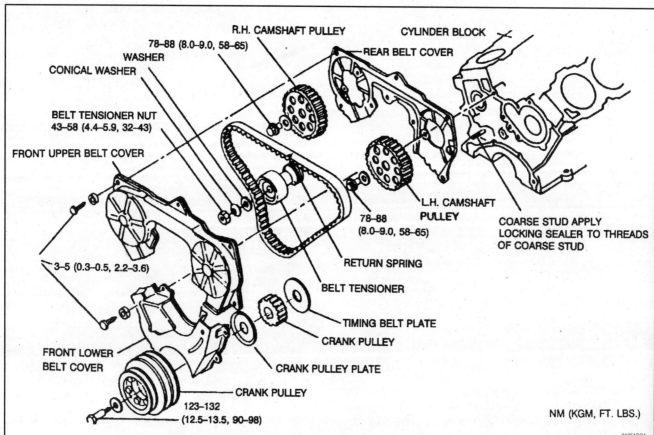

Exploded view of the timing belt assembly—1990–94 Nissan 3.0L VG30E engine

NISSAN—TIMING BELTS

6. Remove the suction pipe bracket and disconnect the lower coolant hose from the suction pipe.
7. Remove the spark plugs.
8. Set No. 1 cylinder at TDC of the compression stroke.
9. Remove the crankshaft pulley.
10. Remove compressor drive belt idler bracket.
11. Remove the front upper and lower belt covers and gaskets.
12. Using chalk or paint, mark the relationship of the timing belt to the camshaft and the camshaft sprockets. Also mark the timing belt's direction of rotation. Align the punch mark on the left hand camshaft pulley with the mark on the upper rear timing belt cover. Align the punchmark on the crankshaft with the notch on the oil pump housing. Temporarily install the crankshaft pulley bolt to allow for crankshaft rotation.
13. Loosen the timing belt tensioner and return spring then remove the timing belt. Check that the tensioner spring turns smoothly and check the tensioner spring for wear.

To install:

14. Before installing the timing belt confirm that No. 1 cylinder is at TDC on its compression stroke. Install tensioner and tensioner spring. If stud is removed, apply locking sealant to threads before installing.
15. Swing the tensioner fully clockwise with hexagon wrench and temporarily tighten locknut.
16. Point the arrow on the timing belt toward the front belt cover. Align the white lines on the timing belt with the punch marks on all 3 pulleys.

➡ There are 133 total timing belt teeth. If timing belt is installed correctly, there will be 40 teeth between left hand and right hand camshaft sprocket timing marks. There will be 43 teeth between left hand camshaft sprocket and crankshaft sprocket timing marks.

17. Loosen tensioner locknut, keeping tensioner steady with an Allen wrench.
18. Swing tensioner 70–80 degrees clockwise with the Allen wrench and temporarily tighten locknut.
19. Install the spark plugs. Turn crankshaft clockwise 2–3 times, then slowly set No. 1 cylinder at TDC on its compression stroke.
20. Push middle of timing belt between right hand camshaft sprocket and tensioner pulley with a force of 22 ft. lbs.
21. Loosen tensioner locknut, keeping tensioner steady with the Allen wrench.
22. Insert a 0.138 in. (0.35mm) thick and 0.5 in. (12.7mm) wide feeler gauge between the bottom of tensioner pulley and timing belt. Turn crankshaft clockwise and position gauge completely between tensioner pulley and timing belt. The timing belt will move about 2.5 teeth.
23. Tighten tensioner locknut, keeping tensioner steady with the Allen wrench.
24. Turn crankshaft clockwise or counterclockwise and remove the gauge.
25. Rotate the engine 3 times, then set No. 1 at TDC on its compression stroke.
26. Timing belt deflection is 0.512–0.571 in. (13.0–14.5mm) at 22 lbs. of pressure. If it is out of specified range, readjust the timing belt.
27. Install the upper and lower timing belt covers with new gaskets. Tighten the covers bolts to 2–4 ft. lbs. (3–5 Nm).

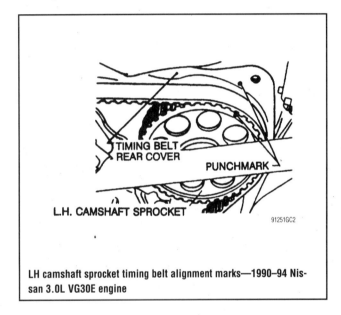

LH camshaft sprocket timing belt alignment marks—1990–94 Nissan 3.0L VG30E engine

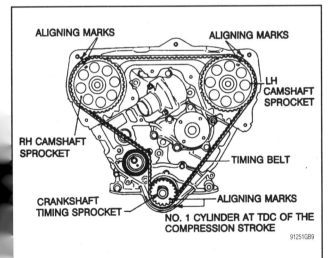

Camshaft and crankshaft alignment mark locations for timing belt service—1990–94 Nissan 3.0L VG30E engine

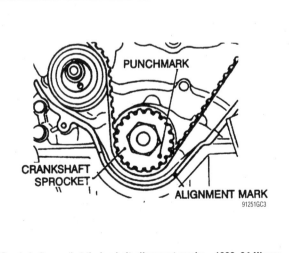

Crankshaft sprocket timing belt alignment marks—1990–94 Nissan 3.0L VG30E engine

164 TIMING BELTS—NISSAN

28. Install the crankshaft pulley. Tighten the pulley bolt to 90–98 ft. lbs. (123–132 Nm).
29. Install the compressor drive belt idler bracket.
30. Connect the lower coolant hose to the suction pipe and install the suction pipe bracket.
31. Install and tension the drive belts.

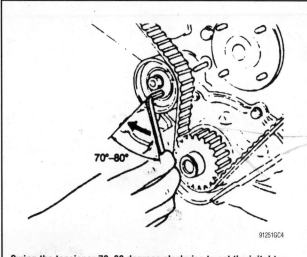

Swing the tensioner 70–80 degrees clockwise to set the initial tension of the timing belt—1990–94 Nissan 3.0L VG30E engine

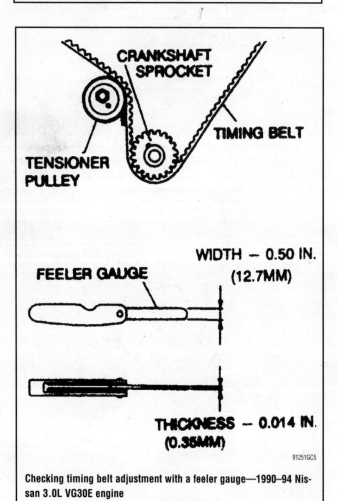

Checking timing belt adjustment with a feeler gauge—1990–94 Nissan 3.0L VG30E engine

32. Install the radiator fan and shroud.
33. Install the engine undercovers.
34. Fill the cooling system to the proper level.
35. Connect the negative battery.

1990–94 MODELS—MAXIMA

1. Disconnect the negative battery cable.
2. Raise and support the front of the vehicle safely.
3. Remove the engine undercovers.
4. Drain the cooling system.
5. Remove the front right side wheel.
6. Remove the engine side cover.
7. Remove the alternator, power steering and air conditioning compressor drive belts from the engine. When removing the power steering drive belt, loosen the idler pulley from the right side wheel housing.
8. Remove the upper radiator and water inlet hoses.
9. Remove the water pump pulley.
10. Remove the idler bracket of the compressor drive belt.
11. Remove the crankshaft pulley with a suitable puller.
12. Remove the upper and lower timing belt covers and gaskets.
13. Rotate the engine with a socket wrench on the crankshaft pulley bolt to align the punchmark on the left hand camshaft pulley with the mark on the upper rear timing belt cover; align the punchmark on the crankshaft with the notch on the oil pump housing; temporarily install the crankshaft pulley bolt to allow for crankshaft rotation.
14. Use a hex wrench to turn the belt tensioner clockwise and tighten the tensioner locknut just enough to hold the tensioner in position. Then, remove the timing belt.

To install:

15. Before installing the timing belt, confirm that No. 1 cylinder is at TDC on its compression stroke. Install tensioner and tensioner spring. If stud is removed apply locking sealant to threads before installing.
16. Swing tensioner fully clockwise with hexagon wrench and temporarily tighten locknut.
17. Point the arrow on the timing belt toward the front belt cover. Align the white lines on the timing belt with the punch marks on all 3 pulleys.

➡ There are 133 total timing belt teeth. If timing belt is installed correctly there will be 40 teeth between left hand and right hand camshaft sprocket timing marks. There will be 43 teeth between left hand camshaft sprocket and crankshaft sprocket timing marks.

18. Loosen tensioner locknut, keeping tensioner steady with a hexagon wrench.
19. Swing tensioner 70–80 degrees clockwise with hexagon wrench and temporarily tighten locknut.
20. Turn crankshaft clockwise 2–3 times, then slowly set No. 1 cylinder at TDC of the compression stroke.
21. Push middle of timing belt between right hand camshaft sprocket and tensioner pulley with a force of 22 lbs. (10 Kg).
22. Loosen tensioner locknut, keeping tensioner steady with a hexagon wrench.
23. Insert a 0.138 in. (0.35mm) thick and 0.5 in. (12.7mm) wide feeler gauge between the bottom of tensioner pulley and timing belt. Turn crankshaft clockwise and position gauge completely between tensioner pulley and timing belt. The timing belt will move about 2.5 teeth.

NISSAN—TIMING BELTS

24. Tighten tensioner locknut, keeping tensioner steady with a hexagon wrench.
25. Turn crankshaft clockwise or counterclockwise and remove the gauge.
26. Rotate the engine 3 times, then set No. 1 at TDC on its compression stroke. Check timing belt tension, the reference valve is 0.51–0.059 in. (13–15mm) deflection with a 22 lbs. (10 Kg) force applied.
27. Install the upper and lower timing belt covers with new gaskets.
28. Install the crankshaft pulley. Tighten the pulley bolt to 90–98 ft. lbs. (123–132 Nm).
29. Install the compressor drive belt idler bracket.
30. Install the water pump pulley and tighten the nuts to 12–15 ft. lbs. (16–21 Nm). Install the upper radiator and water inlet hoses.
31. Install the drive belts.
32. Install the engine side cover.
33. Mount the front right wheel.
34. Install the engine undercovers.
35. Lower the vehicle.
36. Fill the cooling system and connect the negative battery cable.

1995–99 MODELS

1. Remove the engine undercover.
2. Remove the radiator shroud, the fan and the pulleys.
3. Drain the coolant from the radiator and remove the water pump hose.

✱✱ CAUTION

When draining the coolant, keep in mind that cats and dogs are attracted by the ethylene glycol antifreeze, and are quite likely to drink any that is left in an uncovered container or in puddles on the ground. This will prove fatal in sufficient quantity. Always drain the coolant into a sealable container. Coolant should be reused unless it is contaminated or several years old.

4. Remove the radiator.
5. Remove the power steering, A/C compressor and alternator drive belts.
6. Remove the spark plugs.
7. Remove the distributor protector (dust shield).
8. Remove the A/C compressor drive belt idler pulley and bracket.
9. Remove the fresh air intake tube at the cylinder head cover.
10. Disconnect the radiator hose at the thermostat housing.
11. Remove the crankshaft pulley bolt, then pull off the pulley with a suitable puller.
12. Remove the bolts, then remove the front upper and lower timing belt covers.
13. Set the No. 1 piston at TDC of its compression stroke. Align the punchmark on the left camshaft sprocket with the punchmark on the timing belt upper rear cover. Align the punchmark on the crankshaft sprocket with the notch on the oil pump housing. Temporarily install the crank pulley bolt so the crankshaft can be rotated if necessary.
14. Loosen the timing belt tensioner and return spring, then remove the timing belt.

To install:

✱✱ CAUTION

Before installing the timing belt, confirm that the No. 1 cylinder is set at the TDC of the compression stroke.

15. Remove both cylinder head covers and loosen all rocker arm shaft retaining bolts.

➡ The rocker arm shaft bolts MUST be loosened so that the correct belt tension can be obtained.

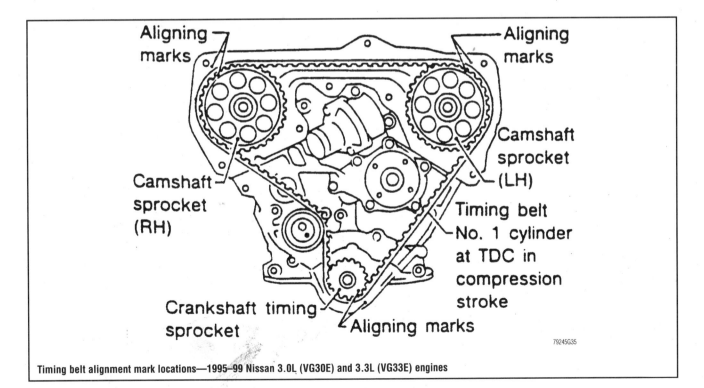

Timing belt alignment mark locations—1995–99 Nissan 3.0L (VG30E) and 3.3L (VG33E) engines

16. Install the tensioner and the return spring. Using a hexagon wrench, turn the tensioner clockwise and temporarily tighten the locknut.
17. Be sure that the timing belt is clean and free from oil or water.
18. When installing the timing belt align the white lines on the belt with the punchmarks on the camshaft and crankshaft sprockets. Have the arrow on the timing belt pointing toward the front belt covers.

➡ **A good way (although rather tedious!) to check for proper timing belt installation is to count the number of belt teeth between the timing marks. There are 133 teeth on the belt; there should be 40 teeth between the timing marks on the left and right side camshaft sprockets, and 43 teeth between the timing marks on the left side camshaft sprocket and the crankshaft sprocket.**

19. While keeping the tensioner steady, loosen the locknut with a hex wrench.
20. Turn the tensioner approximately 70–80 degrees clockwise with the wrench, then tighten the locknut.

※※ WARNING

If any binding is felt when adjusting the timing belt tension by turning the crankshaft, STOP turning the engine, because the pistons may be hitting the valves.

21. Turn the crankshaft in a clockwise direction several times, then **slowly** set the No. 1 piston to TDC of the compression stroke.
22. Apply 22 lbs. of pressure (push it in!) to the center span of the timing belt between the right side camshaft sprocket and the tensioner pulley, then loosen the tensioner locknut.
23. Using a 0.0138 in. (0.35mm) thick feeler gauge (the actual width of the blade **must** be ½ in. or 13mm!), turn the crankshaft clockwise (**slowly!**). The timing belt should move approximately 2½ teeth. Tighten the tensioner locknut, turn the crankshaft slightly and remove the feeler gauge.
24. Slowly rotate the crankshaft clockwise several more times, then set the No. 1 piston to TDC of the compression stroke.
25. Position the two timing covers on the block, then tighten the mounting bolts to 24 ft. lbs. (35 Nm).
26. Press the crankshaft pulley onto the shaft, then tighten the bolt to 90–98 ft. lbs. (123–132 Nm).
27. Connect the radiator hose to the thermostat housing.
28. Reconnect the fresh air intake tube at the cylinder head cover.
29. Install the A/C compressor drive belt idler pulley and bracket.
30. Install the distributor protector (dust shield).
31. Install the spark plugs.
32. Install the power steering, A/C compressor and alternator drive belts.
33. Install the radiator.
34. Reconnect the water pump hose and fill the engine with coolant. Install the fan shroud and pulleys.
35. Install the engine undercover.
36. Start the engine and check for any leaks.

Quest

On this vehicle, right side refers to the "rear" components (near the firewall) and left side refers to the "front" components (near the radiator).

1. If the timing belt is to be removed, it is good practice to turn the crankshaft until the engine is at Top Dead Center (TDC) of the number one cylinder, compression stroke (firing position), before beginning work. This should align all timing marks and serve as a reference for all work that follows. After verifying that the engine is at TDC for the number one cylinder, do not crank the engine or allow the crankshaft or camshaft sprockets to be turned otherwise engine timing will be lost.
2. Drain the cooling system.
3. Disconnect the negative battery cable.
4. Remove the alternator drive belt, water pump and power steering pump belt and the A/C compressor belt (if equipped), using the recommended drive belt removal procedure.
5. If equipped with A/C, remove the three A/C compressor drive belt idler pulley bolts and remove the idler pulley.
6. Remove the upper radiator hose bracket bolt. Remove the upper hose with the bracket from the vehicle.
7. Remove the water bypass hose from between the thermostat housing and the lower water hose connection.
8. Remove the main wiring harness from the upper engine front cover.
9. Remove the eight upper engine front cover bolts and remove the upper cover.
10. Raise and safely support the vehicle.
11. Remove the right side front wheel and tire assembly.
12. Remove the four right side engine and transmission splash shield bolts and two screws, and remove the right side outer engine and transaxle splash shield.
13. Use a strap wrench to hold the water pump pulley. Remove the four pulley bolts, and the water pump pulley.
14. Use a strap wrench to hold the crankshaft pulley. Remove the center pulley bolt, and the crankshaft pulley using a harmonic bal-

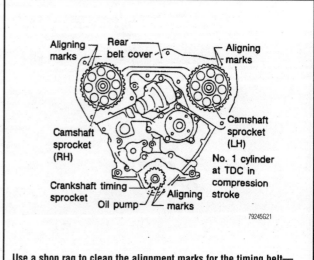

Use a shop rag to clean the alignment marks for the timing belt—Nissan 3.0L (VG30E) Quest engine

NISSAN—TIMING BELTS

ancer (damper) puller to draw the pulley from the front of the crankshaft.

15. Remove the five lower engine front cover bolts, then remove the lower engine front cover.

16. Be sure that the timing marks between the crankshaft sprocket and the oil pump housing line up.

17. If the timing belt is to be reused, mark an arrow on the belt indicating the direction of rotation. The directional arrow is necessary to ensure that the timing belt, if it to be reused, can reinstalled in the same direction.

18. Loosen the timing belt tensioner nut and slip the timing belt off of the sprockets.

19. If necessary, the camshaft sprockets can be removed. A special spanner tool is designed to hold the sprocket to keep it from turning while the center bolt is being loosened. Use care if using substitutes.

➥ **The sprockets are not interchangeable.**

20. If necessary, the crankshaft sprocket can be removed. The outer timing belt guide (looks like a large washer) and the crankshaft sprocket simply pull off the front of the crankshaft.

➥ **Be careful, there are two crankshaft keys. Use care not to loose them.**

To install:

21. Clean all parts well. If removed, inspect the crankshaft sprocket for warping or abnormal wear. Check the sprocket teeth for wear, deformation, chipping or other damage. Replace as necessary. Clean the sprocket mounting surface to ease installation. Install the key. Slip the sprocket onto the crankshaft. Tap it in place with a suitably-sized socket.

22. If removed, inspect the camshaft sprockets for damage and wear. Replace as required. The sprockets should be marked **L3** to designate the front, or left side camshaft and **R3** to designate the rear, or right side camshaft. Use care to install the sprockets properly. A special spanner tool is designed to hold the sprocket to keep it from turning while the center bolt is being tightened. Use care if using a substitute. Tighten the camshaft sprocket center bolts to 58–65 ft. lbs. (78–88 Nm). Verify that the timing marks on the camshaft sprockets and the timing marks on the rear cover (called the seal plate) are aligned.

23. Use an Allen wrench to turn the timing belt tensioner clockwise until the belt tensioner spring is fully extended. Temporarily tighten the tensioner nut to 32–43 ft. lbs. (43–58 Nm).

24. If a new timing belt is to be installed, look for a printed arrow on the belt. Be sure the arrow is pointing away from the engine. If the original timing belt is to be reused, be sure that the directional arrow that was marked at disassembly is facing the correct direction.

25. A new Original Equipment Manufacture (OEM) timing belt should have three white timing marks on it that indicate the correct timing positions of the camshafts and the crankshaft. These marks are to help ensure that the engine is properly timed. When the engine is properly timed, each white timing mark on the timing belt will be aligned with the corresponding camshaft and crankshaft timing mark on the sprocket. Because the white timing marks are not evenly spaced, the technician needs to use care in installing the belt. There should be 40 timing belt teeth between the timing marks on the front and rear camshaft sprockets and 43 teeth between the timing mark on the front camshaft sprocket and the timing mark on the crankshaft sprocket.

26. Verify that the camshaft timing marks are aligned with the timing marks on the rear cover (seal plate) and that the crankshaft sprocket timing mark is aligned with the timing mark on the oil pump housing.

27. Install the timing belt starting at the crankshaft sprocket and moving around the camshaft sprockets following a counterclockwise path. Do not allow any slack in the timing belt between the sprockets. After all of the timing marks are matched up with the timing belt installed, slip the timing belt onto the belt tensioner.

28. While holding the timing belt tensioner with an Allen wrench, loosen the tensioner nut. Allow the tensioner to put pressure on the timing belt. Use an Allen wrench to turn the timing belt tensioner 70–80 degrees clockwise and tighten the timing belt tensioner nut to 32–43 ft. lbs. (43–58 Nm).

✲✲ WARNING

If any binding is felt when adjusting the timing belt tension by turning the crankshaft, STOP turning the engine, because the pistons may be hitting the valves.

29. Rotate the crankshaft clockwise twice and align the number one piston to TDC on the compression stroke (firing position).

30. Apply 22 lbs. (10kg) of force on the timing belt between the rear camshaft sprocket and the timing belt tensioner. An assistant may be needed. While holding the timing belt tensioner steady with an Allen wrench, loosen the timing belt tensioner nut. Remove the Allen wrench and adjust the timing belt tensioner using the following procedure:

 a. Install a 0.0138 in. (0.35mm) thick and 0.500 in. (12.7mm) wide feeler gauge where the timing belt just starts to go around the tensioner (approximately the four o'clock position, looking at the tensioner).

 b. Turn the crankshaft sprocket clockwise, which should force the feeler gauge between the timing belt and the tensioner, up to a position on the tensioner of about 1 o'clock.

 c. Tighten the timing belt tensioner nut to 32–43 ft. lbs. (43–58 Nm).

 d. Turn the crankshaft clockwise to rotate the feeler gauge out from between the timing belt tensioner and the timing belt.

31. Rotate the crankshaft clockwise twice, and once again align the number one piston to TDC on the compression stroke (firing position).

32. Apply 22lbs. (10kg) of force on the timing belt between the front and rear camshaft sprockets. Measure the amount of belt deflection. Belt deflection should be between 0.51–0.59 in. (13–15mm). If belt deflection is out of specification, repeat Steps 29 through 33. If the timing belt deflection cannot be adjusted into specification, the timing belt will have to be replaced.

33. Position the lower engine front cover and install the five lower cover bolts. Do not overtighten. Tighten to 27–44 inch lbs. (3–5 Nm).

34. Install the outer timing belt guide next to the crankshaft sprocket with the dished side facing away from the cylinder block. Install the crankshaft pulley. Use a strap wrench to keep the crankshaft pulley from turning and tighten the center bolt to 90–98 ft. lbs. (123–132 Nm).

35. Position the water pump pulley on the pump. Install the four bolts. Use a strap wrench to keep the water pump pulley from turning and tighten the four water pump pulley bolts to 12–15 ft. lbs. (16–21 Nm).

TIMING BELTS—NISSAN, SAAB

36. Position the right side outer engine and transaxle splash shield, and secure with the four bolts and two screws.
37. Install the right side front wheel and tire assembly. Tighten the lug nuts to 72–87 ft. lbs. (98–118 Nm).
38. Lower the vehicle.
39. Position the upper engine timing belt front cover, and tighten the eight bolts to 27–44 inch lbs. (3–5 Nm).
40. Install the main wiring harness on the upper engine front cover.
41. Position the water bypass hose between the thermostat housing and water connection. Install the upper radiator hose between the radiator and the water hose connection. Secure the hoses with clamps. Install the upper radiator hose bracket. Tighten the bracket bolt to 34–58 ft. lbs. (46–65 Nm).
42. If equipped, position the A/C compressor drive belt idler pulley and install the three bolts. Tighten to 15 ft. lbs. (21 Nm).
43. Install and adjust the alternator drive belt, the water pump and power steering pump drive belt and the A/C compressor drive belt (of equipped).
44. Connect the battery cable.
45. Fill the cooling system.
46. Start the engine and allow it to warm to operating temperature. Check and adjust the ignition timing. Road test to verify correct engine operation.

Saab

2.5L (B258L) ENGINE

⁂ WARNING

To avoid damage to the valves, DO NOT rotate the camshafts once the timing belt is removed. The crankshaft may only be turned between 0° and 60° BTDC when the camshafts are locked in position with the appropriate locking tool.

1. Disconnect the negative battery cable.
2. Remove the air cleaner complete with hoses.
3. Using a 15mm wrench turn the serpentine drive belt tensioner towards the front of the vehicle, remove the belt from the water pump pulley and slowly release the belt tensioner.
4. Remove the belt tensioner and the power steering pump pulley.
5. Remove the water pump pulley. To do this you will have to carefully press the engine towards the left in the vehicle, using the engine bracket as a fulcrum.
6. Remove the timing cover.
7. Remove the right front wheel and the cover in the wheel well.
8. Remove the 6 bolts holding the crankshaft pulley and remove the pulley. Do not remove the center bolt.
9. Put the No. 1 cylinder in the top dead center position (on the compression stroke).
10. The timing marks on the crankshaft and camshafts should be in alignment with their respective marks on the engine. Insert Camshaft Locking Tool 83-94-926 and install Crankshaft Locking Tool 83-94-868 (or their equivalents).
11. If reusing the timing belt, mark the direction of its rotation. To help with refitting, the belt can be marked at the camshaft timing marks and also at the crankshaft timing mark.
12. Remove the tensioning roller and the two adjusting rollers and remove the timing belt.
13. Release tension from and remove the timing belt. Loosen the timing belt adjuster bolts.
14. Rotate the crankshaft back to 60° BTDC, to prevent damage to the valves.
15. Remove the bracket with the upper timing belt adjuster and tensioner rollers.

To install:
16. Install the bracket with the upper timing belt adjuster and tensioner pulleys.
17. Install and locking tools 83-94-926 (or equivalent) between the camshaft sprockets to lock the camshafts of both heads in position.
18. Rotate the crankshaft forward to just before 0° TDC and install the crankshaft locking tool on the crankshaft. Carefully rotate the engine until the arm of the tool is against the water pump flange. Be sure the crankshaft is at 0° TDC and all timing marks are aligned. Remove the locking tool.
19. Install the timing belt, noting the direction of rotation marked at disassembly.
    a. Adjust the tensioner lightly by hand counterclockwise to keep the timing belt from falling off. Be sure the crankshaft is a 0° TDC and all timing marks are aligned.
    b. Install the camshaft locking tool.
    c. Install tool (83-93-985) with a cut piece from an old timing belt, to measure belt tension.
    d. Snug the center bolts of the adjusting rollers. Turn the lower adjusting roller counterclockwise, until a belt tension of 202–221 ft. lbs. (275–300 Nm) is registered on the torque wrench.
    e. Tighten the adjusting roller center bolts to 30 ft. lbs. (40 Nm).

➡ **This adjustment of the timing belt is only preparatory and should, therefore, not be used as a final check.**

20. Adjust the tensioner pulley until the marks are aligned. Tighten the tensioner pulley to 15 ft. lbs. (20 Nm). Remove the camshaft locking tool on camshaft sprockets 1 and 2. Adjust the upper adjusting roller until sprocket No. 2 moves 0.04–0.08 in. (1–2mm) clockwise. Tighten the upper adjusting roller to 30 ft. lbs. (40 Nm) and remove the upper locking tool.
21. Rotate the engine two complete revolutions to just before 0° TDC and install the locking tool on the crankshaft. Carefully turn the crankshaft until the arm of the locking tool is against the water pump flange and tighten the locking tool. Set the camshaft locking tool into position on the front of the camshaft sprockets. Be sure that the timing marks on the camshaft sprockets are aligned with the marks on the tool and that the edge of the timing belt is flush with the edge of the camshaft sprockets.

➡ **Also check that the alignment marks on the tensioner pulley are still aligned.**

22. Make sure the valve cover O-rings are still in position and clean. Lubricate the O-rings with soapy water, apply 81-52-381 (or equivalent) sealer at the corners of the large end bearing caps and install the valve covers.
23. Install the spark plug wires.
24. Install the crankshaft pulley and tighten the retaining bolts to 15 ft. lbs. (20 Nm).
25. Install the timing belt cover tighten the bolts to 6 ft. lbs. (8 Nm).

SAAB—TIMING BELTS

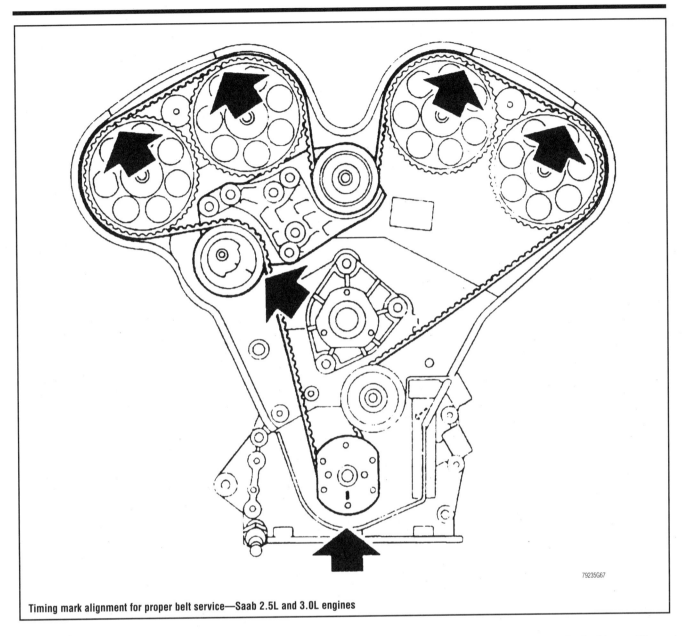

Timing mark alignment for proper belt service—Saab 2.5L and 3.0L engines

26. Install the water pump pulley. To do this you will have to carefully press the engine towards the left in the vehicle, using the engine bracket as a fulcrum.
27. Install the power steering pump pulley and tighten the retaining bolts to 6 ft. lbs. (8 Nm).
28. Using a 15mm wrench turn the serpentine drive belt tensioner towards the front of the vehicle, position the belt and slowly release the belt tensioner.
29. Install the air cleaner complete with hoses.
30. Connect the negative battery cable.
31. Start the engine and check the ignition timing.

3.0L (B308L) ENGINE

※※ WARNING

To avoid damage to the valves, DO NOT rotate the camshafts once the timing belt is removed from the engine. The crankshaft may only be turned between 0° and 60° BTDC when the camshafts are locked in position with the appropriate locking tool.

1. Disconnect the negative battery cable.
2. Lift up the power steering reservoir.
3. Disconnect the connection on the torque arm engine mount and remove the torque arm.
4. Remove the power steering line clamp from the torque arm engine mount and remove the engine mount.
5. Disconnect the hose from the coolant expansion tank and remove the upper alternator air intake.
6. Remove the right front wheel and the cover in the wheel well.
7. Loosen the power steering pump, water pump pulley and 6 outer crankshaft pulley bolts.
8. Using a 15mm wrench turn the serpentine drive belt tensioner towards the front of the vehicle, remove the belt from the water pump pulley and slowly release the belt tensioner. Remove the belt tensioner.
9. Remove the power steering pump and water pump pulley.
10. Remove the timing cover.

170 TIMING BELTS—SAAB, SUBARU

➡ **When removing the crankshaft pulley, remove the 6 outer bolts only, DO NOT remove the center bolt.**

11. Remove the crankshaft pulley.
12. Rotate the crankshaft to TDC of No. 1 cylinder.
13. The timing marks on the crankshaft and camshafts should be in alignment with their respective marks on the engine. Install camshaft locking tools (such as Saab tools KM-800-1 for camshaft sprockets 1 and 2 and KM-800-2 for sprockets 3 and 4) and a flywheel locking tool (such as Saab tool 83-94-868).
14. Mark the direction of rotation of the timing belt for reassembly.
15. Release tension from and remove the timing belt. Loosen the timing belt adjuster bolts.
16. Rotate the crankshaft back to 60° BTDC, to prevent damage to the valves.
17. Remove the bracket with the upper timing belt adjuster and tensioner rollers.

To install:
18. Remove the flywheel locking tool and install the flywheel inspection cover.
19. Install the bracket with the upper timing belt adjuster and tensioner pulleys.
20. Install both camshaft locking tools.
21. Rotate the crankshaft forward to just before 0° TDC and install the crankshaft locking tool on the crankshaft. Carefully rotate the engine until the arm of the tool is against the water pump flange. Be sure the crankshaft is at 0° TDC and all timing marks are aligned. Remove the locking tool.
22. If reusing the belt, fit the timing belt according to its marked direction of rotation and timing marks. Adjust the tensioning roller loosely by hand to prevent the belt from slipping out of the cogs. Always adjust counterclockwise.
23. Measure the belt tension with Saab tool 83-93-985, or equivalent.
24. Snug the center bolts of the adjusting rollers. Turn the lower adjusting roller counterclockwise, until a belt tension of 202–220 ft. lbs. (275–300 Nm) is reached. Tighten the adjusting roller center bolts to 30 ft. lbs. (40 Nm).

➡ **This is a preliminary adjustment of the belt tension and must not be used as a check when the belt is finally adjusted.**

25. Continue to carry out the adjustment by means of the tensioning roller, mark against mark. Remove the locking tool for camshaft sprockets 1 and 2. Carry out the final adjustment with the upper center adjusting roller until camshaft sprocket No. 2 moves 0.04–0.08 in. (1–2mm) forward.
26. Remove the locking tool for camshaft sprockets 3 and 4 and also remove the crankshaft locking tool.
27. Tighten the tensioning roller to 15 ft. lbs. (20 Nm). Tighten the upper adjusting roller to 30 ft. lbs. (40 Nm) and tighten the lower adjusting roller to 15 ft. lbs. (20 Nm).
28. Rotate the engine two complete revolutions to just before 0° TDC and install the locking tool on the crankshaft. Carefully turn the crankshaft until the arm of the locking tool is against the water pump flange and tighten the locking tool. Set Saab tool KM-800-20 (or equivalent) into position. Be sure that the timing marks on the camshaft sprockets are aligned with the marks on the tool and that the edge of the timing belt is flush with the edge of the camshaft sprockets.

➡ **Also check that the alignment marks on the tensioner pulley are still aligned.**

29. Install the crankshaft pulley and tighten the retaining bolts to 15 ft. lbs. (20 Nm).
30. Install the timing belt cover tighten the bolts to 6 ft. lbs. (8 Nm).
31. Install the water pump pulley. Tighten water pump pulley bolts to 6 ft. lbs. (8 Nm).
32. Install power steering pump.
33. Install the power steering pump pulley and tighten the retaining bolts to 6 ft. lbs. (8 Nm).
34. Using a 15mm wrench turn the serpentine drive belt tensioner towards the front of the vehicle, position the belt and slowly release the belt tensioner.
35. Install the upper alternator air intake.
36. Install the torque arm engine mount and attach the power steering line clamp.
37. Install the torque arm and bolt the connection to the torque arm engine mount.
38. Connect the upper hose to the coolant expansion tank and set the power steering reservoir back into position.
39. Connect the negative battery cable.
40. Start the engine and check the ignition timing.

Subaru

1.2L ENGINE

1. Disconnect the negative battery cable.
2. Position the crankshaft with No. 3 cylinder at TDC.
3. Loosen the alternator-to-engine bolts, relax the drive belt tension and remove the drive belt from the front of the engine.
4. Using a socket wrench (through the hole in the right fender) and the crank/camshaft pulley wrench tool 499205500 or equivalent (to hold the crankshaft pulley), remove the crankshaft pulley-to-crankshaft bolt and the pulley from the crankshaft.
5. Remove the timing belt cover-to-engine bolts and the cover from the engine.

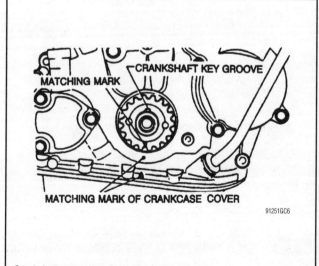

Crankshaft gear alignment—Subaru 1.2L engine

SUBARU—TIMING BELTS   171

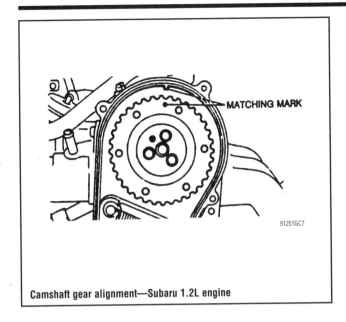

Camshaft gear alignment—Subaru 1.2L engine

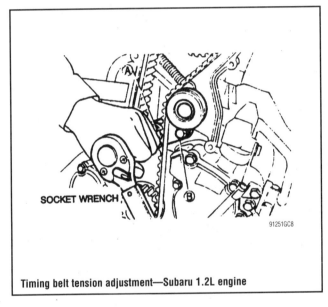

Timing belt tension adjustment—Subaru 1.2L engine

6. Remove the crankshaft bolts and pulley. Remove the outer front timing belt cover.

7. Loosen the tensioner bolt and position it in the direction that loosens the belt. Tighten the tensioner bolt in that position.

8. Remove the camshaft drive pulley plate. Mark the timing belt, if to be used again, in the direction of rotation for reinstallation. Remove the belt from the sprockets.

9. If necessary, remove the tensioner and spring. Remove the camshaft pulley. Remove the inner belt cover and cover mount, only as required.

To install:

10. When installing the timing belt, rotate and align the matchmark of the camshaft driven pulley 0.120 in. (3mm) diameter hole with the matchmark of the cam belt side cover.

11. Align the matchmark of the camshaft drive pulley and the crankshaft cover. Install the camshaft drive belt.

12. Ensure that each rocker arm can be moved. Loosen the tensioner bolt ½ turn.

13. Tighten the tensioner bolt below the adjusting wheel first. Tighten the other bolt. Check to be sure all sprocket and housing matching marks are in agreement.

14. Install the camshaft drive pulley plate.

15. Install the cam belt cover.

16. Install the crankshaft pulley and bolts. Tighten the crankshaft bolts to 58–72 ft. lbs. (78–98 Nm).

17. Check and adjust the valve clearance.

18. Install the accessory drive belts and tension to specification.

19. Connect the negative battery cable. Start the engine and allow it to reach operating temperature. Test drive the vehicle and check for leaks.

1.8L ENGINE

1990–94 Models

1. Disconnect the negative battery cable.
2. Loosen the water pump pulley nut/bolts and the alternator-to-engine bolts, the remove the drive belt.
3. Disconnect the electrical connector from the oil pressure switch.
4. Remove the oil level gauge guide with the gauge.
5. Remove the timing hole cover from the top of the flywheel housing.
6. Using the flywheel stopper tool 498277000 or equivalent (MT), or the driveplate stopper tool 498407000 or equivalent (AT), insert it through the timing hole (in the flywheel housing) and lock the flywheel.
7. Remove the crankshaft pulley bolt and using a puller, remove the crankshaft pulley.
8. If equipped with a turbocharger, remove the belt cover plate.
9. Remove the left side, the right side and the front timing belt cover.
10. Loosen the timing belt tensioner mounting bolts ½ turn and slacken the timing belt. Tighten the mounting bolts.
11. Mark the rotating direction of the timing belt, then remove the belt.
12. Perform the same procedure for the No. 2 timing belt. Remove the crankshaft sprockets.
13. Remove both tensioners together with the tensioner springs.
14. Remove the belt idler. Remove the camshaft sprockets.
15. Remove the No. 2 belt covers.

To install:

16. Inspect the timing belt for breaks, cracks and wear. Replace as required.
17. Check the belt tensioner and idler for smooth rotation. Replace if noisy or excessive play is noticed.
18. Install the left hand belt cover seal No. 3 to the cylinder block.
19. Install the left hand belt cover seal, left hand belt cover seal No. 4, and belt cover mount to the right rear belt cover, then install the assembly on the cylinder block. Tighten to 3–4 ft. lbs. (4–5 Nm).
20. Install the left hand belt cover seal No. 2 and belt cover mounts to left hand belt cover No. 2, then install to the cylinder head and camshaft case. Tighten to 3–4 ft. lbs. (4–5 Nm).
21. Install the right hand belt cover seal, belt cover seal No. 2 and belt cover mounts to the right hand belt cover No. 2, then

172 TIMING BELTS—SUBARU

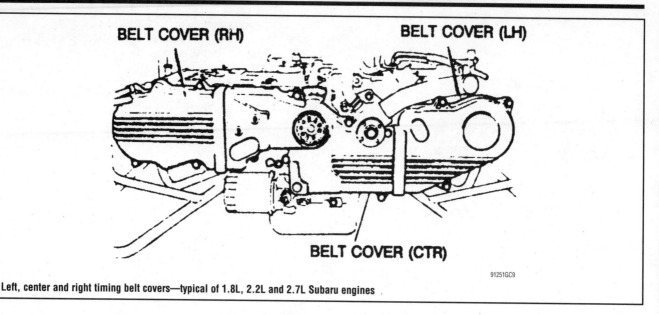

Left, center and right timing belt covers—typical of 1.8L, 2.2L and 2.7L Subaru engines

1. Right hand timing belt
2. Left hand timing belt
3. Crankshaft sprocket
4. Camshaft sprocket
5. Right hand tensioner
6. Left hand tensioner
7. Oil pump sprocket
8. Idler

Timing belt routing—1990–94 Subaru 1.8L and 2.7L engines

install to the cylinder head and camshaft case. Tighten to 3–4 ft. lbs. (4–5 Nm).

22. Install the camshaft sprockets to the right and left camshafts. Tighten the bolts gradually in 2–3 steps to 6–7 ft. lbs. (9–10 Nm).

23. Attach the tensioner spring to the tensioner, then install to the right side of the cylinder block. Tighten the bolts temporarily by hand.

24. Attach the tensioner spring to the bolt, tighten the right side bolt and then loosen ½ turn.

25. Push down tensioner until it stops, then temporarily tighten the left bolt.

26. Install the left side tensioner in the same manner.

27. Install the belt idler to the cylinder block using care not to turn the seal. Tighten to 29–35 ft. lbs. (39–47 Nm).

28. Install sprockets to the crankshaft. Install the crankshaft pulley and tighten the bolt temporarily.

29. Align the center of the three lines scribed on the flywheel with the timing mark on the flywheel housing.

30. Align the timing mark on the left hand camshaft sprocket with the notch on the belt cover.

31. Attach timing belt No. 2 to the crankshaft sprocket No. 2, oil pump sprocket, belt idler, and camshaft sprocket in that order. Avoid downward slackening of the belt.

SUBARU—TIMING BELTS 173

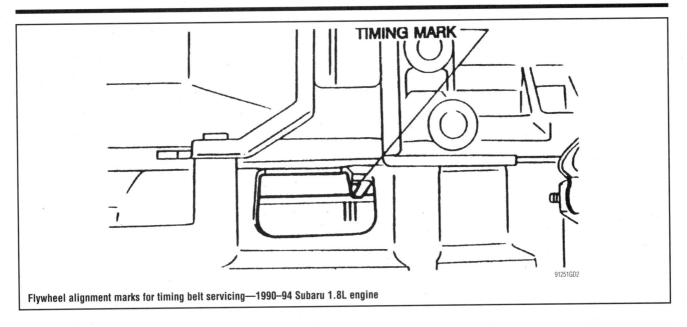

Flywheel alignment marks for timing belt servicing—1990–94 Subaru 1.8L engine

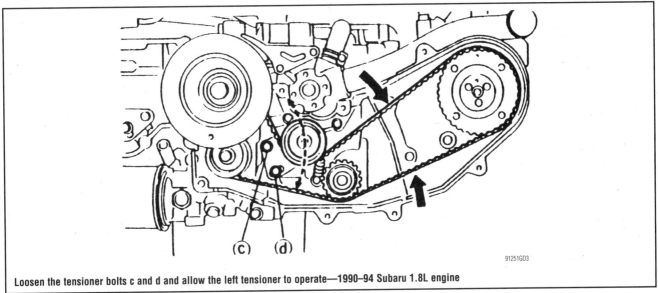

Loosen the tensioner bolts c and d and allow the left tensioner to operate—1990–94 Subaru 1.8L engine

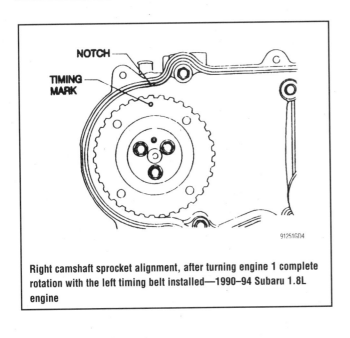

Right camshaft sprocket alignment, after turning engine 1 complete rotation with the left timing belt installed—1990–94 Subaru 1.8L engine

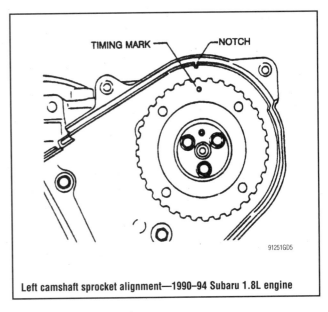

Left camshaft sprocket alignment—1990–94 Subaru 1.8L engine

174 TIMING BELTS—SUBARU

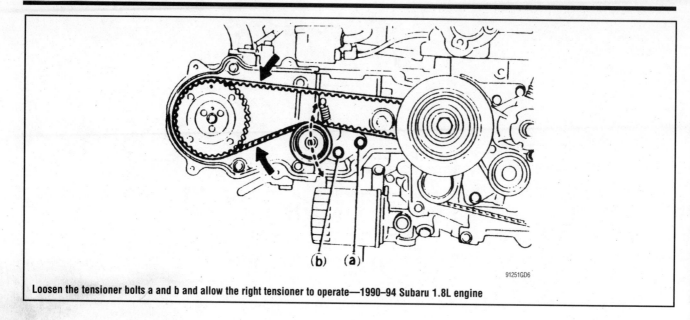

Loosen the tensioner bolts a and b and allow the right tensioner to operate—1990–94 Subaru 1.8L engine

32. Loosen tensioner No. 2 lower bolt ½ turn to apply tension. Push timing belt by hand to ensure smooth movement of tensioner.
33. Apply 25 ft. lbs. (new belt) or 18 ft. lbs. (used belt) torque to the camshaft sprocket in counterclockwise direction. While applying torque tighten tensioner No. 2 lower bolt temporarily, then tighten upper bolt temporarily.
34. Tighten the lower bolt, then the upper bolt to 13–15 ft. lbs. (17–20 Nm) in that order.
35. Check that the flywheel timing mark and left hand camshaft sprocket mark are in their proper positions.
36. Turn the crankshaft 1 turn clockwise from the position where timing belt No. 2 was installed, and align the center of the 3 lines on the flywheel with the timing mark on the flywheel housing.
37. Align the timing mark on the right hand camshaft sprocket with the notch in the belt cover.
38. Attach the timing belt to the crankshaft sprocket and camshaft sprocket, avoiding slackening of the belt on the upper side.
39. Loosen the tensioner ½ turn to apply tension to the belt. Push the belt by hand to ensure smooth operation.
40. Apply 25 ft. lbs. (new belt) or 18 ft. lbs. (used belt) torque to the camshaft sprocket in counterclockwise direction. While applying torque tighten tensioner left bolt temporarily, then tighten right bolt temporarily.
41. Tighten the left bolt, then the right bolt to 13–15 ft. lbs. (17–20 Nm) in that order.
42. Check that the flywheel timing mark and left hand camshaft sprocket mark are in their proper positions.
43. Remove the crankshaft pulley.
44. Install the right front belt cover seals and belt cover plug. Install the belt covers to the cylinder block.
45. On turbocharged engines, install the belt cover plate.
46. Install the crankshaft pulley and tighten to 66–79 ft. lbs. (89–107 Nm).
47. Install the water pump pulley and tighten to 6–7 ft. lbs. (9–10 Nm). Install the pulley cover, oil level guide and gauge and oil pressure switch connector.
48. Install and properly tension the accessory drive belt.

1995–99 Models

1. Disconnect the negative battery cable.
2. Loosen the water pump pulley nut/bolts and the alternator-to-engine bolts, the remove the drive belt.
3. Disconnect the electrical connector from the oil pressure switch.
4. Remove the oil level gauge guide with the gauge.
5. Remove the timing hole cover from the top of the flywheel housing.
6. Using the flywheel stopper tool 498277000 or equivalent (MT), or the driveplate stopper tool 498407000 or equivalent (AT), insert it through the timing hole (in the flywheel housing) and lock the flywheel.
7. Remove the crankshaft pulley bolt and using a puller, remove the crankshaft pulley.
8. If equipped with a turbocharger, remove the belt cover plate.
9. Remove the left side, the right side and the front timing belt cover.
10. Loosen the timing belt tensioner mounting bolts ½ turn and slacken the timing belt. Tighten the mounting bolts.
11. Mark the rotating direction of the No. 1 timing belt, then remove the belt.
12. Perform the same procedure for the No. 2 timing belt. Remove the crankshaft sprockets.
13. Remove both tensioners together with the tensioner springs.
14. Remove the belt idler. Remove the camshaft sprockets.
15. Remove the No. 2 belt covers.

To install:

16. Inspect the timing belt for breaks, cracks and wear. Replace as required.
17. Check the belt tensioner and idler for smooth rotation. Replace if noisy or excessive play is noticed.
18. Install the driver's side belt cover seal No. 3 to the cylinder block.
19. Install the driver's side belt cover seal, driver's side belt cover seal No. 4, and belt cover mount to the right rear belt cover,

SUBARU—TIMING BELTS

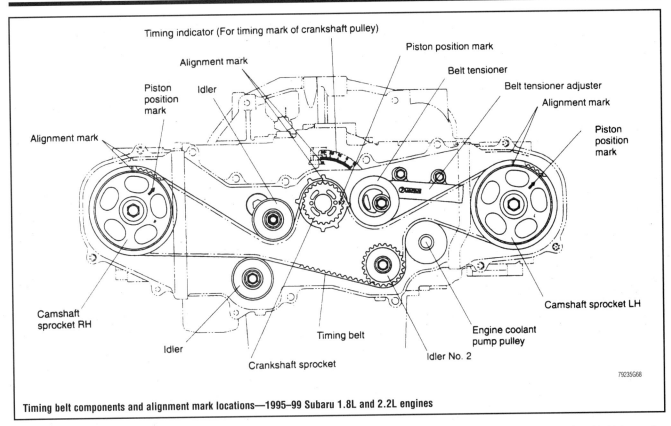

Timing belt components and alignment mark locations—1995-99 Subaru 1.8L and 2.2L engines

then install the assembly on the cylinder block. Tighten to 34 ft. lbs. (45 Nm).

20. Install the driver's side belt cover seal No. 2 and belt cover mounts to driver's side belt cover No. 2, then install to the cylinder head and camshaft case. Tighten to 34 ft. lbs. (45 Nm).

21. Install the passenger's side belt cover seal, belt cover seal No. 2 and belt cover mounts to the passenger's side belt cover No. 2, then install to the cylinder head and camshaft case. Tighten to 34 ft. lbs. (45 Nm).

22. Install the camshaft sprockets to the right and left camshafts. Tighten the bolts gradually in two or three steps to 67 ft. lbs. (91 Nm).

23. Attach the tensioner spring to the tensioner, then install to the right side of the cylinder block. Tighten the bolts temporarily by hand.

24. Attach the tensioner spring to the bolt, tighten the right side bolt, then loosen it ½ turn.

25. Push down the tensioner until it stops, then temporarily tighten the left bolt.

26. Install the left side tensioner in the same manner.

27. Install the belt idler to the cylinder block using care not to turn the seal. Tighten to 29–35 ft. lbs. (39–47 Nm).

28. Install the sprockets on the crankshaft. Install the crankshaft pulley and tighten the bolt temporarily.

29. Align the center of the three lines scribed on the flywheel with the timing mark on the flywheel housing.

30. Align the timing mark on the driver's side camshaft sprocket with the notch on the belt cover.

31. Attach timing belt No. 2 to the crankshaft sprocket No. 2, oil pump sprocket, belt idler, and camshaft sprocket in that order. Avoid downward slackening of the belt.

32. Loosen tensioner No. 2 lower bolt ½ turn to apply tension. Push timing belt by hand to ensure smooth movement of tensioner.

33. Apply 25 ft. lbs. (new belt) or 18 ft. lbs. (used belt) torque to the camshaft sprocket in counterclockwise direction. While applying torque tighten tensioner No. 2 lower bolt temporarily, then tighten upper bolt temporarily.

34. Tighten the lower bolt, then the upper bolt to 13–15 ft. lbs. (17–20Nm) in that order.

35. Check that the flywheel timing mark and driver's side camshaft sprocket mark are in their proper positions.

36. Turn the crankshaft one turn clockwise from the position where timing belt No. 2 was installed, and align the center of the three lines on the flywheel with the timing mark on the flywheel housing.

37. Align the timing mark on the passenger's side camshaft sprocket with the notch in the belt cover.

38. Attach the timing belt to the crankshaft sprocket and camshaft sprocket, avoiding slackening of the belt on the upper side.

39. Loosen the tensioner ½ turn to apply tension to the belt. Push the belt by hand to ensure smooth operation.

40. Apply 25 ft. lbs. (34 Nm) for new belts, or 18 ft. lbs. (24 Nm) for used belts, tighten to the camshaft sprocket in counterclockwise direction. While applying torque, tighten the tensioner left bolt temporarily, then tighten right bolt temporarily.

41. Tighten the left bolt, then the right bolt to 13–15 ft. lbs. (17–20 Nm) in that order.

42. Check that the flywheel timing mark and driver's side camshaft sprocket mark are in their proper positions.

43. Remove the crankshaft pulley.

44. Install the right front belt cover seals and belt cover plug. Install the belt covers to the cylinder block.

45. On turbo-charged engines, install the belt cover plate.

46. Install the crankshaft pulley and tighten to 66–79 ft. lbs. (89–107 Nm).

176 TIMING BELTS—SUBARU

47. Install the water pump pulley and tighten to 67 ft. lbs. (91 Nm). Install the pulley cover, oil level guide and gauge and oil pressure switch connector.
48. Install and properly tension the accessory drive belt.

2.2L ENGINE

1990–94 Models

1. Remove the accessory drive belt.
2. Remove the crankshaft pulley bolt and crankshaft pulley.
3. Remove the cam belt covers. Align the camshaft sprockets so each sprocket notch aligns with the cam cover notches. Align the crankshaft sprocket top tooth notch, located at the rear of the tooth, with the notch on the crank angle sensor boss. Mark the 3 alignment points as well as the direction of cam belt rotation.
4. Loosen the tensioner adjusting bolts and remove the bottom 3 idlers, the cam belt and the cam belt tensioner. The cam sprockets can then be removed with a modified camshaft sprocket wrench tool.

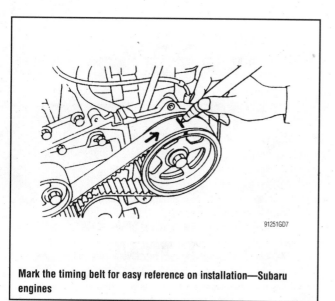

Mark the timing belt for easy reference on installation—Subaru engines

5. If the sprockets are removed, note the reference sensor at the rear of the left cam sprocket.

To install:

6. Install the crankshaft sprocket and all of the idlers except for the lower right. Compress the hydraulic tensioner in a vise slowly and temporarily secure the plunger with a pin. Install the tensioner and the pulley.
7. After the cam belt components are installed, align the crankshaft sprocket notch on the rear sprocket tooth with the crank angle sensor boss. This places the sprocket notch in the 12 o'clock position.
8. Align the camshaft sprockets with the notches in the cam belt cover. As the directional marked belt is installed, align the marks on the belt with the crankshaft sprocket and the left camshaft sprocket. Install the lower right idler.
9. Load the tensioner by pushing it towards the crankshaft with a prybar and tighten the bolts. Remove the tensioner retention pin and the belt tension is automatically set. Rock the crankshaft back and forth 1 time to distribute the belt tension.
10. Verify the correctness of the timing by noting that the notches on the 2 cam pulleys and the notch on the crankshaft pulley all point to the 12 o'clock position when the belt is properly installed.
11. Complete the engine component assembly by installing the cam belt covers, the crankshaft pulley bolt and pulley and the remaining components.

1995–99 Models

The engine uses a single cam belt drive system with a serpentine type belt. The left side of the engine uses a hydraulic cam belt tensioner which is self-adjusting.

➡ **It is recommended that the timing belt be replaced at least every 60,000 miles (96,618 km).**

1. Disconnect the negative battery cable.
2. Position the No. 1 piston to TDC of its compression stroke.
3. Remove the engine drive belts.
4. Remove the timing belt covers.
5. Align the camshaft sprockets so each sprocket notch aligns with the cam cover notches. Align the crankshaft sprocket top tooth

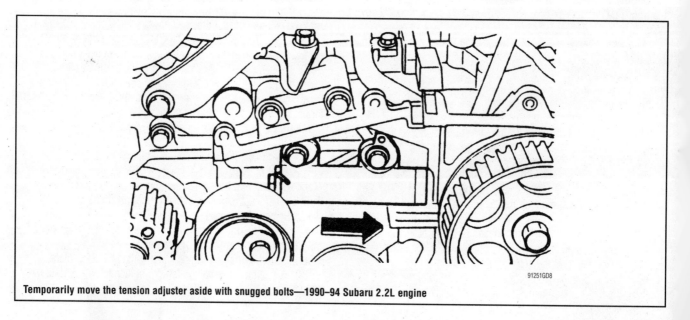

Temporarily move the tension adjuster aside with snugged bolts—1990–94 Subaru 2.2L engine

SUBARU—TIMING BELTS

notch, located at the rear of the tooth, with the notch on the crank angle sensor boss. Mark the three alignment points as well as the direction of cam belt rotation.

6. Loosen the tensioner adjusting bolts. Remove the bottom three idlers, the cam belt and the cam belt tensioner. The cam sprockets can, then be removed with a modified camshaft sprocket wrench tool.

7. Remove the sprockets, if necessary. Note the reference sensor at the rear of the left cam sprocket.

To install:

8. Install the sprockets, if removed and tighten the retaining bolts to 47–54 ft. lbs. (64–74 Nm).

9. Install the crankshaft sprocket and the non-adjustable right side idler. Do not install the tensioner idler at this time.

10. Compress the hydraulic tensioner in a vise slowly and temporarily secure the plunger with a pin or suitable Allen wrench. Install the tensioner and the pulley with the adjustable idler pulley. Temporarily tighten the tensioner while the tensioner is pushed to the right.

11. Align the crankshaft sprocket notch on the rear sprocket tooth with the crank angle sensor boss. This places the sprocket notch in the 12 o'clock position.

12. Align the camshaft sprockets with the notches in the cam rear belt cover. This places the sprocket notch in the 12 o'clock position for each camshaft.

13. Install the timing belt with the directional mark and alignment marks properly positioned (if the belt was reused).

14. Loosen the tensioner retaining bolts and slide the tensioner to the left. Tighten the mounting bolts.

15. After verifying the timing marks are correct, remove the stopper pin from the tensioner.

16. Verify the correctness of the timing by noting that the notches on the 2 cam pulleys and the notch on the crankshaft pulley all point to the 12 o'clock position when the belt is properly installed.

17. Complete the engine component assembly by installing the cam belt covers, the crankshaft pulley bolt and pulley and the remaining components.

18. Connect the negative battery cable.

2.5L ENGINE

Except Forester

The engine uses a single cam belt drive system with a serpentine type belt. The left side of the engine uses a hydraulic cam belt tensioner which is self-adjusting.

➡ **It is recommended that the timing belt be replaced at least every 60,000 miles (96,618 km).**

1. Remove all necessary components for access to the left, right and center timing belt covers, then remove the covers.

2. Align the camshaft sprockets so each sprocket notch aligns with the rear cover notches. Align the crankshaft sprocket top tooth notch, located at the rear of the tooth with the notch. The crankshaft notch will be at 12 o'clock and the key way will be at 6 o'clock.

➡ **Mark the sprocket alignment points as well as the direction of cam belt rotation for reinstallation purposes if the belt is to be reused.**

3. Loosen the tensioner adjusting bolts.
4. Remove the lower timing belt idler.
5. Remove the timing belt from the pulleys.

※※ WARNING

After the timing belt is removed, DO NOT rotate the camshaft sprockets or the crankshaft. Severe internal damage will result from the valve and/or piston contact.

6. Remove the timing belt tensioner and the timing belt tension adjuster.

To install:

➡ **Inspect the timing belt and tensioner for wear or damage and replace as necessary.**

7. Inspect the timing belt tensioner as follows:
   a. When compressing the pushrod of the tensioner with a force of 33 lbs. (147 N), the tensioner should not sink.

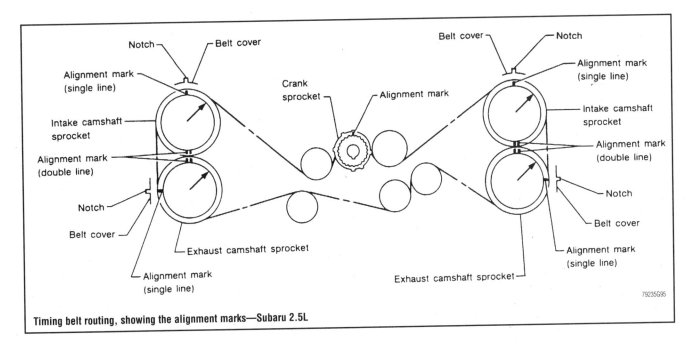

Timing belt routing, showing the alignment marks—Subaru 2.5L

178 TIMING BELTS—SUBARU

b. When compressing the pushrod of the tensioner with a force of 33–110 lbs. (147–490 N), the tensioner should not sink within 8.5 seconds.

c. Measure the extension of the rod beyond the body of the tensioner for a length of 0.606–0.646 in. (15.4–16.4mm). If not within specifications, replace the tensioner.

➡ **Check the idler sprockets for smooth operation. Replace as necessary.**

8. Using a press, compress the tensioner gradually, taking three minutes or more, and insert a 0.059 in. (1.5mm) pin to secure the rod.

9. Install the tensioner and the pulley with the adjustable idler pulley. Temporarily tighten the tensioner while the tensioner is pushed to the right.

10. Align the crankshaft sprocket notch on the rear sprocket tooth with the crank angle sensor boss. This places the sprocket notch in the 12 o'clock position and the key way at the 6 o'clock position.

11. Align the camshaft sprockets with the notches in the cam rear belt cover. This places the sprocket notch in the 12 o'clock position for each camshaft.

12. Install the timing belt in a clockwise direction starting at the crankshaft with the directional mark and alignment marks properly positioned (if the belt was reused).

13. Install the lower timing belt idler and tighten the mounting bolt to 29 ft. lbs. (39 Nm).

✱✱ WARNING

Be sure all the timing marks are properly aligned.

14. Loosen the tensioner retaining bolts and slide the tensioner to the left. Tighten the mounting bolts to 18 ft. lbs. (25 Nm).

15. After verifying the timing marks are correct, remove the stopper pin from the tensioner and recheck the timing marks.

16. Install the center, right, then the center timing belt covers. Tighten the bolts to 44 inch lbs. (5 Nm).

17. Install the remaining components in the reverse order of the removal procedure. When installing the crankshaft sprocket, be sure to tighten the mounting bolt to 94 ft. lbs. (127 Nm).

Forester

When servicing the timing belt heed the following:
- The intake and exhaust camshafts for the 2.5L DOHC engine can be rotated independently when the timing belt is removed. If the intake and exhaust valves are lifted off of their seats simultaneously, their heads will contact each other, possibly causing damage.
- When the timing belt is removed, the camshafts are positioned so that none of the valves are lifted off of their seats, resulting in a "zero-lift" position.
- The left-hand cylinder head camshafts must be rotated from the "zero-lift" position as little as possible when orienting it for timing belt installation, otherwise possible valve head interference may occur.
- Never allow the camshafts to rotate in the direction shown in the accompanying illustration, which would cause both the intake and exhaust valves to lift simultaneously, causing interference.

1. Remove all necessary components to gain access to the timing belt.

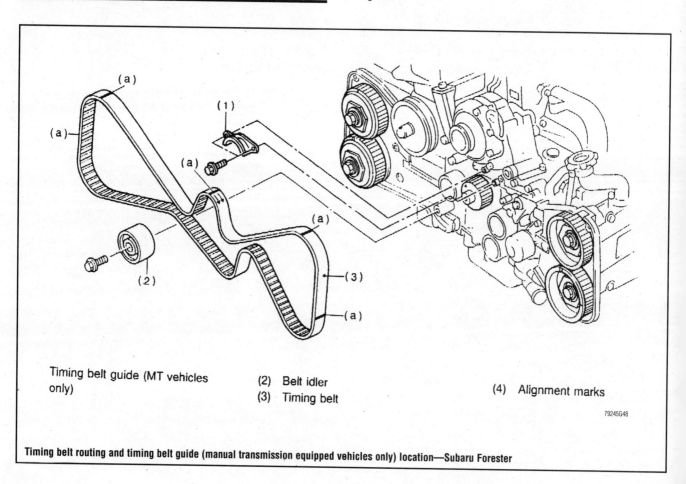

Timing belt guide (MT vehicles only)   (2) Belt idler   (3) Timing belt   (4) Alignment marks

Timing belt routing and timing belt guide (manual transmission equipped vehicles only) location—Subaru Forester

SUBARU—TIMING BELTS 179

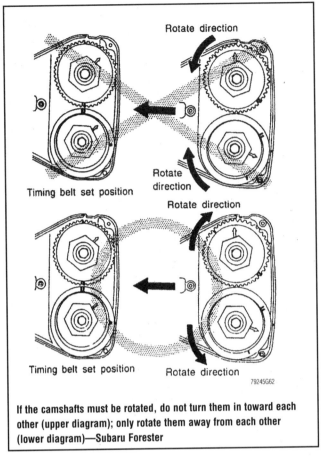

If the camshafts must be rotated, do not turn them in toward each other (upper diagram); only rotate them away from each other (lower diagram)—Subaru Forester

crankshaft until the crankshaft sprocket, left-hand exhaust camshaft sprocket, left-hand intake camshaft sprocket, right-hand intake camshaft sprocket and right-hand exhaust camshaft sprocket timing mark notches are aligned with the respective marks on the belt cover and engine block.

  b. Make alignment and/or arrow marks on the timing belt in relation to the sprockets as indicated in the accompanying illustration.
  - Z1—54.5 tooth length
  - Z2—51 tooth length
  - Z3—28 tooth length

4. Loosen the center bolt from the timing belt idler pulley, then remove the idler pulley from the engine block.

✼✼ WARNING

After removing the timing belt, DO NOT rotate the camshafts. Damage to the valves may occur.

2. On models equipped with manual transmissions, loosen the two timing belt guide mounting bolts, then separate the guide from the engine block.

3. If the directional arrow and alignment marks on the timing belt are faded, and the belt is to be reused, remark the belt with white paint or a grease pencil as follows:

  a. Using a specific socket (Subaru part No. ST 499987500 Crankshaft Socket) installed on the crankshaft sprocket, rotate the

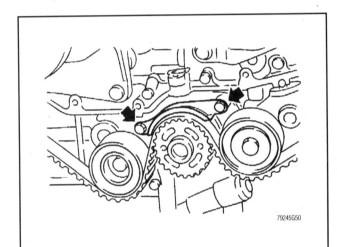

On models equipped with a manual transmission, loosen the two timing belt guide bolts and separate the guide from the engine block—Subaru Forester

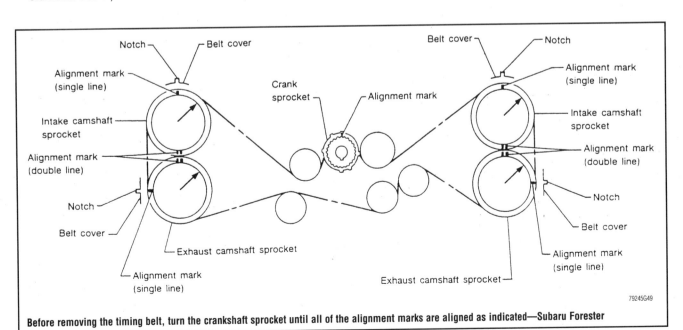

Before removing the timing belt, turn the crankshaft sprocket until all of the alignment marks are aligned as indicated—Subaru Forester

180 TIMING BELTS—SUBARU

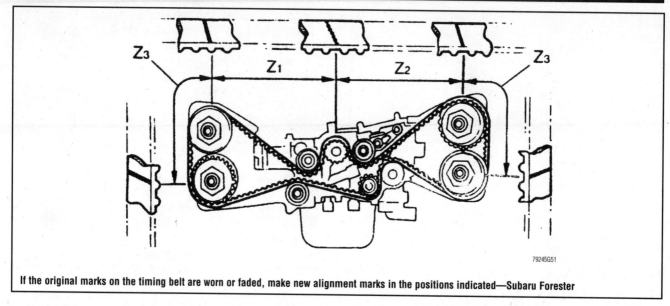

If the original marks on the timing belt are worn or faded, make new alignment marks in the positions indicated—Subaru Forester

5. Carefully remove the timing belt from all of the sprockets.
6. Remove the automatic belt tension adjuster assembly as follows:
   a. Remove the two timing belt idler pulleys, as indicated in the accompanying illustration.
   b. Loosen the automatic tension adjuster assembly mounting bolts, then separate the adjuster assembly from the engine block.

To install:

✱✱ WARNING

Do not allow oil, grease, or coolant to come in contact with the timing belt. If this occurs, quickly and thoroughly remove all traces of the compound. Also, never bend the timing belt sharply; the minimum bending radius is 2.36 in. (60mm).

7. Inspect the camshaft and crankshaft sprocket teeth for abnormal or excessive wear or scratches. Ensure there is no free-play between the sprocket and the key. Inspect the crankshaft sprocket sensor notch for damage or contamination with debris or dirt.

➡ When preparing the automatic tension adjuster assembly for installation, adhere to the following points:

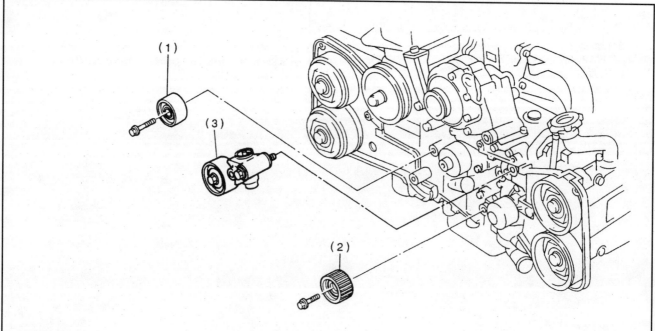

(1) Belt idler
(2) Belt idler No. 2
(3) Automatic belt tension adjuster ASSY

It is necessary to remove the automatic adjuster assembly and reset the pushrod for timing belt installation—Subaru Forester

SUBARU—TIMING BELTS

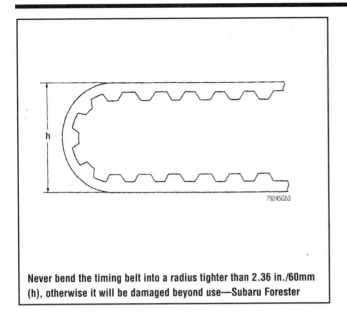

Never bend the timing belt into a radius tighter than 2.36 in./60mm (h), otherwise it will be damaged beyond use—Subaru Forester

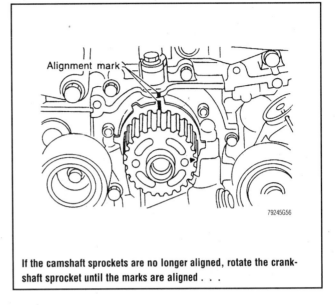

If the camshaft sprockets are no longer aligned, rotate the crankshaft sprocket until the marks are aligned . . .

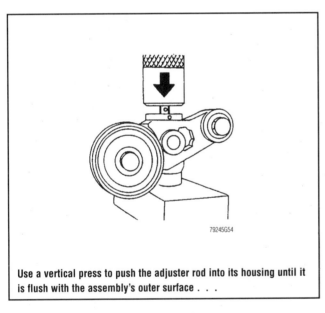

Use a vertical press to push the adjuster rod into its housing until it is flush with the assembly's outer surface . . .

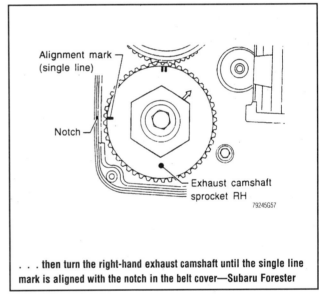

. . . then turn the right-hand exhaust camshaft until the single line mark is aligned with the notch in the belt cover—Subaru Forester

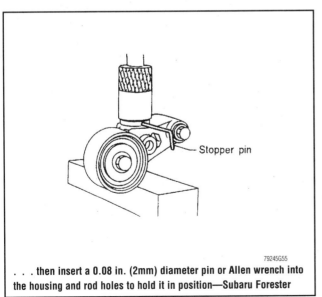

. . . then insert a 0.08 in. (2mm) diameter pin or Allen wrench into the housing and rod holes to hold it in position—Subaru Forester

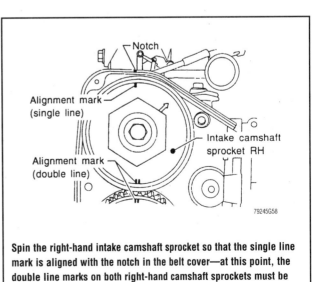

Spin the right-hand intake camshaft sprocket so that the single line mark is aligned with the notch in the belt cover—at this point, the double line marks on both right-hand camshaft sprockets must be aligned

182 TIMING BELTS—SUBARU

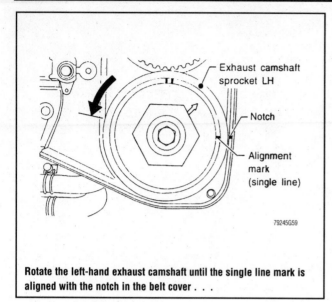

Rotate the left-hand exhaust camshaft until the single line mark is aligned with the notch in the belt cover . . .

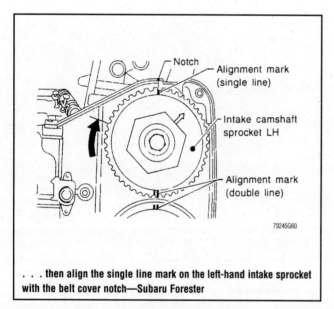

. . . then align the single line mark on the left-hand intake sprocket with the belt cover notch—Subaru Forester

- Always use a vertical press, rather than a horizontal press or vise, to depress the adjuster assembly rod
- Depress the adjuster rod in a vertical position ONLY
- Depress the adjuster rod slowly (taking more than 3 minutes) with a force of 66 lbs. (30 kg)
- Do not allow the press force to exceed 2205 lbs. (1000 kg)
- Press the adjuster rod in as far as the end surface of the cylinder—do not press the rod into the cylinder, which may cause damage to the assembly
- Do not release the press force from the rod until the stopper pin is completely inserted in the cylinder

8. Prepare the automatic timing belt tension adjuster assembly for installation as follows:

   a. Position the adjuster assembly in a vertical press.

   b. Slowly depress the adjuster rod with a force of 66 lbs. (30 kg) until the hole in the rod is aligned with the hole in the adjuster cylinder housing.

   c. Insert a 0.08 in. (2mm) diameter stopper pin or Allen wrench through the hole in the cylinder housing and rod, then slowly release the press force from the adjuster rod.

9. Install the adjuster assembly onto the engine block.
10. Install timing belt idler pulley No. 2 on the engine block.
11. Install the timing belt idler pulley No. 1 on the engine block.
12. If the camshaft and crankshaft timing marks are no longer aligned, perform the following:

   a. Position the crankshaft sprocket so that its mark is aligned with the mark on the oil pump cover on the engine block.

   b. Align the single line mark on the right-hand exhaust camshaft sprocket with the notch on the belt cover.

   c. Rotate the right-hand intake camshaft so that the single line mark is aligned with the notch on the belt cover.

➡ **At this point, the double line marks on both right-hand camshaft sprockets should be aligned.**

   d. Turn the left-hand exhaust (lower) camshaft counterclockwise (as viewed from the front of the engine) until the single line mark is aligned with the notch on the belt cover.

   e. Position the single line mark on the left-hand intake camshaft sprocket so that it is aligned with the notch on the belt cover. When rotating the camshaft, do so only in a clockwise direction (as viewed from the front of the engine).

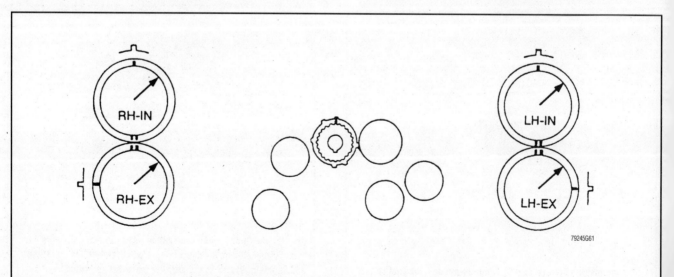

After orienting all five sprockets, the alignment marks should be positioned as shown—Subaru Forester

SUBARU—TIMING BELTS

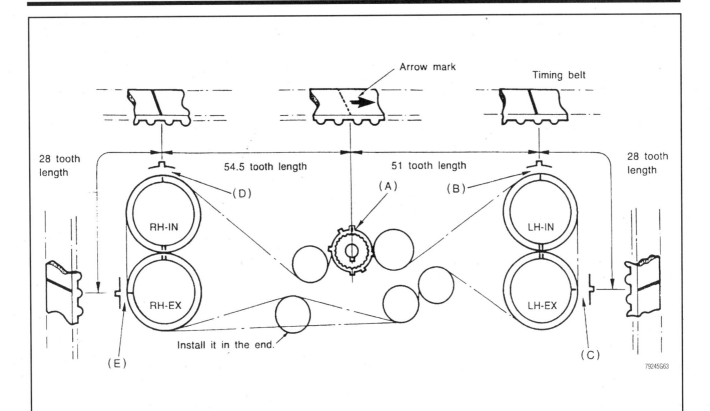

When installing the timing belt, be sure to route it in the proper order (a through e), and ensure that all of the matchmarks are properly aligned—Subaru Forester

➥**At this point, the double line marks on both left-hand camshaft sprockets should be aligned.**

   f. Ensure the timing marks are aligned as shown in the accompanying illustration. If they are not, repeat Substeps 12a through 12e until they are properly aligned.

13. Install the timing belt around the camshaft, crankshaft and idler pulleys so that the positioning marks on the timing belt are aligned with the marks on the sprockets as follows:

   a. Position the timing belt on the crankshaft sprocket so that the marks are aligned.

   b. Route the belt down and under the left-hand, upper idler pulley, then up and around the left-hand intake camshaft sprocket, ensuring the camshaft sprocket mark is aligned with the mark on the belt.

   c. Route the belt down and around the left-hand exhaust camshaft sprocket, making sure the marks are properly aligned, then up and over the first lower idler pulley and down and around the second lower idler pulley.

   d. While holding the timing belt on the inner, left-hand, lower idler pulley, route the other side of the timing belt (from the crankshaft sprocket) down and under the right-hand upper idler pulley.

   e. Route the timing belt up and around the right-hand intake camshaft sprocket so that the belt and sprocket marks are aligned.

   f. Position the belt down and around the right-hand exhaust camshaft sprocket, ensuring the positioning marks are aligned.

14. Install the right-hand lower idler pulley so that the timing belt is routed over the top side of it.

➥**Once the belt is completely installed on all of the pulleys and sprockets, ensure that the positioning marks are still all aligned.**

15. After ensuring all of the marks are still aligned, use a pair of pliers to withdraw the stopper pin or Allen wrench from the adjuster assembly housing.

16. On models with manual transmissions, perform the following:

   a. Install the timing belt guide by temporarily tightening the mounting bolts.

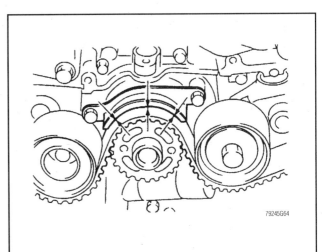

On models equipped with manual transmissions, ensure the timing belt-to-guide clearance (arrows) is correct before tightening the mounting bolts—Subaru Forester

184 TIMING BELTS—SUBARU

b. Position the timing belt guide so that there is 0.019–0.059 in. (0.5–1.5mm) clearance between the timing belt and the belt guide.

c. Tighten the guide mounting bolts securely, then double check the guide clearance.

17. Install the timing belt covers and all remaining engine components.

2.7L ENGINE

1. Disconnect the negative battery cable.
2. Loosen the water pump pulley nut/bolts and the alternator-to-engine bolts, the remove the drive belt.
3. Disconnect the electrical connector from the oil pressure switch.
4. Remove the oil level gauge guide with the gauge.
5. Remove the timing hole cover from the top of the flywheel housing.
6. Using the flywheel stopper tool 498277000 or equivalent (MT), or the driveplate stopper tool 498407000 or equivalent (AT), insert it through the timing hole (in the flywheel housing) and lock the flywheel.
7. Remove the crankshaft pulley bolt and using a puller, remove the crankshaft pulley.
8. If equipped with a turbocharger, remove the belt cover plate.
9. Remove the left side, the right side and the front timing belt cover.
10. Remove the crankshaft pulley.
11. Remove the timing belt covers.
12. Loosen the timing belt tensioner mounting bolts ½ turn and slacken the timing belt. Tighten the mounting bolts.
13. Mark the rotating direction of the timing belt, then remove the right hand belt.
14. Remove the right hand tensioner. Remove the crankshaft sprocket.

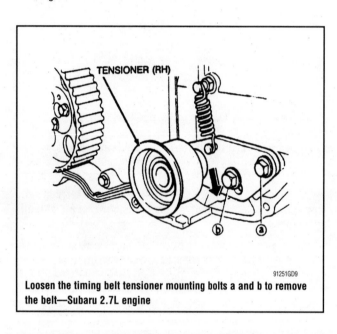

Loosen the timing belt tensioner mounting bolts a and b to remove the belt—Subaru 2.7L engine

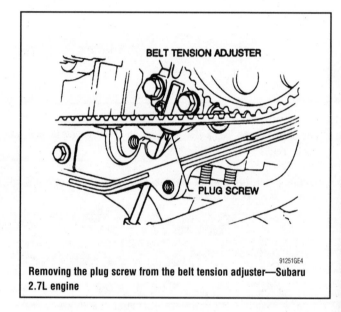

Removing the plug screw from the belt tension adjuster—Subaru 2.7L engine

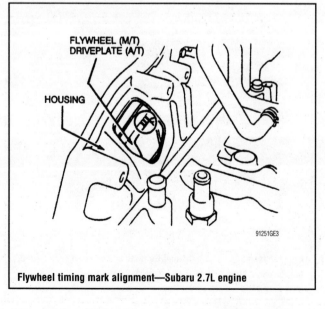

Flywheel timing mark alignment—Subaru 2.7L engine

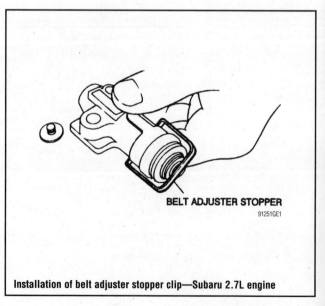

Installation of belt adjuster stopper clip—Subaru 2.7L engine

SUBARU—TIMING BELTS

15. Remove the idler pulley and rubber plug.
16. Remove the plug screw from the left belt tension adjuster lower side.
17. Insert a suitable tool into the hose in the bottom of the belt tension adjuster and turn the screw clockwise to loosen the belt tension. Install a belt adjuster stopper (13082AA000). Remove the left belt tensioner.
18. Remove the left timing belt after marking the rotating direction.
19. Remove the crankshaft sprocket No. 2 and idler pulley. Remove the belt tension adjuster.
20. Remove the camshaft sprockets.
21. Remove the belt cover No. 2 from both sides.

To install:
22. Inspect the timing belt for breaks, cracks and wear. Replace as required.
23. Check the belt tensioner and idler for smooth rotation. Replace if noisy or excessive play is noticed.
24. Install belt cover No. 2 on both sides. Tighten to 3–4 ft. lbs. (4–5 Nm).
25. Install camshaft sprockets on both sides and tighten to 8–9 ft. lbs. (11–13 Nm).
26. Remove the plug screw from the belt tension adjuster lower side. Insert a suitable tool into the hole in the bottom of the tension adjuster and turn the screw clockwise to compress the rubber boot. Install a belt adjuster stopper (13082AA000).
27. Using a syringe, add engine oil through the air vent hole on top of the rubber boot until it overflows. Install the plug screw.
28. Install the belt tension adjuster and tighten to 17–20 ft. lbs. (23–26 Nm).
29. Install the plug rubber and idler pulley. Tighten the pulley to 29–35 ft. lbs. (39–47 Nm).
30. Install crankshaft sprocket with no dowel pin to the crankshaft.
31. Align the center of the 3 lines scribed on the flywheel with the timing mark on the flywheel housing.
32. Align the timing mark on the left hand camshaft sprocket with the notch on the belt cover.
33. Install the timing belt from the crankshaft side and take care not to loosen it.
34. Install the left tensioner and check for smooth operation. Tighten the tensioner to 29–35 ft. lbs. (39–47 Nm).
35. Remove the belt adjuster stopper from the belt tension adjuster.
36. Check that the end of the left tensioner arm contacts the top of the belt tension adjuster.
37. Make sure the flywheel timing mark and left hand camshaft sprocket timing mark are in the proper positions.
38. Turn the crankshaft 1 turn clockwise from the position where the left timing belt was installed and align the center of the 3 lines scribed on the flywheel with the timing mark on the flywheel housing.
39. Align the mark on the right hand camshaft sprocket with the notch in the belt cover.
40. Temporarily tighten both belt tensioner bolts while forcing the tensioner against spring pressure (downward).
41. Install the crankshaft sprocket.
42. Install the timing belt from the crankshaft side and take care no to loosen it.
43. Loosen the left tensioner bolt by ½ turn to apply tension to the belt.

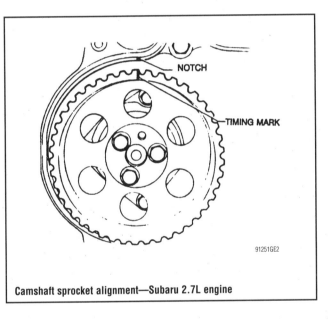

Camshaft sprocket alignment—Subaru 2.7L engine

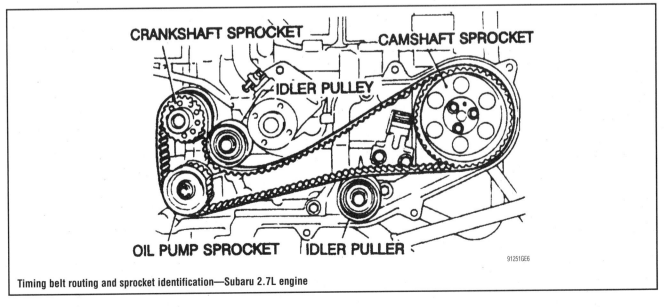

Timing belt routing and sprocket identification—Subaru 2.7L engine

186 TIMING BELTS—SUBARU

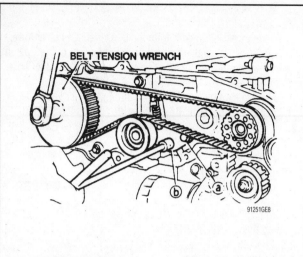

Loosen the tensioner bolts a and b and allow the left tensioner to operate—Subaru 2.7L engine

44. Apply 33–55 ft. lbs. torque to the camshaft sprocket in a counterclockwise direction and tighten the tensioner bolts temporarily.
45. Tighten the left, then right tensioner bolts to 17–20 ft. lbs. (23–26 Nm).
46. Make sure the flywheel timing mark and left hand camshaft sprocket timing mark are in the proper positions.
47. Install the center belt cover and crankshaft pulley. Tighten pulley bolt to 66–79 ft. lbs. (89–107 Nm).
48. Install the oil level gauge guide, water pipe, water pump pulley and belt covers.
49. Install and properly tension the accessory drive belt.
50. Connect the negative battery cable.

3.3L ENGINE

1. Disconnect the negative battery cable.
2. Remove the timing belt covers.
3. Matchmark the timing belt to the sprocket, cover and block marks as follows:

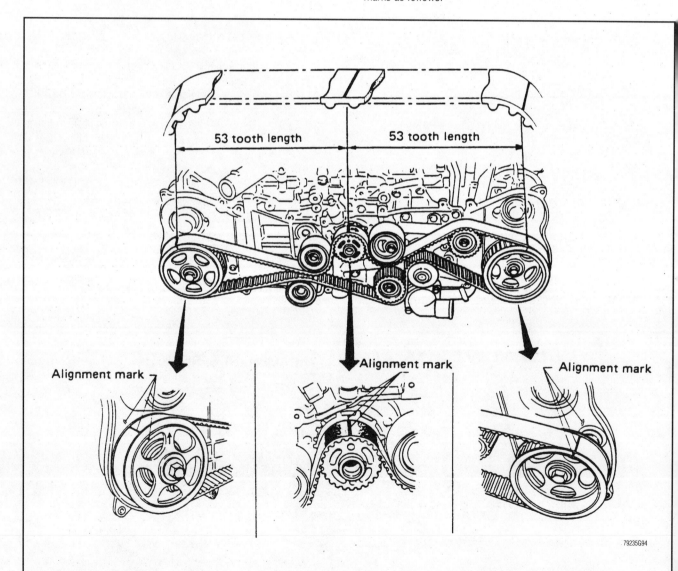

To ensure proper installation of the timing belt, be sure the proper number of belt teeth are between each sprockets, as indicated—Subaru 3.3L engine

SUBARU, SUZUKI—TIMING BELTS

a. Turn the crankshaft to align the timing marks on the crankshaft sprocket with the mark on the block.
b. With the crankshaft marks aligned be sure the left and right camshaft sprocket marks are lined up with marks on the timing covers.
c. If all the marks are in line, use white paint to mark the direction of rotation of the belt as well as mark the spots on the belt where it crosses over the timing marks on the pulleys.
4. Loosen the belt tensioner bolts.
5. Remove belt idler pulley No. 1.
6. Remove belt idler pulley No. 2.
7. Remove the timing belt.
8. Remove the tensioner pulley bolt and remove the tensioner pulley.
9. Remove the two bolts and the tensioner assembly.

To install:
10. Insert a 0.059 in. (1.5mm) diameter stopper pin into place while pushing the tension adjuster rod into the tensioner body.
11. Install the tensioner and tighten the bolts to 18 ft. lbs. (24 Nm), while the tensioner is pushed all the way to the right.
12. Install the tensioner pulley and mounting bolt. DO NOT tighten the idler pulley bolt completely.
13. Be sure the crankshaft and both camshaft sprockets are still lined up with their respective timing marks.
14. Install the timing belt onto the sprockets with the direction of rotation arrow in the correct direction and the timing marks on the belt in line with the marks on the sprockets.
15. Install the number 1 and 2 idler pulleys and tighten the mounting bolts to 29 ft. lbs. (39 Nm).
16. Loosen the tensioner pulley bolt and the tensioner assembly mounting bolts. Slide the tensioner assembly all the way to the left and tighten the bolts to 18 ft. lbs. (24 Nm).
17. Check again that all the timing marks are still in alignment. If they are remove the stopper pin from the tensioner assembly.
18. Install the timing belt covers.
19. Connect the negative battery cable.

Suzuki

1.3L ENGINE

Cars

1. Disconnect the negative battery cable.
2. Remove the air cleaner assembly with the MAF sensor and outlet hose.
3. Remove the air cleaner bracket.
4. Raise and safely support the vehicle.
5. Remove the right side fender apron clips by pushing the center pin.

➡ Do not push the center pin too far in, or it will fall off into the fender.

6. If equipped, remove the power steering and air conditioning accessory belt.
7. Loosen the water pump pulley bolts.
8. Remove the alternator drive belt.
9. Remove the water pump pulley.

10. Remove the crankshaft pulley by performing the following:
a. If equipped with a manual transaxle, insert a suitable flat bladed tool into the hole in the bell housing next to the exhaust pipe. This will lock the crankshaft in place.
b. If equipped with a automatic transaxle, hold a suitable flat bladed tool in line with the oil pan and insert the flat bladed tool into the teeth of the drive plate.
c. Loosen the crankshaft pulley bolts. Some vehicles may be equipped with bolts requiring a 5 mm hexagon tool.
d. Remove the crankshaft timing belt pulley bolt with special tool 09919–16020 or a 17mm socket.
e. Remove the pulley from the crankshaft.
f. Install the crankshaft bolt.
g. Remove the flat bladed tool that was used to lock the crankshaft in place.

➡ On single overhead cam models, to remove the crankshaft pulley with the engine assembly mounted on the body, it is necessary to remove the crankshaft timing belt pulley bolt. If the engine assembly is dismounted, the bolt does not need to be removed.

11. Loosen the right engine mounting bolt and push the air cleaner bracket to the right.
12. Remove the upper and lower timing belt outside covers.
13. Set the camshaft timing belt pulley(s) to their timing marks. The crankshaft mark and camshaft marks are straight up.
14. On single overhead cam models. remove the resonator and the timing belt outside cover.
15. Remove the tensioner stud and loosen the tensioner bolt.

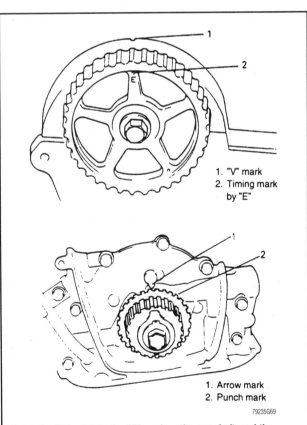

1. "V" mark
2. Timing mark by "E"

1. Arrow mark
2. Punch mark

Match the "V" notch to the "E" mark on the camshaft, and the punch and arrow on the crankshaft to properly position the engine for belt service—Suzuki 1.3L and 1.6L engines

188 TIMING BELTS—SUZUKI

16. Remove the tensioner spring and damper, then remove the timing belt.

> ✱✱ **WARNING**
>
> **After the timing belt is removed never turn the camshafts or the crankshaft. Interference may occur between the pistons and the valves causing component damage.**

17. Remove the tensioner and the tensioner plate.

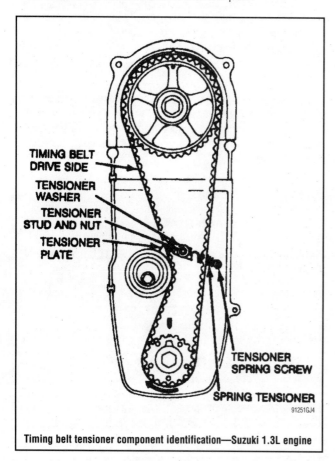

Timing belt tensioner component identification—Suzuki 1.3L engine

To install:

18. Install the timing belt tensioner plate and tensioner. Only hand-tighten the tensioner bolt.

➡ Be sure that the lug on the tensioner plate is inserted into the hole on the tensioner.

19. Be sure the tensioner plate and the tensioner move uniformly. If they do not move together remove the tensioner and the tensioner plate and reinsert the plate lug into the tensioner hole.
20. Check the camshaft sprockets to verify that they have not moved.

➡ On dual overhead cam models, if the sprockets will not stay aligned due to valve spring tensions they can be positioned by using four 8mm bolts and two flanged nuts. Install two bolts into the holes on the head in between the camshafts. Take and install the nuts with the flanges on two other bolts, the flange must face away from the head of the bolt. Position the head of the second pair of bolts on the threaded section of the first bolts, the nut should be facing up. The nut can be positioned into a groove on the appropriate camshaft sprocket so the sprocket alignment can be adjusted. Turn the flanged nut without having the sprocket resting on the nut so the sprocket is not damaged.

21. Check the crankshaft alignment by verifying that the punch mark on the timing belt pulley(s) is aligned with the arrow on the oil pump case.
22. On dual overhead cam models, install the timing belt on the three pulleys in such a way that there is no slack in the belt. Install the spring and the spring damper and hand-tighten the tensioner stud.

➡ If installed, remove the bolts used to secure the camshaft pulleys prior to rotating the engine.

23. On single overhead cam models, remove the cylinder head cover.

➡ This is to permit the free rotation of the camshaft. When installing the timing belt to the pulleys, the belt should be correctly tensioned by the tensioner spring force. If the camshaft does not rotate freely the belt will not be correctly tensioned.

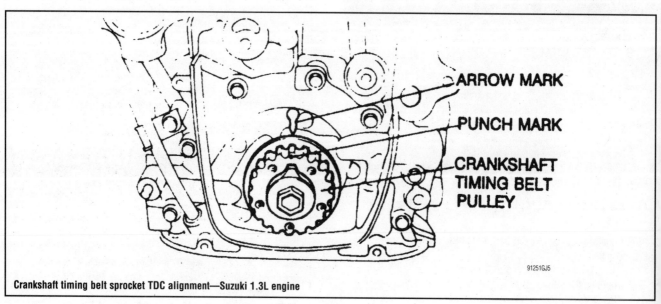

Crankshaft timing belt sprocket TDC alignment—Suzuki 1.3L engine

SUZUKI—TIMING BELTS

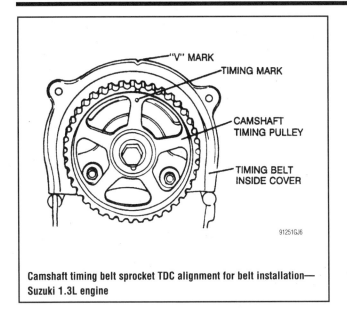

Camshaft timing belt sprocket TDC alignment for belt installation—Suzuki 1.3L engine

24. On single overhead cam models, with the timing marks aligned, hold the tensioner plate up by hand and install the timing belt on the pulleys so there is no slack on the drive side of the belt.

25. Turn the crankshaft two rotations clockwise. Confirm that the three sets of timing marks are still properly aligned.

26. If the belt is free of slack and the alignment marks are correct tighten the tensioner stud to 7–8 ft. lbs. (9–12 Nm). Tighten the tensioner bolt to 17–21 ft. lbs. (24–30 Nm).

27. Install the timing belt upper and lower outside covers. Tighten the timing cover bolts to 7–8 ft. lbs. (9–12 Nm).

28. Install all remaining components in the reverse order of the removal procedure.

Trucks

1. Disconnect the negative battery cable.
2. Remove the radiator cooling fan and fan shroud.
3. If equipped, disconnect the air conditioning compressor drive belt, properly discharge the air conditioning system and remove the air conditioning compressor flexible suction hose.
4. Loosen the alternator mounting bolts and remove the water pump drive belt and pulley.
5. Remove the crankshaft mounting bolts and disconnect the crankshaft pulley.

➡ **The crankshaft drive belt pulley can be removed without loosening the center crankshaft bolt.**

6. Disconnect the timing belt cover mounting bolts and remove the timing belt cover. Loosen but do not remove the tensioner bolt.
7. Disconnect the air intake case from the intake manifold.
8. Loosen the timing belt tensioner adjusting bolt and pivot nut. Hold pressure on the tensioner to loosen the timing belt and remove the timing belt from the camshaft and crankshaft pulleys.
9. Remove the timing belt tensioner, tensioner plate and tensioner spring.

To install:

10. Install the timing belt tensioner, plate and spring. Hand tighten the tensioner bolt and stud only at this time.
11. Remove the cylinder head cover and loosen all the valve adjusting screws to permit free rotation of the camshaft.

12. Turn the camshaft pulley clockwise and align the timing marks.
13. Turn the crankshaft clockwise, using a 17mm wrench to crank the timing belt pulley bolt.
14. Align the punch mark on the timing belt pulley with the arrow mark on the oil pump.
15. With the 4 marks aligned, remove any slack from the drive side of the belt. Tighten the tensioner bolt to 17.5–21.5 ft. lbs.
16. To allow the belt to be free of any slack, turn the crankshaft clockwise 2 full rotations. Confirm that the 4 marks are aligned.
17. Install the timing cover and tighten the bolts to 7.0–8.5 ft. lbs.
18. Adjust the valve lash and install the cylinder head cover, using a new gasket.
19. Connect the air intake case to the intake manifold.
20. Install the crankshaft pulley and install the mounting bolts. Tighten to 7.0–8.5 ft. lbs.
21. Install the water pump pulley, water pump drive belt and tighten the alternator mounting bolts.
22. If equipped, install the air conditioning compressor flexible suction hose and install the air conditioning compressor belt.
23. Install the radiator cooling fan and fan shroud.
24. Connect the negative battery cable.
25. If equipped, properly recharge the air conditioning system. Run the engine and check for any leaks.

1.6L 8-VALVE ENGINE

Cars

1. Disconnect the negative battery cable.
2. If equipped with A/C or power steering, remove the drive belt, the A/C compressor, and the power steering pump. Do not disconnect any lines from either the power steering pump or A/C compressor.

➡ **Suspend the A/C compressor and power steering pump. Do not allow the components hang from the lines.**

3. Raise the engine slightly with a floor jack and a block of wood. Place the wood between the jack and the engine.
4. Remove the air cleaner case with the air cleaner outlet hose.
5. Remove the engine right mounting bracket and stiffener by removing the nut and bolts.
6. Loosen the water pump drive belt and pulley.
7. Remove the crankshaft pulley mounting bolts and remove the crankshaft pulley.

➡ **The crankshaft drive belt pulley can be removed without loosening the center crankshaft bolt.**

8. Disconnect and remove the timing belt cover bolts and cover. Loosen but do not remove the tensioner bolt.

✱✱ CAUTION

After timing belt is removed, never turn the camshaft and crankshaft independently. This engine is an interference engine and if the camshaft or crankshaft is turned beyond a certain point, damage to the valves could occur.

9. Loosen the timing belt tensioner adjusting bolt and pivot nut. Hold pressure on the tensioner to loosen the timing belt and remove the timing belt from the camshaft and crankshaft sprockets.

190 TIMING BELTS—SUZUKI

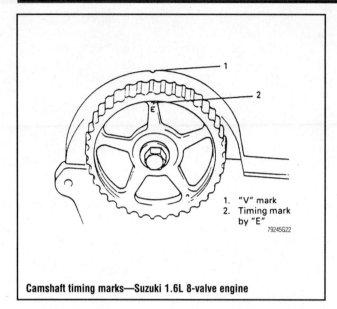

1. "V" mark
2. Timing mark by "E"

Camshaft timing marks—Suzuki 1.6L 8-valve engine

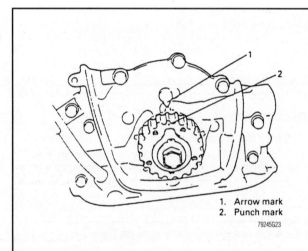

1. Arrow mark
2. Punch mark

Align the punch mark with the arrow for proper timing belt installation—Suzuki 1.6L 8-valve engine

18. Install the crankshaft pulley and replace the mounting bolts. Tighten to 10–13 ft. lbs. (14–18 Nm).
19. Install the water pump pulley and water pump drive belt.
20. Install the engine right mounting bracket and stiffener. Tighten the nuts and bolts to 40 ft. lbs. (55 Nm).
21. Install the remaining components.
22. Adjust all the drive belts.
23. Connect the negative battery cable.
24. Run the engine and check the timing.

Trucks

➡ **Do not rotate the crankshaft counterclockwise or attempt to rotate the crankshaft by turning the camshaft sprocket.**

1. Remove the timing belt cover.
2. Remove rocker arm cover.
3. If the timing belt is not already marked with a directional arrow, use white paint, a grease pencil or correction fluid to do so.
4. Disconnect one end of the tensioner spring. Loosen the timing belt tensioner bolt and stud, then, using your finger, press the tensioner plate up and remove the timing belt from the crankshaft and camshaft sprockets.
5. Remove the timing belt tensioner, tensioner plate and spring from the engine.
6. Insert a metal rod through the hole in the camshaft to lock the camshaft from rotating. Loosen the camshaft sprocket retaining bolt, then pull the camshaft sprocket off of the end of the camshaft.
7. Remove the crankshaft timing belt sprocket by loosening the center bolt, while preventing the crankshaft from rotating. To hold the crankshaft from turning, you can use Suzuki Tool 09927–56010 (or equivalent), or a large prybar inserted in the transmission housing slot and the flywheel teeth. Pull the sprocket off of the end of the crankshaft. Be sure to retain the crankshaft sprocket key and belt guide for assembly.
8. If necessary, remove the timing belt inside cover from the cylinder head.

To install:
9. If necessary, install the timing belt inside cover.
10. Slide the timing belt guide on the crankshaft so that the concave side faces the oil pump, then install the sprocket key in the groove in the crankshaft.
11. Slide the pulley onto the crankshaft, and install the center retaining bolt. Tighten the center bolt to 58–65 ft. lbs. (80–90 Nm). To hold the crankshaft from turning, you can use Suzuki Tool 09927–56010 (or equivalent), or a large prybar inserted in the transmission housing slot and the flywheel teeth.
12. Install the timing belt camshaft sprocket, ensuring that the slot in the sprocket engages the camshaft (pulley) pin; this ensures that the sprocket is properly positioned on the end of the camshaft. Secure the camshaft with the metal rod used during removal, then tighten the sprocket bolt to 41–46 ft. lbs. (56–64 Nm).
13. Assemble the timing belt tensioner plate and the tensioner, making sure that the lug of the tensioner plate engages the tensioner.
14. Install the timing belt tensioner, tensioner plate and spring on the engine. Tighten the mounting bolt and stud only finger-tight at this time. Ensure that when the tensioner is moved in a counterclockwise direction, the tensioner moves in the same direction. If the tensioner does not move, remove it and the tensioner plate to reassemble them properly.

10. Remove the timing belt tensioner, tensioner plate and tensioner spring.

To install:
11. Install the timing belt tensioner, plate and spring. Hand-tighten the tensioner bolt and stud only at this time.
12. Turn the camshaft sprocket clockwise and align the timing marks.
13. Turn the crankshaft clockwise, using a 17mm wrench to crank the timing belt sprocket bolt.
14. Align the punch mark on the timing belt sprocket with the arrow mark on the oil pump.
15. With the 4 marks aligned, remove any slack from the drive side of the belt. Tighten the tensioner bolt to 16–20 ft. lbs. (22–28 Nm).
16. To allow the belt to be free of any slack, turn the crankshaft clockwise 2 full rotations. Confirm that the 4 marks are aligned.
17. Install the timing cover and tighten the bolts to 7–8 ft. lbs. (9–12 Nm).

SUZUKI—TIMING BELTS

15. Loosen all rocker arm valve lash locknuts and adjusting screws. This will permit movement of the camshaft without any rocker arm associated drag, which is essential for proper timing belt tensioning. If the camshaft does not rotate freely (free of rocker arm drag), the belt will not be properly tensioned.

※※ WARNING

If any binding is felt when adjusting the timing belt tension by turning the crankshaft, STOP turning the engine, because the pistons may be hitting the valves.

16. Rotate the camshaft sprocket clockwise until the timing mark on the sprocket and the V mark on the timing belt inside cover are aligned.
17. Using a 17mm wrench, or socket and breaker bar, on the crankshaft sprocket center bolt, turn the crankshaft clockwise until the punch mark on the sprocket is aligned with the arrow mark on the oil pump.
18. With the camshaft and crankshaft marks properly aligned, push the tensioner up with your finger and install the timing belt on the two sprockets, ensuring that the drive side of the belt is free of all slack. Release your finger from the tensioner. Be sure to install the timing belt so that the directional arrow is pointing in the appropriate direction.

→**In this position, the No. 4 cylinder is at Top Dead Center (TDC) on the compression stroke.**

19. Rotate the crankshaft clockwise two full revolutions, then tighten the tensioner stud to 80–106 inch lbs. (9–12 Nm). Then, tighten the tensioner bolt to 18–21 ft. lbs. (24–30 Nm).
20. Ensure that all four timing marks are still aligned as before; if they are not, remove the timing belt, and install and tension it again.
21. Install the timing belt cover and all related components.

1.6L 16-VALVE ENGINE

The 1.6L 16-valve engine is known as an interference motor, because it is fabricated with such close tolerances between the pistons and valves that, if the timing belt is incorrectly positioned, jumps teeth on one of the sprockets or breaks, the valve and pistons will come into contact. This can cause severe internal engine damage

→**Do not rotate the crankshaft counterclockwise or attempt to rotate the crankshaft by turning the camshaft sprocket.**

1. Remove the timing belt cover.
2. If the timing belt is not already marked with a directional arrow, use white paint, a grease pencil or correction fluid to do so.
3. Rotate the crankshaft clockwise until the timing mark on the camshaft sprocket and the V mark on the timing belt inside cover are aligned, and the punch mark on the crankshaft sprocket is aligned with the mark on the engine.

※※ WARNING

Do not rotate the crankshaft or camshaft once the timing belt is removed, because the valves and pistons can come into contact, which may cause internal engine damage.

4. Disconnect one end of the tensioner spring. Loosen the timing belt tensioner bolt and stud, then, using your finger, press the tensioner plate up and remove the timing belt from the crankshaft and camshaft sprockets.

5. Remove the timing belt tensioner, tensioner plate and spring from the engine.
6. Install Suzuki Tool 09917–68220, or equivalent, onto the camshaft sprocket to hold the camshaft from rotating. Loosen the camshaft sprocket retaining bolt, then pull the camshaft sprocket off of the end of the camshaft.
7. Remove the crankshaft timing belt sprocket by loosening the center bolt, while preventing the crankshaft from rotating. To hold the crankshaft from turning, you can use Suzuki Tool 09927–56010 (or equivalent), or a large prybar inserted in the transmission housing slot and the flywheel teeth. Pull the sprocket off of the end of the crankshaft. Be sure to retain the crankshaft sprocket key and belt guide for assembly.
8. If necessary, remove the timing belt inside cover from the cylinder head.

To install:

9. If necessary, install the timing belt inside cover.
10. Slide the timing belt guide on the crankshaft so that the concave side faces the oil pump, then install the sprocket key in the groove in the crankshaft.
11. Slide the pulley onto the crankshaft, and install the center retaining bolt. Tighten the center bolt to 80 ft. lbs. (110 Nm). To hold the crankshaft from turning, you can use Suzuki Tool 09927–56010 (or equivalent), or a large prybar inserted in the transmission housing slot and the flywheel teeth.
12. Install the timing belt camshaft sprocket, ensuring that the slot in the sprocket engages the camshaft (pulley) pin; this ensures that the sprocket is properly positioned on the end of the camshaft. Secure the camshaft with the holding tool used during removal, then tighten the sprocket bolt to 44 ft. lbs. (60 Nm).
13. Assemble the timing belt tensioner plate and the tensioner, making sure that the lug of the tensioner plate engages the tensioner.

※※ WARNING

If any binding is felt when adjusting the timing belt tension by turning the crankshaft, STOP turning the engine, because the pistons may be hitting the valves.

14. Install the timing belt tensioner, tensioner plate and spring on the engine. Tighten the mounting bolt and stud only finger-tight

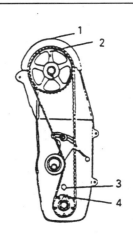

1. "V" mark on cylinder head cover
2. Timing mark by "E" on camshaft timing belt pulley
3. Arrow mark on oil pump case
4. Punch mark on crankshaft timing belt pulley

Rotate the crankshaft clockwise until the camshaft and crankshaft timing marks are aligned—Suzuki 1.6L 16-valve engine

192 TIMING BELTS—SUZUKI, TOYOTA

at this time. Ensure that when the tensioner is moved in a counter-clockwise direction, the tensioner moves in the same direction. If the tensioner does not move, remove it and the tensioner plate to reassemble them properly.

15. Loosen all rocker arm valve lash locknuts and adjusting screws. This will permit movement of the camshaft without any rocker arm associated drag, which is essential for proper timing belt tensioning. If the camshaft does not rotate freely (free of rocker arm drag), the belt will not be properly tensioned.

16. Rotate the camshaft sprocket clockwise until the timing mark on the sprocket and the V mark on the timing belt inside cover are aligned.

17. Using a wrench, or socket and breaker bar, on the crankshaft sprocket center bolt, turn the crankshaft clockwise until the punch mark on the sprocket is aligned with the arrow mark on the oil pump.

18. With the camshaft and crankshaft marks properly aligned, push the tensioner up with your finger and install the timing belt on the two sprockets, ensuring that the drive side of the belt is free of all slack. Release your finger from the tensioner. Be sure to install the timing belt so that the directional arrow is pointing in the appropriate direction.

➡ **In this position, the No. 4 cylinder is at Top Dead Center (TDC) on the compression stroke.**

19. Rotate the crankshaft clockwise two full revolutions, then tighten the tensioner stud to 97 inch lbs. (11 Nm). Then, tighten the tensioner bolt to 18 ft. lbs. (24 Nm).

20. Ensure that all four timing marks are still aligned as before; if they are not, remove the timing belt, and install and tension it again.

21. Install the timing belt cover and all related components.

Toyota

1.5L (3E-E & 3E) ENGINES

1. Disconnect the negative battery cable.
2. Remove the air cleaner assembly. Disconnect the accelerator and throttle cables.
3. Remove all drive belts.
4. If equipped with cruise control remove the actuator and bracket assembly.
5. Raise and support the vehicle safely.
6. Remove the right tire and wheel assembly. Remove the right side engine undercover. Remove the right side engine mount insulator.
7. Remove the cylinder head cover if necessary.
8. Set the No. 1 piston to TDC of the compression stroke and remove the crankshaft pulley using the proper tools.
9. Remove the engine front cover retaining bolts. Remove both front covers from the engine.
10. Remove the timing belt guide. Remove the timing belt and the No. 1 idler pulley. If using the old belt matchmark it in the direction of engine rotation. Matchmark the pulleys.

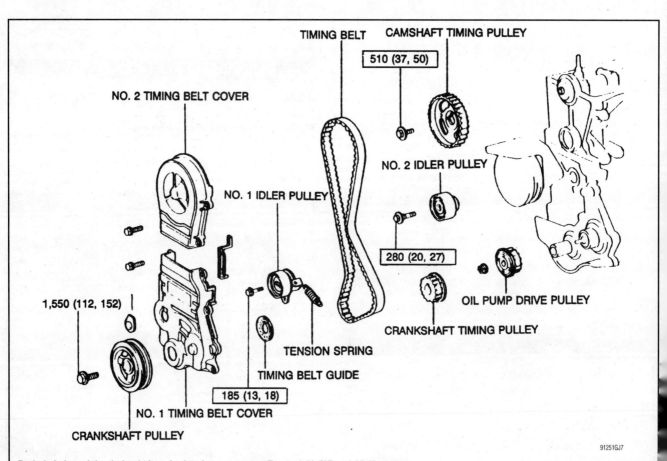

Exploded view of the timing belt and related components—Toyota 1.5L (3E and 3E-E) engines

TOYOTA—TIMING BELTS

11. Remove the tension spring. Remove the No. 2 idler pulley. Remove the crankshaft pulley, camshaft pulley and oil pump pulley using the proper tools.

To install:

12. Inspect the belt for defects. Replace as required. Inspect the idler pulleys and springs. Replace defective components as required.

13. Align and install the oil pump pulley. Tighten the retaining bolt to 20 ft. lbs. (26 Nm).

14. To install the camshaft timing pulley, align the camshaft knock pin with the No. 1 bearing cap mark. Align the knock pin hole on the 3E mark side with the camshaft knock pin hole. Tighten the retaining bolt to 37 ft. lbs. (50 Nm).

15. Install the crankshaft timing pulley and align the TDC marks on the oil pump body and the crankshaft timing pulley. Install the No. 1 idler pulley. Pry the idler pulley toward the left as far as it will go and temporarily tighten the retaining bolt.

16. Install the No. 2 idler pulley and tighten the retaining bolt to 20 ft. lbs. (27 Nm). Install the timing belt. If reusing the old belt, align it with the marks made during the removal procedure.

17. Inspect the valve timing and the belt tension by loosening the No. 1 idler pulley set bolt. Temporarily install the crankshaft pulley bolt and turn the crankshaft 2 complete revolutions in the clockwise direction.

18. Check that each pulley aligns with the proper markings. Tighten the No. 1 idler pulley bolt to 13 ft. lbs. (18 Nm). Check for proper belt tension. Install the belt guide.

19. Install the timing belt covers. Align and install the crankshaft pulley. Tighten the retaining bolt to 112 ft. lbs. (152 Nm).

20. Installation of the remaining components is the reverse of the removal procedure.

1.5L (5E-FE) ENGINE

1990–94 Models

1. Disconnect the negative battery cable.
2. Remove the right engine undercover.

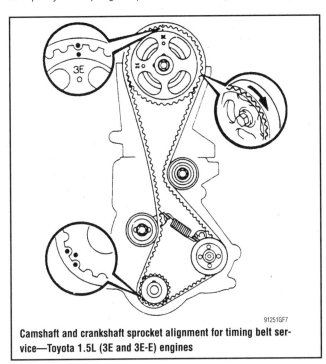

Camshaft and crankshaft sprocket alignment for timing belt service—Toyota 1.5L (3E and 3E-E) engines

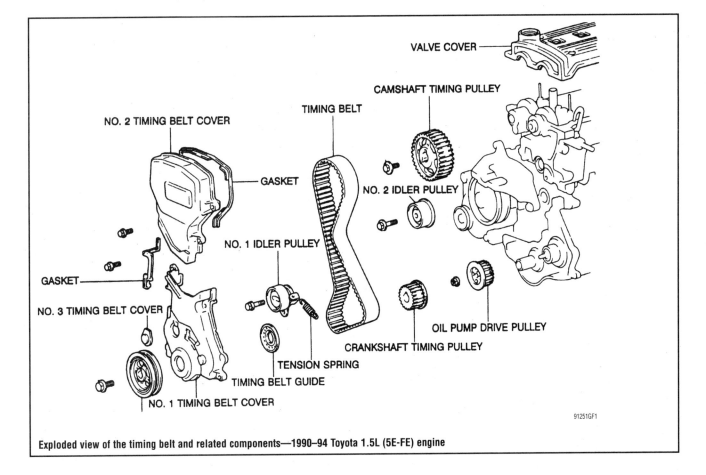

Exploded view of the timing belt and related components—1990–94 Toyota 1.5L (5E-FE) engine

3. If equipped with power steering, remove the power steering pump and bracket. If equipped with air conditioning and without power steering, remove the idler pulley bracket.
4. Tag and disconnect all electrical wire and vacuum hoses that interfere with removal of the timing belt cover.
5. Remove the No. 2 timing belt cover by removing the 4 bolts.
6. Remove the alternator belt, then the No. 3 timing belt cover from the No. 1 timing belt cover.
7. Remove the crankshaft pulley bolt and the pulley, then remove the No. 1 cover.
8. Place matchmarks on the timing belt and marks that indicate the direction of travel of the belt.
9. Remove the tension spring.
10. Loosen the idler pulley bolt and push the pulley to the left as far as it will go, then temporarily tighten the bolt.
11. Remove the belt.
12. Remove the crankshaft timing sprocket, the camshaft timing sprocket and the oil pump pulley sprockets.

To install:
13. Install the oil pump sprocket, torquing the bolt to 27 ft. lbs. (36 Nm).
14. Install the camshaft timing sprocket and tighten the bolt to 37 ft. lbs. (50 Nm).
15. Install the crankshaft sprocket, aligning the slot with the Woodruff key. Then using the crankshaft bolt, turn the crankshaft until the timing marks on the sprocket and oil pump body align. This is the setting at TDC before the marks on the belt cover can be seen.
16. Install the belt on the crankshaft gear (using matchmarks if reinstalling old belt) and install the guide and lower cover.
17. Set the No. 1 cylinder to TDC on the compression stroke.
18. Turn the camshaft and align the hole of the camshaft timing pulley with the timing mark of the bearing cap. The matchmarks if using the old belt should line up. Place the belt over all pulleys.
19. Loosen the adjuster pulley bolt until the pulley is moved slightly by the spring tension.
20. Turn the crankshaft pulley 2 revolutions from TDC to TDC.

➡ **Always rotate the crankshaft clockwise.**

21. Check that the pulleys align with the reference marks. If not, reinstall the belt.

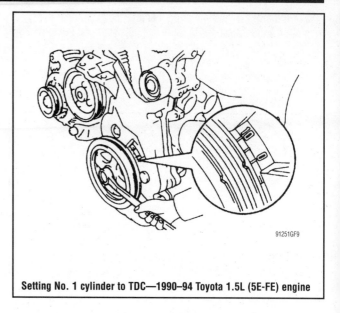

Setting No. 1 cylinder to TDC—1990–94 Toyota 1.5L (5E-FE) engine

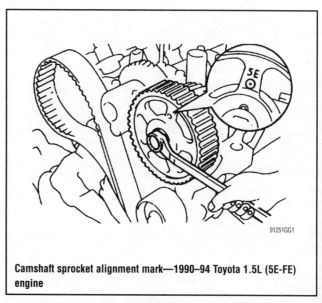

Camshaft sprocket alignment mark—1990–94 Toyota 1.5L (5E-FE) engine

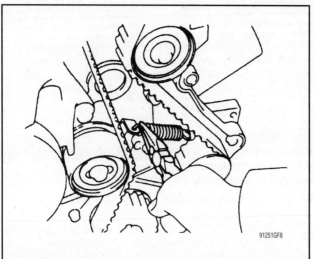

Removing the timing belt tensioning spring with a pair of needle nose pliers—1990–94 Toyota 1.5L (5E-FE) engine

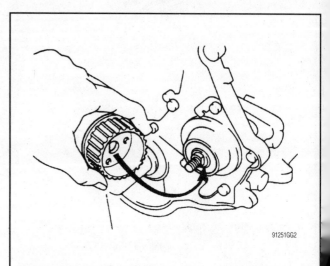

When installing oil pump pulley, be sure to align the flat on the stud to the hole in the pulley—1990–94 Toyota 1.5L (5E-FE) engine

TOYOTA—TIMING BELTS 195

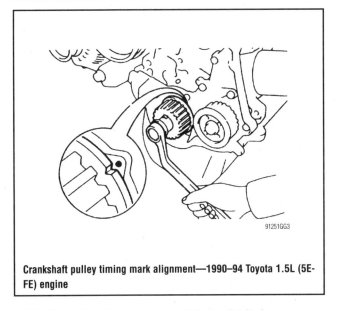

Crankshaft pulley timing mark alignment—1990–94 Toyota 1.5L (5E-FE) engine

22. Tighten the adjuster pulley to 13 ft. lbs. (18 Nm).
23. Install the crankshaft pulley bolt and the pulley, then install the No. 1 cover. Tighten the crankshaft pulley bolt to 112 ft. lbs. (152 Nm).
24. Install the No. 3 timing belt cover, the alternator belt and the No. 2 timing belt cover with its gasket and 4 bolts.
25. Reinstall the remaining components, start and check for unusual noises.

1995–99 Models

1. Disconnect the negative battery cable.
2. Remove the right engine undercover.
3. If equipped with power steering, remove the power steering pump and bracket. If equipped with air conditioning and without power steering, remove the idler pulley bracket.
4. Tag and disconnect all electrical wire and vacuum hoses that interfere with removal of the timing belt cover.

※※ CAUTION

On vehicles equipped with an air bag, be sure to disconnect the negative battery cable and wait at least 90 seconds before proceeding.

5. Remove the No. 2 timing belt cover.
6. Rotate the engine clockwise until the crankshaft pulley is aligned with the **0** mark on the No. 1 timing belt cover. Verify the hole in the camshaft timing pulley is aligned with the timing mark on the No. 1 bearing cap. If not as specified, rotate the crankshaft an additional 360 degrees.
7. For vehicles with A/C and/or power steering, remove the four bolts to the No. 2 crankshaft pulley. Then, remove the No. 2 crankshaft pulley.
8. Using Toyota Tools SST 09213–14010 and SST 09330–00021 (or their equivalents), remove the No. 1 crankshaft pulley bolt.
9. Using a crankshaft pulley/damper puller (such as Toyota Tool SST 09950–50010), remove the No. 1 crankshaft pulley from the crankshaft.
10. Remove the No. 3 timing belt cover (plug).
11. Remove the No. 1 timing belt cover and timing belt guide.

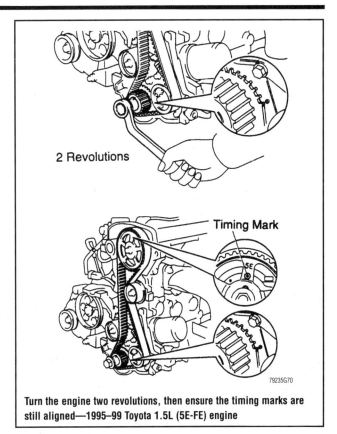

Turn the engine two revolutions, then ensure the timing marks are still aligned—1995–99 Toyota 1.5L (5E-FE) engine

12. Place matchmarks on the timing belt on both sides of the cam and crankshaft gear timing marks. Also, place an arrow on the top surface of the belt to indicate the direction of travel.
13. Using pliers, remove the tension spring.
14. Loosen the No. 1 idler pulley bolt and push the pulley to the left as far as it will go, then temporarily tighten the bolt.
15. Remove the belt.

To install:

16. For vehicle with a distributor (distributor ignition), use the crankshaft bolt to turn the crankshaft until the timing marks on the sprocket and oil pump body align. This is method is used to set the piston at TDC before the marks on the belt cover can be seen.
17. For vehicles with a crankshaft position sensor (distributorless ignition), use the crankshaft bolt to turn the crankshaft until the rotor side of the crankshaft position sensor faces inward.
18. Turn the camshaft and align the hole of the camshaft timing pulley with the timing mark of the bearing cap. The matchmarks, if using the old belt should line up. Place the belt over the crankshaft pulley and the idler pulleys.
19. Install the belt on the crankshaft gear (using the matchmarks if reinstalling the old belt) and install the timing belt guide with flange out.
20. Loosen the No. 1 idler pulley bolt until the pulley is moved slightly by the spring tension.
21. Turn the crankshaft pulley two revolutions from TDC to TDC.

➥**Always rotate the crankshaft clockwise.**

22. Check that the pulleys align with the reference marks. If not, reinstall the belt.
23. When the timing is verified, tighten the adjuster pulley (No. 1 idler pulley) to 13 ft. lbs. (18 Nm).

196 TIMING BELTS—TOYOTA

24. Install No. 1 and No. 3 lower timing belt covers.
25. Install the No. 1 crankshaft pulley and tighten the pulley bolt to 112 ft. lbs. (152 Nm).
26. Tighten the four No. 2 crankshaft pulley bolts to 14 ft. lbs. (19 Nm).
27. Install the No. 2 timing belt cover with the four bolts.
28. Install the remaining components in the reverse order of the removal procedure. When installing the right-hand engine mounting insulator, tighten the bracket bolt to 47 ft. lbs. (64 Nm) and the through-bolt to 54 ft. lbs. (73 Nm).

1.6L (4A-FE) & 1.8L (7A-FE) ENGINES

1990–94 Models

1. Raise and support the vehicle safely. Remove the right wheel and undercover. Remove the air cleaner.
2. Remove the drive belts. Remove the power steering pump and the air conditioning compressor, with brackets, and position them aside. Leave the hydraulic and refrigerant lines connected.
3. Remove the spark plugs and the cylinder head cover. Rotate the crankshaft pulley so the **0** mark is in alignment with the groove in the No. 1 front cover. Check that the lifters on the No. 1 cylinder are loose. If not, turn the crankshaft 1 complete revolution (360 degrees).
4. Position a floor jack under the engine and remove the right side engine mounting insulator.
5. Remove the water pump and crankshaft pulleys. The crankshaft pulley will require a 2-armed puller.
6. Loosen the 9 bolts and remove the No. 1, No. 2 and No. 3 front covers. Remove the timing belt guide.

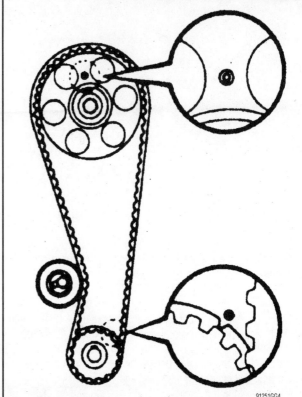

Crankshaft and camshaft timing belt sprocket alignment for timing belt replacement—1990-94 Toyota 1.6L (4A-FE) and 1.8L (7A-FE) engines

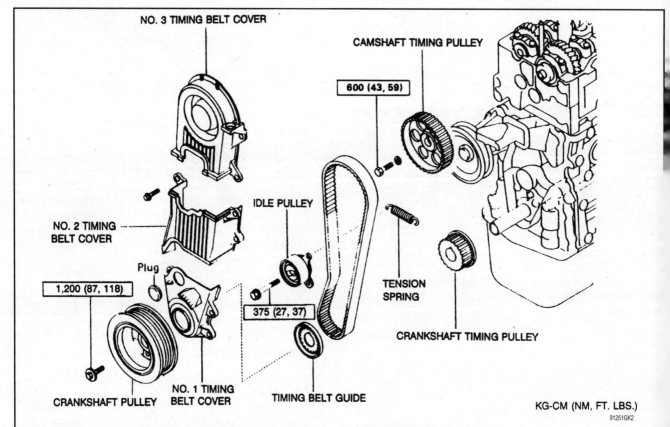

Exploded view of the timing belt and related components—1990-94 Toyota 1.6L (4A-FE) and 1.8L (7A-FE) engines

TOYOTA—TIMING BELTS

7. Loosen the bolt on the idler pulley, push it to the left as far as it will go and then retighten it. If reusing the timing belt, draw an arrow in the direction of engine revolution (clockwise) and then matchmark the belt to the pulleys.

8. Remove the timing belt. Remove the idler pulley bolt, the pulley and the tension spring.

9. Remove the crankshaft timing pulley.

10. Lock the camshaft and remove the camshaft timing pulley.

To install:

11. Install the camshaft timing pulley so it aligns with the knock pin on the exhaust camshaft. Tighten the pulley to 43 ft. lbs. (59 Nm). Align the mark on the No. 1 camshaft bearing cap with the center of the small hole in the pulley.

12. Install the crankshaft timing pulley so the marks on the pulley and the oil pump body are in alignment.

13. Install the idler pulley and tension spring, move to the left as far as it will go and tighten temporarily.

14. Align the matchmarks made during removal and then install the timing belt on the camshaft pulley. Loosen the idler pulley set bolt. Make sure the timing belt contact at the crankshaft pulley does not shift.

15. Rotate the crankshaft clockwise 2 revolutions from TDC to TDC. Make sure each pulley aligns with the marks made previously. If the marks are not in alignment, the valve timing is wrong. Shift the timing belt contact slightly and then repeat Steps 14–15.

16. Tighten the set bolt on the timing belt idler pulley to 27 ft. lbs. (37 Nm). Measure the timing belt deflection at the SIDE point, looking for 0.20–0.24 in. (5–6mm) of deflection at 4.4 pounds of pressure. If the deflection is not correct, readjust the idler pulley.

17. Installation of the remaining components is the reverse of the removal procedure.

1995–99 Models

1. Disconnect the negative battery cable.
2. Remove the air cleaner assembly and disconnect the accelerator and throttle cables.
3. Remove all drive belts and the washer tank.
4. Remove the alternator, alternator bracket and right engine mounting stay, if clearance is needed. If equipped with cruise control, remove the actuator and bracket assembly.
5. Raise and support the vehicle safely.
6. Remove the right tire and wheel assembly. Remove the right side engine undercover. Remove the right side engine mount insulator.
7. Remove the valve cover on the 4A-FE and 7A-FE engines, if necessary.
8. Set the No. 1 piston to TDC of the compression stroke and remove the crankshaft pulley using the proper tools.
9. Remove the engine front cover retaining bolts. Remove both front covers from the engine.

✴✴ CAUTION

On vehicles equipped with an air bag, be sure to disconnect the negative battery cable and wait at least 90 seconds before proceeding.

10. Turn the crankshaft to align the timing mark on crankshaft pulley at **0**, setting the piston in No. 1 cylinder at Top Dead Center (TDC) on the compression stroke. Check that the valve lifters on the No. 1 cylinder are loose. If not, turn crankshaft pulley 1 complete revolution (360 degrees).

11. Remove the nine bolts and timing belt covers from the engine.
12. Slide the timing belt guide from crankshaft.
13. Set the camshaft and crankshaft timing sprockets to align the marks.

➡**Do not turn crankshaft or camshaft independently after removal of timing belt. Binding or damage to engine components could result. If the timing belt is to be reused, mark timing belt with arrow showing direction of engine revolution. Put matchmarks where timing belt meets with crankshaft timing sprocket and camshaft timing sprocket to ensure installation in the same position.**

14. Remove the timing belt tensioner bolt, tensioner, and the tension spring.
15. Remove the timing belt from camshaft and crankshaft timing sprockets.

✴✴ WARNING

Do not bend, twist or turn the timing belt inside out. Do not allow the belt to come in contact with oil, coolant or steam.

To install:

➡**Inspect the camshaft timing sprocket to ensure mark is still aligned as indicated.**

16. Reinstall the timing belt tensioner and the tension spring. Pry the tensioner to the left as far as it will go and temporarily tighten the retaining bolt.

17. Install the timing belt. If reinstalling the old belt, observe the matchmarks made during removal. Be sure the belt is fully and squarely seated on the sprockets.

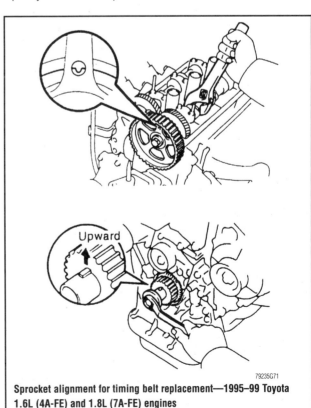

Sprocket alignment for timing belt replacement—1995–99 Toyota 1.6L (4A-FE) and 1.8L (7A-FE) engines

198 TIMING BELTS—TOYOTA

18. Loosen the retaining bolt for the timing belt tensioner and allow it to tension the belt.
19. Temporarily install the crankshaft pulley bolt and turn the crank clockwise 2 full revolutions from TDC to TDC. Insure that each timing mark realigns exactly.
20. Tighten the timing belt tensioner bolt to 27 ft. lbs. (37 Nm).
21. Measure the timing belt deflection at the **SIDE** point, looking for ¼ in. (5–6mm) of deflection at 4.4 lbs. (2 kg) of pressure. If the deflection is not correct, adjust with the timing belt tensioner.
22. Install the timing belt guide, with the cup side facing outward, onto the crankshaft and install the timing belt covers from the lowest to the highest. Tighten the nine cover bolts to 62 inch lbs. (7 Nm).
23. Install all applicable remaining components. During assembly, be sure to tighten the crankshaft pulley bolt to 87 ft. lbs. (118 Nm), the mounting bracket-to-engine mount bolt to 47 ft. lbs. (64 Nm), the mounting bracket-to-engine mount nuts to 38 ft. lbs. (52 Nm), the engine mount-to-body bolt **A** to 19 ft. lbs. (25 Nm), the engine mount-to-body bolts **B** to 19 ft. lbs. (25 Nm), the engine mount to body bolt **C** to 19 ft. lbs. (25 Nm), if equipped with cruise control.

1.6L (4A-GE) ENGINE

1. Disconnect the negative battery cable.
2. Raise the vehicle and safely support it on jackstands.
3. Remove the right front wheel.
4. Remove the splash shield from under the car.
5. Drain the coolant into clean containers. Close the draincocks when the system is empty.
6. Lower the car to the ground. Disconnect the accelerator cable and, if equipped, the cruise control cable.
7. Remove the cruise control actuator, if equipped.
8. Carefully remove the ignition coil.
9. Disconnect the radiator hose at the water outlet.
10. Remove the power steering drive belt and the alternator drive belt.
11. Remove the spark plugs.
12. Rotate the crankshaft clockwise and set the engine to TDC/compression on No. 1 cylinder. Align the crankshaft marks at zero; look through the oil filler hole and make sure the small hole in the end of the camshaft can be seen.

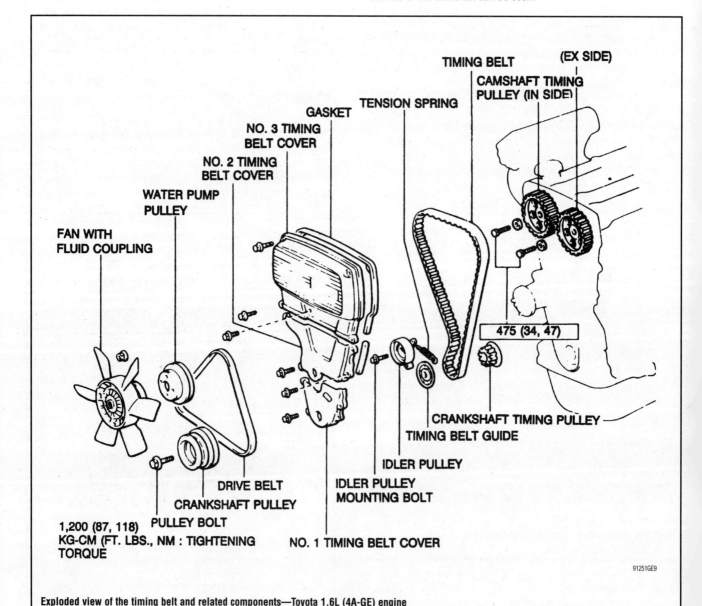

Exploded view of the timing belt and related components—Toyota 1.6L (4A-GE) engine

13. Raise and safely support the vehicle. Disconnect the center engine mount.
14. Lower the vehicle to the ground.
15. Support the engine either from above or below. Disconnect the right engine mount from the engine.
16. Raise the engine and remove the mount.
17. Remove the water pump pulley.
18. Remove the crankshaft pulley.
19. Remove the 10 bolts and remove the timing belt covers with their gaskets.

➡ **The bolts are different lengths; they must be returned to their correct location at reassembly. Label or diagram the bolts during removal.**

20. Remove the timing belt guide from the crankshaft pulley.
21. Loosen the timing belt idler pulley, move it to the left (to take tension off the belt) and tighten its bolt.
22. Make matchmarks on the belt and all pulleys showing the exact placement of the belt. Mark an arrow on the belt showing its direction of rotation.
23. Carefully slip the timing belt off the pulleys.

➡ **Do not disturb the position of the camshafts or the crankshaft during removal.**

24. Remove the idler pulley bolt, pulley and return spring.
25. Remove the PCV hose and the valve cover.
26. Use an adjustable wrench to counterhold the camshaft. Be careful not to damage the cylinder head. Loosen the center bolt in each camshaft pulley and remove the pulley. Label the pulleys and keep them clean.
27. Check the timing belt carefully for any signs of cracking or deterioration. Pay particular attention to the area where each tooth or cog attaches to the backing of the belt. If the belt shows signs of damage, check the contact faces of the pulleys for possible burrs or scratches.
28. Check the idler pulley by holding it in your hand and spinning it. It should rotate freely and quietly. Any sign of grinding or abnormal noise indicates the pulley should be replaced.
29. Check the free length of the tension spring. Correct length is 43.5mm measured at the inside faces of the hooks. A spring which has stretched during use will not apply the correct tension to the pulley; replace the spring.
30. If you can test the tension of the spring, look for 22 lbs. of tension at 50mm of length. If in doubt, replace the spring.

To install:

31. Align the camshaft knock pin and the pulley. Reinstall the camshaft timing belt pulleys, making sure the pulley fits properly on the shaft and that the timing marks align correctly. Tighten the center bolt on each pulley to 43 ft. lbs. Be careful not to damage the cylinder head during installation.
32. Before reinstalling the belt, double check that the crank and camshafts are exactly in their correct positions. The alignment marks on the pulleys should align with the cast marks on the head and oil pump.
33. Reinstall the valve covers and the PCV hose.
34. Install the timing belt idler pulley and its tensioning spring. Move the idler to the left and temporarily tighten its bolt.
35. Carefully observing the matchmarks made earlier, install the timing belt onto the pulleys.

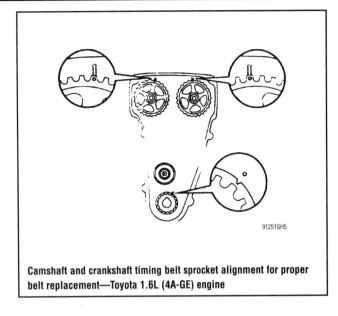

Camshaft and crankshaft timing belt sprocket alignment for proper belt replacement—Toyota 1.6L (4A-GE) engine

36. Slowly release tension on the idler pulley bolt and allow the idler to take up tension on the timing belt. DO NOT allow the idler to slam into the belt; the belt may become damaged.
37. Temporarily install the crankshaft pulley bolt. Turn the engine clockwise through two complete revolutions, stopping at TDC. Check that each pulley aligns with its marks.
38. Check the tension of the timing belt at a top point halfway between the two camshaft sprockets. The correct deflection is 4mm at 4.4 lbs. pressure. If the belt tension is incorrect, readjust it by repeating steps 18–20. If the tension is correct, tighten the idler pulley bolt to 27 ft. lbs.
39. Remove the crankshaft pulley bolt.
40. Install the timing belt guide onto the crankshaft timing pulley.
41. When reinstalling, make certain that the gaskets and their mating surfaces are clean and free from dirt and oil. The gasket itself must be free of cuts and deformations and must fit securely in the grooves of the covers.
42. Reinstall the covers and their gaskets and the 10 bolts in their proper positions.
43. Install the crankshaft pulley, again using the counterholding tool. Tighten the bolt to specifications.
44. Install the water pump pulley.
45. Install the right engine mount. Tighten the through bolt to 58 ft. lbs.
46. Reinstall the spark plugs and their wires.
47. Install the alternator drive belt and the power steering drive belt. Adjust the belts to the correct tension.
48. Connect the radiator hose to the water outlet port.
49. Install the ignition coil.
50. Install the cruise control actuator and the cruise control cable, if equipped.
51. Connect the accelerator cable.
52. Refill the cooling system with the correct amount of antifreeze and water.
53. Connect the negative battery cable.
54. Start the engine and check for leaks. Allow the engine to warm up and check the work areas carefully for seepage.

TIMING BELTS—TOYOTA

55. Install the splash shield under the car. Install the right front wheel.

2.0L (3S-GTE) ENGINES

1. Disconnect the negative battery cable. Raise the vehicle and support safely. Remove the right front tire and wheel assembly. Remove the right side fender liner.
2. Remove the windshield washer and radiator reservoir tanks. Remove the cruise control actuator, if equipped.
3. Remove the power steering belt. Remove the power steering pump and position it aside with the hydraulic lines still attached.
4. Remove the alternator and support bracket. Remove the upper timing belt cover.
5. Set the No. 1 piston to TDC of the compression stroke by aligning the groove on the crankshaft pulley with the **0** mark on the lower timing belt cover. Check that the matchmarks on the 2 camshaft timing pulleys and the rear timing belt cover are aligned. If not, turn the crankshaft 1 complete revolution clockwise (360 degrees).
6. If the timing belt is to be reused, draw a directional arrow on it and matchmark the belt to the 2 camshaft pulleys. Loosen the No. 1 idler pulley bolt and shift the pulley as far left as possible; tighten the set bolt. Remove the timing belt from the 2 camshaft pulleys. Support the belt so the contact of the belt with the remaining pulleys does not shift.
7. Carefully hold the camshafts with an adjustable wrench and remove the camshaft pulley set bolts. Remove the pulleys and their set pins.
8. Remove the crankshaft pulley. Remove the lower timing belt.
9. Remove the timing belt guide and then remove the timing belt from the remaining pulleys. Be sure to matchmark the belt to the pulleys if it is to be reused.
10. Remove the No. 1 idler pulley and the tension spring. Remove the No. 2 idler pulley, the crankshaft timing pulley and the oil pump pulley.

To install:
11. Install the oil pump pulley and tighten it to 21 ft. lbs. (28 Nm). Install the crankshaft timing pulley by sliding it onto the crankshaft over the Woodruff key. Install the No. 2 idler pulley and tighten it to 32 ft. lbs. (43 Nm).
12. Install the No. 1 idler pulley and the tension spring. Move the pulley as far to the left as it will go and tighten it.
13. Install the timing belt on all pulleys except the 2 camshaft pulleys. Make sure the matchmarks made earlier are in alignment.
14. Install the timing belt guide with the cup side out. Install the lower timing belt cover and then install the crankshaft pulley. Tighten it to 80 ft. lbs. (108 Nm).
15. Check that the No. 1 cylinder is at TDC of the compression stroke for the crankshaft. The crankshaft pulley groove should be aligned with the **0** mark on the lower timing belt cover.
16. Check that the No. 1 cylinder is at TDC of the compression stroke for the camshaft.

➡There are 2 types of camshafts, one with 2 holes on the timing pulley contact surface and one with 5 holes on the timing pulley contact surface. All replacements have 5 holes.

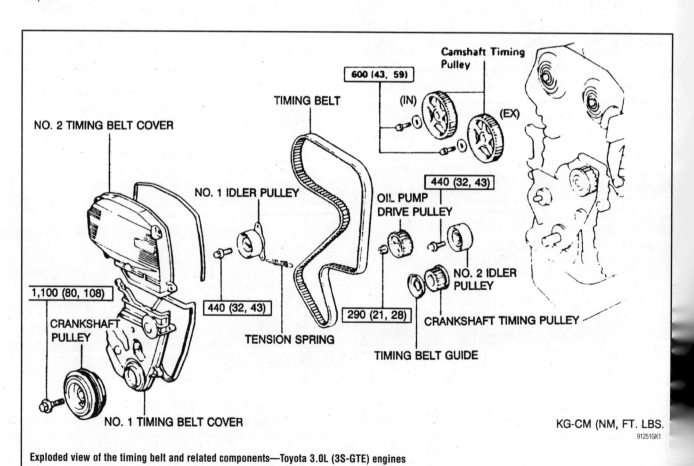

Exploded view of the timing belt and related components—Toyota 3.0L (3S-GTE) engines

TOYOTA—TIMING BELTS 201

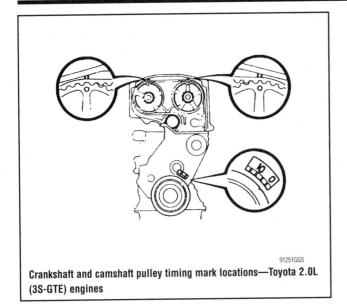

Crankshaft and camshaft pulley timing mark locations—Toyota 2.0L (3S-GTE) engines

2 Hole: Using a wrench, turn the camshafts so the camshaft knock pin aligns with the matchmark on the rear timing belt cover. And the No. 1 cam lobe is pointing outward.

5 Hole: Using a wrench, turn the camshaft so the knock pin aligns with the notch in the No. 1 camshaft bearing cap.

17. Hang the timing belt on the 2 camshaft timing pulleys. Align all matchmarks made during removal. The **S** mark on the pulley should face outward.

18. Align the timing pulley matchmark with the rear timing belt cover matchmark and install the pulleys with the belt.

➡ On 1 hole pulleys, match the camshaft knock pin with the camshaft pulley hole. On 5 hole pulleys, insert the knock pin into whichever pulley and camshaft holes are aligned.

19. Hold the camshaft with an adjustable wrench and tighten the pulley set bolt to 43 ft. lbs. (59 Nm).
20. Rotate the engine 2 complete revolutions. Check that all marks align in the correct location. Tighten the No. 1 idler pulley bolt to 32 ft. lbs. (43 Nm).
21. Installation of the remaining components is the reverse of the removal procedure. Road test the vehicle for proper operation.

2.0L (3S-FE) & 2.2L (5S-FE) ENGINES

Except RAV4

1990–94 MODELS

1. Disconnect the negative battery cable. Raise and support the vehicle safely. Remove the right tire and wheel assembly.
2. If equipped, remove the cruise control actuator and bracket. Remove the drive belts.
3. Remove the alternator and alternator bracket and right engine mounting stay. Raise the engine enough to remove the right side engine mounting insulator and brackets.
4. Remove the spark plugs. Remove the upper timing cover. Position the No. 1 cylinder to TDC on the compression stroke so the groove in the crankshaft pulley is aligned with the **O** mark in the No. 1 front cover. If the hole in the camshaft pulley is not aligned with the mark on the bearing cap, turn the crankshaft 1 complete revolution (360 degrees).

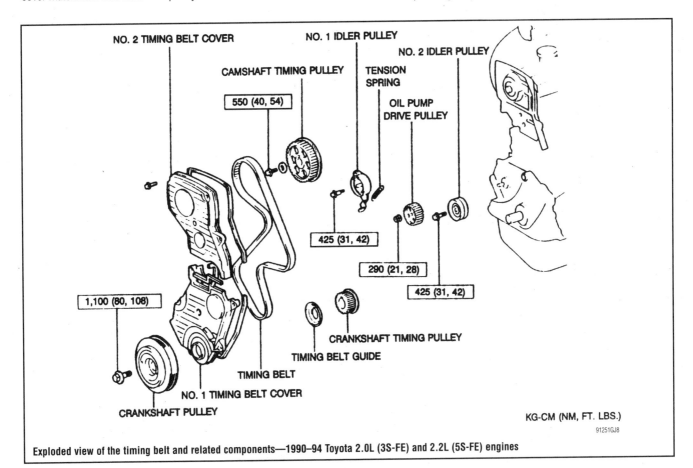

Exploded view of the timing belt and related components—1990–94 Toyota 2.0L (3S-FE) and 2.2L (5S-FE) engines

202 TIMING BELTS—TOYOTA

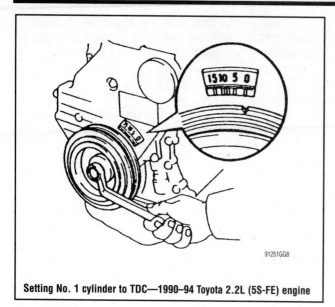

Setting No. 1 cylinder to TDC—1990–94 Toyota 2.2L (5S-FE) engine

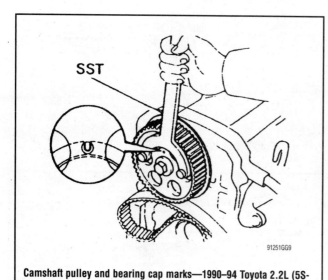

Camshaft pulley and bearing cap marks—1990–94 Toyota 2.2L (5S-FE) engine

5. If reusing the belt place matchmarks on the timing belt and the camshaft pulley. Loosen the mount bolt of the No. 1 idler pulley and position the pulley toward the left as far as it will go. Tighten the bolt. Remove the belt from the camshaft pulley.

6. Remove the camshaft pulley. Remove the crankshaft pulley using the proper removal tool. Remove the lower timing cover.

7. Remove the timing belt and the belt guide. If reusing the belt mark the belt and the crankshaft pulley in the direction of engine rotation.

8. Remove the No. 1 idler pulley and the tension spring. Remove the No. 2 idler pulley. Remove the crankshaft timing pulley. Remove the oil pump pulley.

To install:

9. Inspect the belt for defects. Replace as required. Inspect the idler pulleys and springs. Replace defective components as required.

10. Align the cutouts of the oil pump pulley and shaft. Install the oil pump pulley and tighten the retaining nut to 21 ft. lbs. (28 Nm).

11. To install the crankshaft pulley, align the pulley set key with the key groove of the pulley and slide it in position. Install the No. 2 idler pulley and tighten the bolt to 31 ft. lbs. (42 Nm). Be sure the pulley moves freely.

12. Temporarily install the No. 1 idler pulley and tension spring. Pry the pulley toward the left as far as it will go. Tighten the bolt.

13. Temporarily install the timing belt. If reusing the old belt align the marks made during removal. Install the timing belt guide.

14. Install the No. 1 timing belt cover. Install the crankshaft pulley and tighten the bolt to 80 ft. lbs. (108 Nm).

15. Install the camshaft pulley by aligning the camshaft knock pin with the knock pin groove in the pulley. Install the washer and tighten the retaining bolt to 40 ft. lbs. (54 Nm).

16. With the engine set at TDC on the compression stroke install the timing belt. If reusing the belt, align with the marks made during the removal procedure.

17. Once the belt is installed be sure there is tension between the crankshaft pulley, water pump pulley and camshaft pulley. Loosen the No. 1 idler pulley mount bolt ½ turn. Turn the crankshaft pulley 2 revolutions from TDC to TDC, in the clockwise direction. Tighten the No. 1 idler pulley mount bolt to 31 ft. lbs. (42 Nm).

18. Installation the remaining components is the reverse of the removal procedure.

1995–99 MODELS

1. Disconnect the negative battery cable. Raise and support the vehicle safely. Remove the right tire and wheel assembly.

2. If equipped, remove the cruise control actuator and bracket. Remove the drive belts.

3. Remove the alternator and alternator bracket and right engine mounting stay. Raise the engine enough to remove the right side engine mounting insulator and brackets.

4. Remove the spark plugs. Remove the upper timing cover.

> ※※ **CAUTION**
>
> **On vehicles equipped with an air bag, be sure to disconnect the negative battery cable and wait at least 90 seconds before proceeding.**

5. Remove the No. 2 timing cover.

6. Position the No. 1 cylinder to TDC on the compression stroke by turning the crankshaft pulley and aligning its groove with the timing mark **0** of the No. 1 timing belt cover. Check that the hole of the camshaft timing pulley is aligned with the alignment mark of the bearing cap. If not, turn the crankshaft one revolution (360 degrees).

7. Remove the timing belt from the camshaft timing pulley, as follows:

   a. If reusing the belt, place matchmarks on the timing belt and the camshaft pulley. Loosen the mount bolt of the No. 1 idler pulley and position the pulley toward the left as far as it will go. Tighten the bolt. Remove the belt from the camshaft pulley.

8. Remove the camshaft timing pulley.

   a. Using Toyota tools Nos. 09249–63010 and 09960–10010, or their equivalents, remove the bolt and the camshaft pulley.

9. Remove the crankshaft pulley.

   a. Using Toyota tools Nos. 09213–54015 and 09330–00021, or their equivalents, to hold the crankshaft pulley. Remove the pulley set bolt and remove the pulley using a puller.

TOYOTA—TIMING BELTS 203

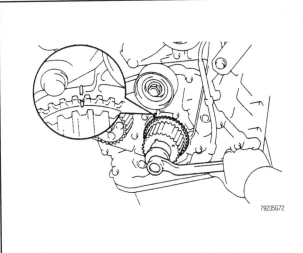

Crankshaft positioning for timing belt removal and installation—1995–99 Toyota 2.2L (5S-FE) engine

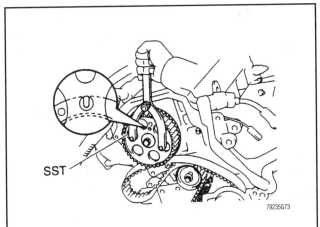

Using a spanner wrench, turn the camshaft into position so that the alignment mark is visible through the hole in the sprocket—1995–99 Toyota 2.2L (5S-FE) engine

10. Remove the No. 1 timing belt cover.
11. Remove the timing belt and the belt guide. If reusing the belt mark the belt and the crankshaft pulley in the direction of engine rotation and matchmark for correct installation.

To install:

12. Install the crankshaft timing pulley, as follows:
    a. Align the timing pulley set key with the key groove of the pulley.
    b. Slide on the timing pulley with the flange side facing inward.
13. Install the No. 2 idler pulley and tighten the bolt to 31 ft. lbs. (42 Nm). Be sure that the pulley moves freely.
14. Temporarily install the No. 1 idler pulley and tension spring. Pry the pulley toward the left as far as it will go. Tighten the bolt. Be sure that the pulley rotates freely.
15. Temporarily install the timing belt, as follows:
    a. Using the crankshaft pulley bolt, turn the crankshaft and align the timing marks of the crankshaft timing pulley and the oil pump body.
    b. If reusing the old belt, align the marks made during removal, and install the belt with the arrow pointing in the direction of the engine revolution.
16. Install the timing belt guide with the cup side facing outward.
17. Install the No. 1 timing belt cover.
18. Install the crankshaft pulley. Align the pulley set key with the key groove of the pulley and slide on the pulley. Tighten the bolt to 80 ft. lbs. (108 Nm).
19. Install the camshaft timing pulley.
    a. Align the camshaft knock pin with the knock pin groove of the pulley and slide on the timing pulley. Tighten the bolt to 40 ft. lbs. (54 Nm).
20. With the No. 1 cylinder set at TDC on the compression stroke install, the timing belt (all timing marks aligned). If reusing the belt, align with the marks made during the removal procedure.
    a. Turn the crankshaft pulley, and align its groove with the timing mark **0** of the No. 1 timing belt cover. Be sure the camshaft sprocket hole is aligned with the mark on the bearing cap.
21. Connect the timing belt to the camshaft timing pulley.
22. Check that the matchmark on the timing belt matches the end of the No. 1 timing belt cover.
23. Once the belt is installed be sure that there is tension between the crankshaft timing pulley and the camshaft pulley.
24. Check the valve timing.
    a. Loosen the No. 1 idler pulley mount bolt ½ turn. Turn the crankshaft pulley two revolutions from TDC in the clockwise direction. Always turn the crankshaft pulley clockwise.
    b. Be sure that the all the timing marks are aligned.
    c. Slowly turn the crankshaft pulley 1 ⅞ revolutions. Align its groove with the mark at 45° BTDC on the No. 1 timing belt cover for the No. 1 cylinder.

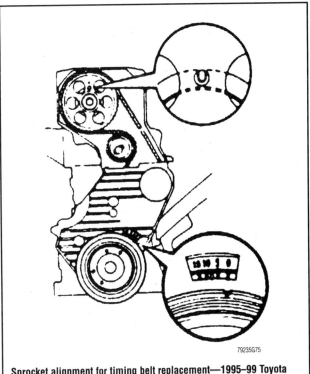

Sprocket alignment for timing belt replacement—1995–99 Toyota 2.2L (5S-FE) engine

TIMING BELTS—TOYOTA

d. Tighten the No. 1 idler pulley mount bolt to 31 ft. lbs. (42 Nm).

25. Install the No. 2 timing belt cover.

   a. Install the upper gasket to the No. 1 timing belt cover.

   b. Disconnect the engine wire protector between the cylinder head cover and the No. 3 timing belt cover.

   c. Install the gasket to the timing belt cover.

   d. Install the belt covers and all remaining components. During assembly, tighten the right engine mount bracket bolts to 38 ft. lbs. (52 Nm), the engine mount insulator bolt to 47 ft. lbs. (64 Nm), the through-bolt to 54 ft. lbs. (78 Nm), the power steering reservoir bracket bolt to 21 ft. lbs. (28 Nm), the power steering reservoir-to-bracket bolt to 27 ft. lbs. (37 Nm) and the nut to 38 ft. lbs. (52 Nm).

RAV4

The timing belt is not adjustable.

1. Disconnect the negative battery cable.

✱✱ CAUTION

To avoid air bag deployment, if equipped, work must be started after approximately 90 seconds or longer from the time the ignition switch is turned to the LOCK position and the negative battery cable is disconnected from the battery.

2. Disconnect the power steering reservoir tank and remove the reservoir bracket.
3. Detach the wiring harness bracket for the Data Link Connector 1(DLC1).
4. Remove the alternator and alternator bracket.
5. If equipped with ABS brakes, remove the ABS actuator.
6. Remove the right front wheel and the fender apron seal.
7. Remove the power steering drive belt.
8. Slightly raise the engine using a block of wood and floor jack under the oil pan to prevent damage.
9. Remove the four bolts, two nuts, and right-hand mounting bracket.
10. Remove the spark plugs.

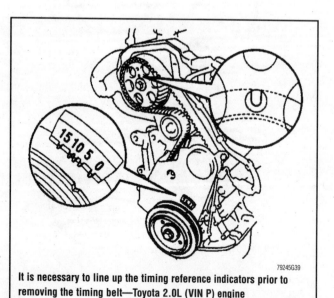

It is necessary to line up the timing reference indicators prior to removing the timing belt—Toyota 2.0L (VIN P) engine

11. Using SST 09213–54015 or equivalent, loosen the crankshaft pulley bolt and remove it by pulling it straight off the crankshaft.
12. Using SST 09249–63010 or equivalent, loosen the retaining bolts and remove the right engine mounting bracket.
13. Remove the upper (No. 2) timing belt cover.
14. Install the crankshaft pulley to the crankshaft and temporarily install the retaining bolt.
15. Turn the crankshaft pulley and align its groove with the timing mark **0** of the No. 1 timing belt cover. Check that the hole of the camshaft timing pulley is aligned with the timing mark of the bearing cap. If not, turn the crankshaft 360° and align the marks.

➡ **If the timing belt is to be reused, matchmark the timing belt to the timing pulleys and timing belt covers so the belt can be reinstalled in its original position. Also, be sure to mark an arrow on the belt to indicate which direction it was turning.**

16. Remove the timing belt from the camshaft timing pulley.
17. Hold the camshaft sprocket with a spanner wrench and remove the mounting bolt. Remove the camshaft pulley.
18. Remove the crankshaft pulley bolt and remove the crankshaft pulley.
19. Remove the No. 1 timing belt cover.
20. Remove the timing belt guide and the timing belt.
21. Remove the No. 1 idler pulley and tension spring.
22. Remove the No. 2 idler pulley.
23. Remove the crankshaft timing pulley.
24. Support the oil pump sprocket with a spanner wrench, then remove the mounting bolt and remove the sprocket.

To install:

25. Install the oil pump pulley. Tighten the nut to 18 ft. lbs. (24 Nm).
26. Install the crankshaft timing pulley. Align the pulley set key with the key groove of the pulley. Slide on the pulley facing the flange side inward.
27. Install the No. 2 idler pulley and tighten the mounting bolt to 31 ft. lbs. (42 Nm). Be sure that the pulley moves smoothly.
28. Install the No. 1 idler pulley with the bolt and the tension spring. Pry the pulley toward the left as far as it will go and tighten the bolt. Make sure that the pulley moves smoothly.
29. Temporarily install the timing belt. Using the crankshaft pulley bolt, turn the crankshaft and position the key groove of the crankshaft timing pulley upward. If reusing the timing belt, align the points marked during removal.
30. Install the timing belt on the crankshaft timing pulley, oil pump pulley, No. 1 idler pulley, water pump pulley and the No. 2 idler pulley.
31. Install the timing belt guide.

➡ **If the old timing belt is being reinstalled, be sure the directional arrow is facing in the original direction and that the belt and sprocket/cover matchmarks are properly aligned.**

32. Install the lower (No. 1) timing belt cover and new gasket with the four bolts.
33. Align the crankshaft pulley set key with the pulley key groove. Temporarily install the crankshaft pulley and bolt.
34. Align the camshaft knock pin with the groove of the pulley, and slide the timing pulley onto the camshaft with the plate washer and set bolt.
35. Tighten the pulley set bolt to 40 ft. lbs. (54 Nm).

TOYOTA—TIMING BELTS

⚠️ WARNING

If any binding is felt when adjusting the timing belt tension by turning the crankshaft, STOP turning the engine, because the pistons may be hitting the valves.

36. Turn the crankshaft pulley and align the **0** mark on the lower (No. 1) timing belt cover.
37. Finish installing the timing belt and check the valve timing, as follows:
    a. If reusing the old timing belt, align the matchmarks that you made previously, and install the timing belt onto the camshaft pulley.
    b. Align the marks on the timing belt with the marks on the camshaft pulley.
    c. Loosen the No. 1 idler pulley set bolt ½ turn.
    d. Turn the crankshaft pulley two complete revolutions TDC to TDC. ALWAYS turn the crankshaft CLOCKWISE. Check that the pulleys are still in alignment with the timing marks.
    e. If the No. 1 idler pulley uses a green tension spring, slowly turn the crankshaft pulley 1⅞ revolutions, and align its groove with the mark at 45° BTDC (for the No. 1 cylinder) of the No. 1 timing belt cover.
    f. Tighten the No. 1 idler pulley set bolt to 31 ft. lbs. (42 Nm).
    g. Be sure there is belt tension between the crankshaft and camshaft timing pulleys.
38. Place the right-hand engine mounting bracket in position but do not install the bolts.
39. Install the upper (No. 2) timing cover with a new gasket(s).
40. Remove the engine crankshaft pulley bolt and pulley.
41. Using SST 09249–63010 or equivalent, install the mounting bolts for the right-hand mounting bracket. Tighten the mounting bolts to 38 ft. lbs. (52 Nm).
42. Align the crankshaft pulley set key with the pulley key groove. Install the pulley. Tighten the pulley bolt to 80 ft. lbs. (108 Nm).
43. Install the spark plugs.
44. Install the right-hand mounting insulator, as follows:
    a. Attach the mounting insulator to the body and mounting bracket with the four bolts and two nuts.
    b. Tighten the three bolts to hold the mounting insulator to the body. Tighten the bolts to 47 ft. lbs. (64 Nm).
    c. Tighten the two nuts and bolt to hold the mounting insulator to the mounting bracket. Tighten the bolt to 27 ft. lbs. (37 Nm) and the nut to 38 ft. lbs. (52 Nm).
45. Install and adjust the power steering pump drive belt.
46. Install the right-hand engine undercover.
47. Install the right front wheel.
48. Lower the engine.
49. If equipped, install the ABS actuator.
50. Install the alternator and alternator bracket.
51. Install the wiring harness bracket for the DLC1.
52. Install the power steering reservoir bracket and reservoir.
53. Connect the negative battery cable.
54. Start the engine and check the timing.

2.5L (2VZ-FE) ENGINE

1. Disconnect the cable from the negative battery terminal.
2. Remove the power steering pump reservoir and position it out of the way. Remove the right fender apron seal and then remove the alternator and power steering belts.
3. Remove the coolant reservoir hose, the washer tank and coolant overflow tank.
4. Remove the right side engine mount stays.
5. Position a piece of wood on a floor jack and then slide the jack under the oil pan. Raise the jack slightly until the pressure is off the engine mounts.
6. Remove the engine control rod.
7. Remove the crankshaft pulley and bolt.
8. Remove the right side engine mounting bracket.
9. Remove the bolts and lift off the upper (No. 2) cover, then the lower cover.
10. Remove the timing belt guide.
11. Remove the timing belt.

➡ **If the timing belt is to be reused, draw a directional arrow on the timing belt in the direction of engine rotation (clockwise) and place matchmarks on the timing belt and crankshaft gear to match the drilled mark on the pulley.**

12. With a 10mm hex wrench, remove the set bolt, plate washer and the No. 1 idler pulley.

To install:
13. Turn the crankshaft until the key groove in the crankshaft timing pulley is facing upward. Slide the timing sprocket on so the flange side faces inward.
14. Apply bolt adhesive to the first few threads of the No. 1 idler pulley set bolt, install the plate washer and pulley and then tighten the bolt to 25 ft. lbs. (34 Nm).
15. Install the timing belt on the crankshaft timing, No. 1 idler and water pump pulleys.

➡ **If the old timing belt is being reinstalled, make sure the directional arrow is facing in the original direction and that the belt and crankshaft gear matchmarks are properly aligned.**

16. Install the lower (No. 1) timing cover and tighten the bolts.
17. Align the crankshaft pulley set key with the key groove on the pulley and slide the pulley on. Tighten the bolt to 181 ft. lbs. (245 Nm).
18. Install the No. 2 idler pulley and tighten the bolt to 29 ft. lbs. (39 Nm). Check that the pulley moves smoothly.
19. Install the left camshaft pulley with the flange side outward. Align the knock pin hole in the camshaft with the knock pin groove on the pulley and then install the pin. Tighten the bolt to 80 ft. lbs. (108 Nm).
20. Set the No. 1 cylinder to TDC again. Turn the right camshaft until the knock pin hole is aligned with the timing mark on the No. 3 belt cover. Turn the left pulley until the marks on the pulley are aligned with the mark on the No. 3 timing cover.
21. Check that the mark on the belt matches with the edge of the lower cover. If not, shift it on the crank pulley until it does. Turn the

206 TIMING BELTS—TOYOTA

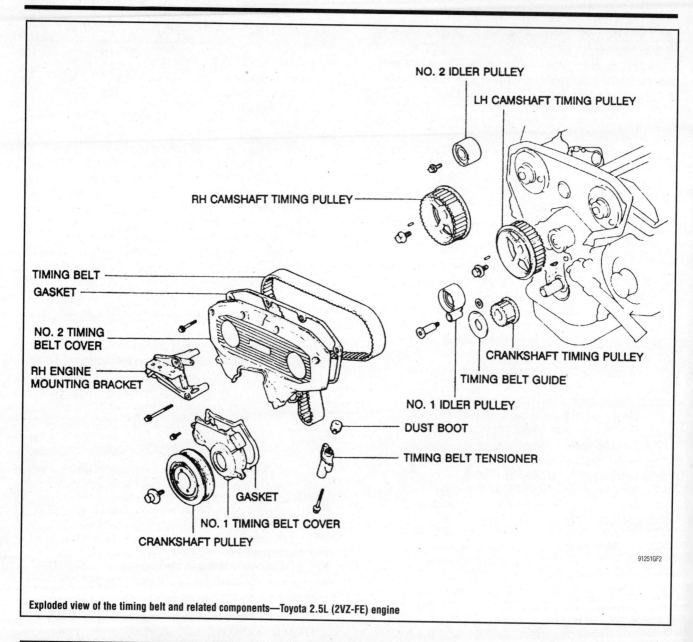

Exploded view of the timing belt and related components—Toyota 2.5L (2VZ-FE) engine

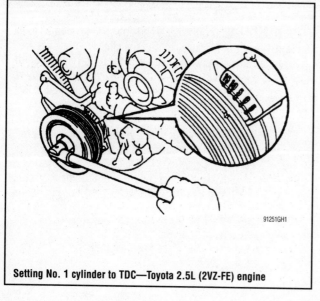

Setting No. 1 cylinder to TDC—Toyota 2.5L (2VZ-FE) engine

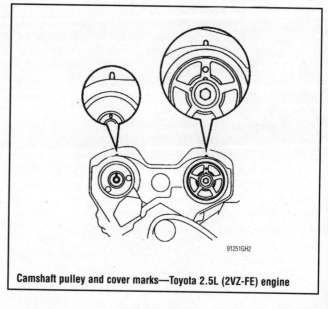

Camshaft pulley and cover marks—Toyota 2.5L (2VZ-FE) engine

TOYOTA—TIMING BELTS

left pulley clockwise a bit and align the mark on the timing belt with the timing mark on the pulley. Slide the belt over the left pulley. Now move the pulley until the marks on it align with the one on the No. 3 cover. There should be tension on the belt between the crankshaft pulley and the left camshaft pulley.

22. Align the installation mark on the timing belt with the mark on the right side camshaft pulley. Hang the belt over the pulley with the flange facing inward. Align the timing marks on the right pulley with the one on the No. 3 cover and slide the pulley onto the end of the camshaft. Move the pulley until the camshaft knock pin hole is aligned with the groove in the pulley and then install the knock pin. Tighten the bolt to 55 ft. lbs. (75 Nm).

23. Position a plate washer between the timing belt tensioner and the a block and then press in the pushrod until the holes are aligned between it and the housing. Slide a 1.5mm Allen wrench through the hole to keep the push rod set. Install the dust boot and then install the tensioner. Tighten the bolts to 20 ft. lbs. (26 Nm). Remove the Allen wrench.

24. Turn the crankshaft clockwise 2 complete revolutions and check that all marks are still in alignment. If not, remove the timing belt and start over again.

25. Install the right engine mount bracket and tighten to 30 ft. lbs. (39 Nm).

26. Position a new gasket and then install the upper No. 2 timing cover.

27. Install the spark plugs.

28. Install the control rod and tighten the bolts to 47 ft. lbs. (64 Nm).

29. Install the right stay and tighten to 23 ft. lbs. (31 Nm).

30. Install and adjust the drive belts.

31. Install the fender apron seal and the wheel.

32. Install the No. 2 stay and tighten the bolt to 55 ft. lbs. (75 Nm) and the nut to 46 ft. lbs. (62 Nm). Install the No. 3 stay and tighten it to 54 ft. lbs. (73 Nm).

33. Install the coolant overflow tank and the washer tank.

34. Install the power steering reservoir tank and the cruise control actuator.

35. Connect the battery cable, start the car and check for any leaks.

3.0L (1MZ-FE) ENGINE

1. With the ignition switch in the **LOCK** position, disconnect the negative battery terminal. If equipped with an air bag system, wait at least 30 seconds or longer before performing any other work.

2. Remove the engine coolant reservoir tank and the alternator belt.

3. Remove the right front wheel and the splash shield.

4. Remove the power steering pump drive belt by loosening the two bolts.

5. Disconnect the two ground wire connectors.

6. Remove the right engine mounting stay.

7. Remove the engine moving control rod and the No. 2 right engine mounting bracket.

8. Remove the No. 2 alternator bracket.

9. Using SST tool 09213–54015 and 09330–00021 or equivalent, remove the crankshaft pulley bolt.

10. Using SST tool 09950–50010 or equivalent (crankshaft pulley remover), remove the crankshaft pulley.

11. Remove the lower timing belt cover by removing four bolts.

12. Remove the No. 2 timing belt cover as follows:

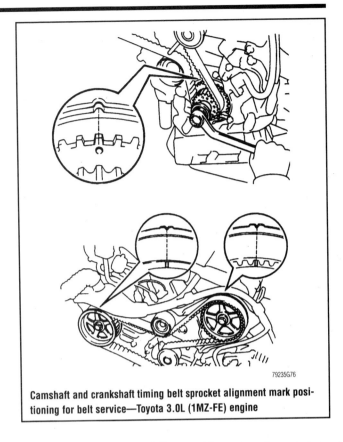

Camshaft and crankshaft timing belt sprocket alignment mark positioning for belt service—Toyota 3.0L (1MZ-FE) engine

a. Remove the bolt and disconnect the engine wire protector from the No. 3 (rear) timing belt cover.

b. Disconnect the engine wire protector clamp from the No. 3 timing belt cover.

c. Remove the five bolts from the No. 2 timing belt cover.

d. Remove the No. 2 cover from the engine.

13. Remove the right engine mounting bracket by removing the nut and two bolts.

14. Remove the crankshaft timing belt guide.

15. Temporarily install the crankshaft pulley bolt.

16. Turn the crankshaft and align the crankshaft timing pulley groove with the oil pump alignment mark. Always turn the engine clockwise.

17. Ensure the timing mark of the camshaft timing pulleys and rear timing belt cover are aligned. If not, turn the engine over an additional 360 degrees (one revolution).

18. Remove the crankshaft pulley bolt.

➥**If the belt is to be reused, align the installation marks on the belt to the marks on the pulleys. If the marks have worn off, make new ones.**

19. Alternately loosen the two timing belt tensioner bolts. Remove the tensioner and dust boot.

20. Remove the timing belt.

To install:

21. Remove any oil or water from the pulleys.

22. Align the front mark of the timing belt with the dot mark of the crankshaft timing pulley.

23. Align the installation marks on the timing belt with the timing marks of the camshaft pulleys.

24. Install the timing belt in the following order:

a. Crankshaft pulley.

b. Water pump pulley.
c. Left camshaft pulley.
d. No. 2 idler pulley.
e. Right camshaft pulley.
f. No. 1 idler pulley.

25. Using a press, slowly press the timing belt tensioner until the holes of the pushrod and housing align. Insert a 0.05 in. (1.27mm) hexagonal Allen wrench through the holes to preserve the setting position.
26. Install the dust boot to the tensioner.
27. Install the tensioner with the two bolts. Alternately tighten the bolts to 20 ft. lbs. (27 Nm). Remove the Allen wrench.
28. Turn the crankshaft clockwise and align the crankshaft timing pulley groove with the oil pump alignment mark.
29. Ensure the camshaft timing marks align with the timing marks on the rear timing belt cover.
30. Install the timing belt guide.
31. Install the right engine mounting bracket and tighten the bolts to 21 ft. lbs. (28 Nm).
32. Install the upper timing belt cover with the five bolts. Tighten the bolts to 74 inch lbs. (8 Nm).
33. Install the engine wire protector clamp to the No. 3 timing belt cover.
34. Install the engine wire protector to the No. 3 timing belt cover with the bolt.
35. Install the lower timing belt cover by installing the four bolts. Tighten the bolts to 74 inch lbs. (8 Nm).
36. Install the crankshaft pulley. Tighten the bolt to 159 ft. lbs. (215 Nm).
37. Install the No. 2 alternator bracket. Tighten the nut to 21 ft. lbs. (28 Nm). Do not tighten the pivot bolt at this time.
38. Install the No. 2 right engine mounting bracket and the moving control rod.
39. Install the right engine mounting stay.
40. Connect the two ground wire connectors.
41. Install and adjust the drive belts.
42. Install the coolant reservoir.
43. Install the right front splash shield and wheel assembly.
44. Connect the negative battery cable.
45. Start the vehicle and check for any leaks.
46. Recheck the ignition timing.

3.0L (2JZ-GTE & 2JZ-GE) ENGINES

1. Disconnect the negative battery cable. Wait at least 90 seconds before performing any other work.
2. Remove the battery and battery tray.
3. Remove the undercovers from the vehicle.
4. Drain the engine coolant and remove the radiator.
5. For vehicles with manual transmissions, remove the drive belt tensioner damper by removing the two nuts.
6. Remove the drive belt.
7. Remove the fan, the fluid coupling, and the water pump pulley.
8. Remove the upper two timing belt covers (Nos. 2 and 3).
9. Remove the drive belt tensioner.
10. Set the No. 1 cylinder to Top Dead Center (TDC) on the compression stroke. Turn the crankshaft pulley clockwise to align the groove with the **0** mark on the lower (No. 1) timing belt cover. Check that the timing marks on the camshaft pulleys are aligned with the marks on the rear belt cover. If the camshaft marks do not align, turn the crankshaft another 360 degrees.

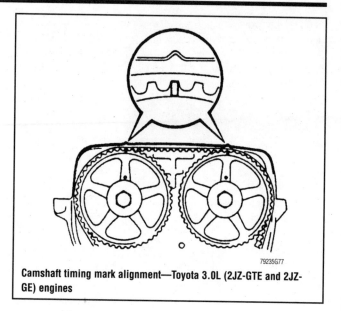

Camshaft timing mark alignment—Toyota 3.0L (2JZ-GTE and 2JZ-GE) engines

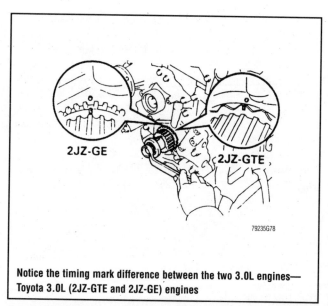

Notice the timing mark difference between the two 3.0L engines—Toyota 3.0L (2JZ-GTE and 2JZ-GE) engines

11. Alternately loosen the two bolts holding the timing belt tensioner. Remove the bolts and remove the tensioner.
12. Remove the timing belt from the camshaft pulleys. If the belt is to be reused, place matchmarks on the belt and gears before removing the belt. Mark the belt with an arrow to show direction of rotation.
13. Using Toyota tool SST 09960–10010 or equivalent, remove the bolts for the camshaft timing gears.
14. Remove the camshaft gears from the engine.
15. If necessary, disconnect the oil cooler tubes from the front of the engine by removing the two bolts and hose clamps.
16. Remove the crankshaft pulley by using Toyota tool 09330–0021, or equivalent, to hold the pulley and using tool 09213–70010, or equivalent, to remove the pulley bolt.
17. Remove the lower (No. 1) timing belt cover and the timing belt guide.
18. Remove the timing belt. If the belt is to be reused, protect it from contact with oil, grease, or fluids.

TOYOTA—TIMING BELTS

To install:

19. Use the crankshaft pulley bolt to turn the crankshaft (clockwise) until the mark on the gear aligns with the oil pump body. Check all the pulleys and gears for cleanliness; remove any grease, oil or coolant. Install the timing belt onto the crankshaft gear and idler pulleys.
20. Install the timing belt guide with the cupped side facing outward.
21. Install the No. 1 timing belt cover.
22. Install the crankshaft pulley. Align the set key with the groove. Hold the pulley with the proper tool and tighten the pulley bolt to 239 ft. lbs. (324 Nm).
23. If equipped with automatic transmission, connect the oil cooler tubes with the clamps and two bolts.
24. Install the camshaft gears as follows:
    a. Align the camshaft knock pin with the groove on the gear and slide on the timing gear.
    b. Temporarily install the timing gear bolt.
    c. Using the same tools as removal, tighten the camshaft gear bolts to 59 ft. lbs. (79 Nm).
    d. Turn the crankshaft pulley and align its groove with the timing mark, **0** on the No. 1 timing belt cover.
    e. Align the timing marks on the camshaft timing gears and the No. 4 timing belt cover.
25. Finish installing the timing belt.
26. Double check that all the timing marks for the crankshaft pulley and the camshaft gears are aligned as they were during disassembly.
27. Set the timing belt tensioner:
    a. Use a press to slowly push in the pushrod on the tensioner. This will require between 220–2200 lbs. (100–1000 kg) of pressure.
    b. Align the holes of the pushrod and housing. Place a 0.06 in. (1.5mm) hex wrench through the holes to keep the pushrod retracted.
    c. Release the press and install the dust boot onto the tensioner.
28. Install the tensioner; alternately tighten the bolts to 20 ft. lbs. (26 Nm).
29. Remove the hex wrench from the tensioner with a pair of pliers.
30. Turn the crankshaft pulley two full turns clockwise. Check that each pulley's timing marks align correctly after the two turns. If any mark does not align, remove the timing belt and reinstall it.
31. Install the drive belt tensioner and tighten the bolts to 15 ft. lbs. (21 Nm).
32. Install the Nos. 2 and 3 timing belt covers.
33. Install the water pump pulley, the fan with fluid coupling, and the drive belt.
34. If equipped with manual transmission, install the drive belt tensioner damper and tighten the nuts to 14 ft. lbs. (20 Nm).
35. Install the radiator and connect the hoses and lines.
36. Install the battery tray and battery. Connect the battery cables.
37. Fill the coolant. Start the engine and check for leaks.
38. Bleed the cooling system and check the ignition timing.
39. Check the fluid levels, including the automatic transmission if so equipped.
40. Install the engine undercover. Road test the vehicle and recheck the coolant level.

3.0L (3VZ-FE) ENGINE

Cars

1. Disconnect the cable from the negative battery terminal.
2. Remove the power steering pump reservoir and position it out of the way. Remove the right fender apron seal and then remove the alternator and power steering belts.
3. Remove the coolant reservoir hose, the washer tank and coolant overflow tank.
4. Remove the right side engine mount stays.
5. Position a piece of wood on a floor jack and then slide the jack under the oil pan. Raise the jack slightly until the pressure is off the engine mounts.
6. Remove the engine control rod.
7. Remove the crankshaft pulley and bolt.
8. Remove the right side engine mounting bracket.
9. Remove the bolts and lift off the upper (No. 2) cover, then the lower cover.
10. Remove the timing belt guide.
11. Remove the timing belt.

➡ **If the timing belt is to be reused, draw a directional arrow on the timing belt in the direction of engine rotation (clockwise) and place matchmarks on the timing belt and crankshaft gear to match the drilled mark on the pulley.**

12. With a 10mm hex wrench, remove the set bolt, plate washer and the No. 1 idler pulley.

To install:

13. Turn the crankshaft until the key groove in the crankshaft timing pulley is facing upward. Slide the timing sprocket on so the flange side faces inward.
14. Apply bolt adhesive to the first few threads of the No. 1 idler pulley set bolt, install the plate washer and pulley and then tighten the bolt to 25 ft. lbs. (34 Nm).
15. Install the timing belt on the crankshaft timing, No. 1 idler and water pump pulleys.

➡ **If the old timing belt is being reinstalled, make sure the directional arrow is facing in the original direction and that the belt and crankshaft gear matchmarks are properly aligned.**

16. Install the lower (No. 1) timing cover and tighten the bolts.
17. Align the crankshaft pulley set key with the key groove on the pulley and slide the pulley on. Tighten the bolt to 181 ft. lbs. (245 Nm).
18. Install the No. 2 idler pulley and tighten the bolt to 29 ft. lbs. (39 Nm). Check that the pulley moves smoothly.
19. Install the left camshaft pulley with the flange side outward. Align the knock pin hole in the camshaft with the knock pin groove on the pulley and then install the pin. Tighten the bolt to 80 ft. lbs. (108 Nm).
20. Set the No. 1 cylinder to TDC again. Turn the right camshaft until the knock pin hole is aligned with the timing mark on the No. 3 belt cover. Turn the left pulley until the marks on the pulley are aligned with the mark on the No. 3 timing cover.
21. Check that the mark on the belt matches with the edge of the lower cover. If not, shift it on the crank pulley until it does. Turn the left pulley clockwise a bit and align the mark on

210 TIMING BELTS—TOYOTA

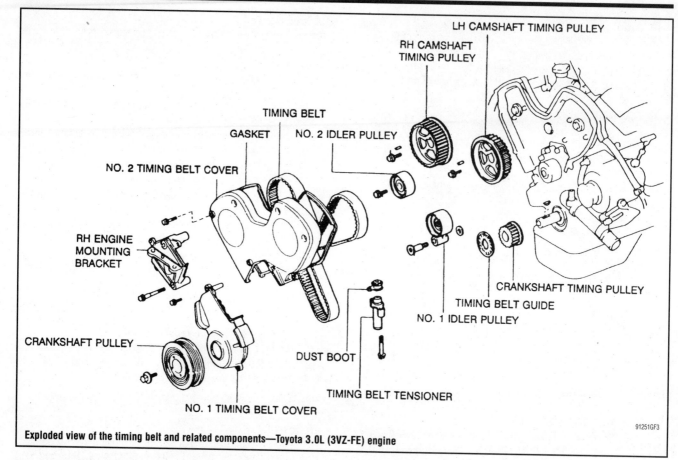

Exploded view of the timing belt and related components—Toyota 3.0L (3VZ-FE) engine

the timing belt with the timing mark on the pulley. Slide the belt over the left pulley. Now move the pulley until the marks on it align with the one on the No. 3 cover. There should be tension on the belt between the crankshaft pulley and the left camshaft pulley.

22. Align the installation mark on the timing belt with the mark on the right side camshaft pulley. Hang the belt over the pulley with the flange facing inward. Align the timing marks on the right pulley with the one on the No. 3 cover and slide the pulley onto the end of the camshaft. Move the pulley until the camshaft knock pin hole is aligned with the groove in the pulley and then install the knock pin. Tighten the bolt to 55 ft. lbs. (75 Nm).

23. Position a plate washer between the timing belt tensioner and the a block and then press in the pushrod until the holes are aligned between it and the housing. Slide a 1.5mm Allen wrench through the hole to keep the push rod set. Install the dust boot and then install the tensioner. Tighten the bolts to 20 ft. lbs. (26 Nm). Remove the Allen wrench.

24. Turn the crankshaft clockwise 2 complete revolutions and check that all marks are still in alignment. If not, remove the timing belt and start over again.

25. Install the right engine mount bracket and tighten to 30 ft. lbs. (39 Nm).

26. Position a new gasket and then install the upper No. 2 timing cover.

27. Install the spark plugs.

28. Install the control rod and tighten the bolts to 47 ft. lbs. (64 Nm).

29. Install the right stay and tighten to 23 ft. lbs. (31 Nm).

30. Install and adjust the drive belts.

31. Install the fender apron seal and the wheel.

32. Install the No. 2 stay and tighten the bolt to 55 ft. lbs. (75 Nm) and the nut to 46 ft. lbs. (62 Nm). Install the No. 3 stay and tighten it to 54 ft. lbs. (73 Nm).

33. Install the coolant overflow tank and the washer tank.

34. Install the power steering reservoir tank and the cruise control actuator.

35. Connect the battery cable, start the car and check for any leaks.

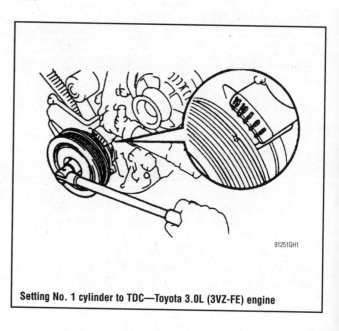

Setting No. 1 cylinder to TDC—Toyota 3.0L (3VZ-FE) engine

TOYOTA—TIMING BELTS 211

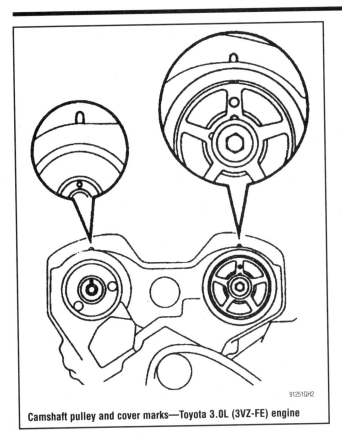

Camshaft pulley and cover marks—Toyota 3.0L (3VZ-FE) engine

Trucks

1. Disconnect the negative battery cable.

❋❋❋ CAUTION

Work must be started after 90 seconds from the time the ignition switch is turned to the LOCK position and the negative battery cable is disconnected.

2. Raise and safely support the vehicle.
3. Remove the engine undercover.

❋❋❋ CAUTION

Never open, service or drain the radiator or cooling system when hot; serious burns can occur from the steam and hot coolant. Also, when draining engine coolant, keep in mind that cats and dogs are attracted to ethylene glycol antifreeze and could drink any that is left in an uncovered container or in puddles on the ground. This will prove fatal in sufficient quantities. Always drain coolant into a sealable container. Coolant should be reused unless it is contaminated or is several years old.

4. Drain the engine coolant.
5. Disconnect the hoses and remove the radiator.
6. Disconnect the No. 2 and No. 3 air hoses from the air pipe.
7. Disconnect the wires from the spark plugs.
8. Remove the spark plugs.
9. Remove the accessory drive belts.
10. Disconnect the power steering pump from the engine.
11. Remove the cooling fan.
12. Remove the water outlet.
13. Disconnect the spark plug wire clamps from the mounting bolts and remove the No. 2 timing belt cover.
14. If reusing the timing belt, check that there are installation marks on the timing belt.
15. Set the No. 1 cylinder at TDC of the compression stroke, as follows:

    a. Turn the crankshaft pulley and align its groove with the timing mark **0** of the No. 1 timing belt cover.

    b. Check that the timing marks of the camshaft timing pulleys and the No. 3 timing belt cover are aligned. If not, turn the crankshaft pulley one revolution (360°).

16. Remove the timing belt tensioner.
17. Remove the fan bracket.
18. Disconnect the timing belt from the camshaft timing pulleys.

➡ **If reusing the timing belt, be sure that you can still read the installation marks. If not, place new installation marks on the timing belt to match the timing marks of the camshaft timing pulleys.**

19. Remove the camshaft timing pulleys, as follows:

    a. Using SST 09960–10010, or equivalent, remove the pulley bolt, the timing pulley, and the knock pin. Remove the two timing pulleys.

20. Remove the crankshaft pulley, as follows:

    a. Using SST 09213–58012 and 09330–00021, or their equivalents, loosen the pulley bolt.

    b. Remove the SST and the pulley bolt. Remove the crankshaft pulley.

21. Remove the No. 1 timing belt cover.
22. Remove the timing belt guide and remove the timing belt.
23. Remove the pivot bolt, the No. 1 idler pulley, and the plate washer.
24. Remove the crankshaft timing pulley.

To install:

25. Install the crankshaft timing pulley, as follows:

    a. Align the timing pulley set key with the key groove of the pulley.

    b. Using SST 09214–60010, or equivalent, and a hammer, tap in the timing pulley with the flange side facing inward.

26. Install the plate washer and the No. 1 idler pulley with the pivot bolt and tighten to 25 ft. lbs. (34 Nm). Check that the pulley bracket moves smoothly.
27. Temporarily install the timing belt, as follows:

    a. Using the crankshaft pulley bolt, turn the crankshaft and align the timing marks of the crankshaft timing pulley and the oil pump body.

    b. Align the installation mark on the timing belt with the dot mark on the crankshaft timing pulley.

    c. Install the timing belt on the crankshaft timing pulley and the water pump pulley.

28. Install the timing belt guide with the cup side facing outward.
29. Install the No. 1 timing belt cover.
30. Install the crankshaft pulley, as follows:

    a. Align the pulley set key with the groove of the crankshaft pulley.

    b. Install the pulley bolt and tighten it to 181 ft. lbs. (245 Nm).

31. Install the left camshaft timing pulley.

    a. Install the knock pin to the camshaft.

    b. Align the knock pin hole of the camshaft with the knock pin groove of the timing pulley.

212 TIMING BELTS—TOYOTA

Turn the crankshaft clockwise to line up the timing marks before removing the timing belt—Toyota T-100 and Tacoma 3.4L (5VZFE) engine

    c. Slide the timing pulley on the camshaft with the flange side facing outward. Tighten the pulley bolt to 80 ft. lbs. (108 Nm).
32. Set the No. 1 cylinder to TDC of the compression stroke, as follows:
    a. Turn the crankshaft pulley, and align its groove with the timing mark **0** of the No. 1 timing belt cover.
    b. Turn the camshaft to align the knock pin hole of the camshaft with the timing mark of the No. 3 timing belt cover.
    c. Turn the camshaft timing pulley and align the timing marks of the camshaft timing pulley and the No. 3 timing belt cover.
33. Connect the timing belt to the left camshaft timing pulley. Check that the installation mark on the timing belt is aligned with the end of the No. 1 timing belt cover.
    a. Using SST 09960–01000, or equivalent, slightly turn the left camshaft timing pulley clockwise. Align the installation mark on the timing belt with the timing mark of the camshaft timing pulley, and hang the timing belt on the left camshaft timing pulley.
    b. Align the timing marks of the left camshaft pulley and the No. 3 timing belt cover.
    c. Check that the timing belt has tension between the crankshaft timing pulley and the left camshaft timing pulley.
34. Install the right camshaft timing pulley and the timing belt, as follows:
    a. Align the installation mark on the timing belt with the timing mark of the right camshaft timing pulley, and hang the timing belt on the right camshaft timing pulley with the flange side facing inward.
    b. Slide the right camshaft timing pulley on the camshaft. Align the timing marks on the right camshaft timing pulley and the No. 3 timing belt cover.
    c. Align the knock pin hole of the camshaft with the knock pin groove of the pulley and install the knock pin. Install the bolt and tighten it to 80 ft. lbs. (108 Nm).
35. Install the fan bracket and tighten the bolts to 30 ft. lbs. (41 Nm).

✲✲ WARNING

If any binding is felt when adjusting the timing belt tension by turning the crankshaft, STOP turning the engine, because the pistons may be hitting the valves.

36. Set the timing belt tensioner, as follows:
    a. Using a press, slowly depress in the pushrod using 200–2,205 lbs. (981–9,807 N) of force.
    b. Align the holes of the pushrod and housing, pass a 1.5 mm hexagon wrench through the holes to keep the setting position of the pushrod.
    c. Release the press and install the dust boot on the tensioner.
37. Install the timing belt tensioner and alternately tighten the bolts to 20 ft. lbs. (28 Nm). Using pliers, remove the 1.5 mm hexagon wrench from the belt tensioner.
38. Check the valve timing, as follows:
    a. Slowly turn the crankshaft pulley two revolutions from TDC to TDC. Always turn the crankshaft pulley clockwise.
    b. Check that each pulley aligns with the timing marks. If the timing marks do not align, remove the timing belt and reinstall it.
39. Install the No. 2 timing belt cover. Connect the four clamps on the spark plug wires to the mounting bolts of the No. 2 timing belt cover.

TOYOTA—TIMING BELTS

40. Remove the old packing material from the water outlet and apply new seal packing before installation. Tighten the nuts to 11 ft. lbs. (15 Nm).
41. Install the alternator drive belt.
42. Install the cooling fan and tighten the nuts to 48 inch lbs. (5.4 Nm).
43. Install the air conditioning drive belt.
44. Install the power steering pump, pump pulley, and the drive belt.
45. Install the spark plugs and tighten them to 13 ft. lbs. (18 Nm).
46. Connect the spark plug wires to the spark plugs.
47. Connect the No. 2 and the No. 3 air hoses to the air pipe.
48. Install the radiator and connect the hoses.
49. Fill the engine with coolant.
50. Connect the negative battery cable.
51. Start the engine and check for leaks.
52. Check the ignition timing and install the engine undercover.

3.0L (7M-GE & 7M-GTE) ENGINES

1. Disconnect the negative battery cable. Drain the cooling system. Remove the radiator. Remove the water outlet.
2. Remove the spark plugs. Remove the drive belts. Remove the No. 3 timing belt cover.
3. Set the No. 1 piston to TDC of the compression stroke. Remove the timing belt from the camshaft sprockets. If reusing the belt, matchmark the belt and the sprockets in the direction of engine rotation.
4. Remove the camshaft pulleys. Remove the crankshaft pulley using the proper removal tools. Remove the power steering air pipe, if equipped.
5. If equipped with air conditioning, remove the compressor and position it aside. Do not disconnect the refrigerant lines.
6. Remove the No. 1 timing belt cover. Remove the timing belt. Remove the idler pulley and the tension spring. Remove the oil pump drive pulley.

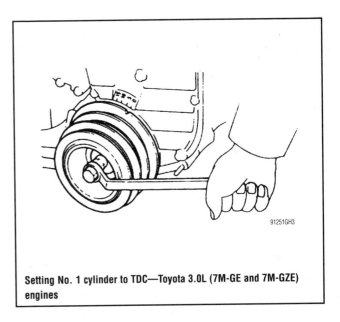

Setting No. 1 cylinder to TDC—Toyota 3.0L (7M-GE and 7M-GZE) engines

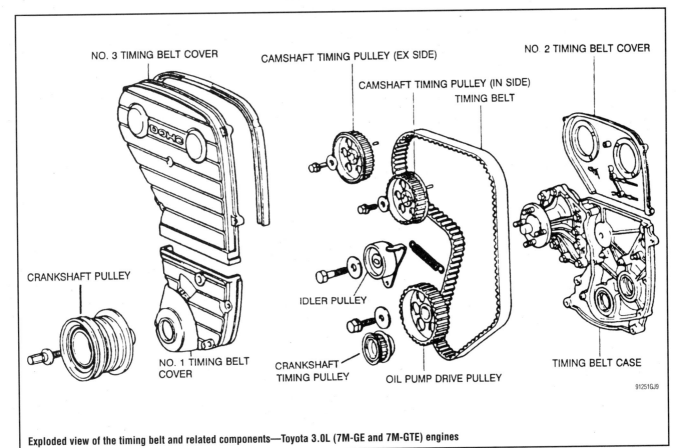

Exploded view of the timing belt and related components—Toyota 3.0L (7M-GE and 7M-GTE) engines

214 TIMING BELTS—TOYOTA

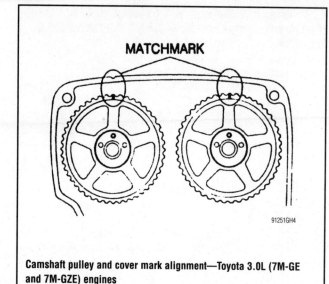

Camshaft pulley and cover mark alignment—Toyota 3.0L (7M-GE and 7M-GZE) engines

To install:

7. Inspect the belt for defects. Replace as required. Inspect the idler pulleys and springs. Replace defective components as required.
8. Install the oil pump drive pulley and retaining bolt. Tighten the bolt to 16 ft. lbs. (22 Nm).
9. Install the crankshaft timing pulley. Temporarily install the idler pulley and tension spring. Tighten the assembly to 36 ft. lbs. (49 Nm). Pry the idler pulley toward the left as far as it will go and temporarily tighten the bolt.
10. Temporarily install the timing belt. If reusing the old belt install it using the marks made during the removal procedure. Install the No. 1 timing belt cover.
11. Align the set key with the key groove and install the crankshaft pulley and tighten the retaining bolt to 195 ft. lbs. (265 Nm).
12. Install the camshaft timing pulleys. Tighten the retaining bolts to 36 ft. lbs. (49 Nm).
13. Loosen the idler pulley bolt. Install the timing belt to the Intake side and the Exhaust side. Tighten the idler pulley bolt to 36 ft. lbs. (49 Nm).
14. Make sure the timing belt tension **A** is equal to the timing belt tension **B**. If not adjust the idler pulley. Turn the engine 2 complete revolutions in the clockwise direction and check to see that everything is aligned properly.
15. Turn both the intake and exhaust camshaft pulleys inward at the same time to loosen the timing belt between the 2 sprockets. Belt deflection should be 4.4–6.6 lbs. If not adjust the idler pulley.
16. Install the remaining components, start the engine and check for leaks and proper operation.

3.4L (5VZ-FE) ENGINE

T-100 and Tacoma

1. Disconnect the negative battery cable.

✻✻ CAUTION

Work must be started after 90 seconds from the time the ignition switch is turned to the LOCK position and the negative battery cable is disconnected.

2. Raise and safely support the vehicle.
3. Remove the engine undercover.
4. Drain the engine coolant.

✻✻ CAUTION

Never open, service or drain the radiator or cooling system when hot; serious burns can occur from the steam and hot coolant. Also, when draining engine coolant,

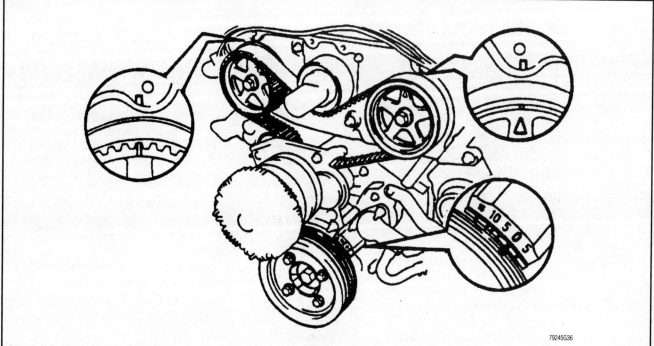

Be sure to properly align the crankshaft timing mark before removing the belt—Toyota 3.0L (3VZ-FE) engine

TOYOTA—TIMING BELTS

keep in mind that cats and dogs are attracted to ethylene glycol antifreeze and could drink any that is left in an uncovered container or in puddles on the ground. This will prove fatal in sufficient quantities. Always drain coolant into a sealable container. Coolant should be reused unless it is contaminated or is several years old.

5. Disconnect the upper radiator hose from the engine.
6. Remove the power steering drive belt.
7. Remove the air conditioning drive belt by loosening the idler pulley nut and the adjusting bolt.
8. Loosen the lockbolt, pivot bolt, and the adjusting bolt and the alternator drive belt.
9. Remove the No. 2 fan shroud by removing the two clips.
10. Remove the fan with the fluid coupling and fan pulleys.
11. Disconnect the power steering pump from the engine and set aside. Do not disconnect the lines from the pump.
12. If equipped with air conditioning, disconnect the compressor from the engine and set aside. Do not disconnect the lines from the compressor.
13. If equipped with air conditioning, disconnect the air conditioning bracket.
14. Remove the No. 2 timing belt cover, as follows:
    a. Detach the camshaft position sensor connector from the No. 2 timing belt cover.
    b. Disconnect the three spark plug wire clamps from the No. 2 timing belt cover.
    c. Remove the six bolts and remove the timing belt cover.
15. Remove the fan bracket, as follows:
    a. Remove the power steering adjusting strut by removing the nut.
    b. Remove the fan bracket by removing the bolt and nut.
16. Set the No. 1 cylinder at TDC of the compression stroke, as follows:
    a. Turn the crankshaft pulley and align its groove with the timing mark **0** of the No. 1 timing belt cover.
    b. Check that the timing marks of the camshaft timing pulleys and the No. 3 timing belt cover are aligned. If not, turn the crankshaft pulley one revolution (360°).

➡ If reusing the timing belt, be sure that you can still read the installation marks. If not, place new installation marks on the timing belt to match the timing marks of the camshaft timing pulleys.

17. Remove the timing belt tensioner by alternately loosening the two bolts.
18. Remove the camshaft timing pulleys, as follows:
    a. Using SST 09960–10010 or equivalent, remove the pulley bolt, the timing pulley and the knock pin. Remove the two timing pulleys with the timing belt.
19. Remove the crankshaft pulley, as follows:
    a. Using SST 09213–54015 and 09330–00021, or their equivalents, loosen the pulley bolt.
    b. Remove the SST tool, the pulley bolt, and the pulley.
20. Remove the starter wire bracket and the No. 1 timing belt cover.
21. Remove the timing belt guide and remove the timing belt.
22. Remove the bolt and the No. 2 idler pulley.
23. Remove the pivot bolt, the No. 1 idler pulley, and the plate washer.
24. Remove the crankshaft gear.

To install:
25. Install the crankshaft timing gear.
    a. Align the timing pulley set key with the key groove of the gear.
    b. Using SST 09214–60010, or equivalent, and a hammer, tap in the timing gear with the flange side facing inward.
26. Install the plate washer and the No. 1 idler pulley with the pivot bolt and tighten it to 26 ft. lbs. (35 Nm). Check that the pulley bracket moves smoothly.
27. Install the No. 2 timing belt idler with the bolt. Tighten the bolt to 30 ft. lbs. (40 Nm). Check that the pulley bracket moves smoothly.
28. Temporarily install the timing belt, as follows:
    a. Using the crankshaft pulley bolt, turn the crankshaft and align the timing marks of the crankshaft timing pulley and the oil pump body.
    b. Align the installation mark on the timing belt with the dot mark of the crankshaft timing pulley.
    c. Install the timing belt on the crankshaft timing pulley, No. 1 idler pulley, and the water pump pulleys.
29. Install the timing belt guide with the cup side facing outward.
30. Install the No. 1 timing belt cover and starter wire bracket. Tighten the timing belt cover bolts to 80 inch lbs. (9 Nm).

❊❊ WARNING

If any binding is felt when adjusting the timing belt tension by turning the crankshaft, STOP turning the engine, because the pistons may be hitting the valves.

31. Install the crankshaft pulley, as follows:
    a. Align the pulley set key with the key groove of the crankshaft pulley.
    b. Install the pulley bolt and tighten it to 184 ft. lbs. (250 Nm).
32. Install the left camshaft timing pulley.
    a. Install the knock pin to the camshaft.
    b. Align the knock pin hose of the camshaft with the knock pin groove of the timing pulley.
    c. Slide the timing belt pulley on the camshaft with the flange side facing outward. Tighten the pulley bolt to 81 ft. lbs. (110 Nm).
33. Set the No. 1 cylinder to TDC of the compression stroke, as follows:
    a. Turn the crankshaft pulley, and align its groove with the timing mark **0** of the No. 1 timing belt cover.
    b. Turn the camshaft to align the knock pin hole of the camshaft with the timing mark of the No. 3 timing belt cover.
    c. Turn the camshaft timing pulley, and align the timing marks of the camshaft timing pulley and the No. 3 timing belt cover.
34. Connect the timing belt to the left camshaft timing pulley, as follows:

➡ **Check that the installation mark on the timing belt is aligned with the end of the No. 1 timing belt cover.**

    a. Using SST 09960–01000 or equivalent, slightly turn the left camshaft timing pulley clockwise. Align the installation mark on the timing belt with the timing mark of the camshaft timing pulley, and hang the timing belt on the left camshaft timing pulley.

    b. Align the timing marks of the left camshaft pulley and the No. 3 timing belt cover.

    c. Check that the timing belt has tension between the crankshaft timing pulley and the left camshaft timing pulley.

35. Install the right camshaft timing pulley and the timing belt, as follows:

    a. Align the installation mark on the timing belt with the timing mark of the right camshaft timing pulley, and hang the timing belt on the right camshaft timing pulley with the flange side facing inward.

    b. Slide the right camshaft timing pulley on the camshaft. Align the timing marks on the right camshaft timing pulley and the No. 3 timing belt cover.

    c. Align the knock pin hole of the camshaft with the knock pin groove of the pulley and install the knock pin. Install the bolt and tighten it to 81 ft. lbs. (110 Nm).

36. Set the timing belt tensioner, as follows:

    a. Using a press, slowly press in the pushrod using 220–2,205 lbs. (981–9,807 N) of force.

    b. Align the holes of the pushrod and housing, pass a 1.5 mm hex wrench through the holes to keep the setting position of the pushrod.

    c. Release the press and install the dust boot on the tensioner.

37. Install the timing belt tensioner and alternately tighten the bolts to 20 ft. lbs. (28 Nm). Using pliers, remove the 1.5 mm hex wrench from the belt tensioner.

38. Check the valve timing, as follows:

    a. Slowly turn the crankshaft pulley two revolutions from TDC to TDC. Always turn the crankshaft pulley clockwise.

    b. Check that each pulley aligns with the timing marks. If the timing marks do not align, remove the timing belt and reinstall it.

39. Install the fan bracket with the bolt and nut.
40. Install the power steering adjusting strut with the nut.
41. Install the No. 2 timing belt cover. Tighten the bolts to 80 inch lbs. (9 Nm). Install the remaining components.
42. Fill the cooling system with coolant.
43. Connect the negative battery cable.
44. Start the engine and check for leaks.

4Runner

1. Disconnect the negative battery cable.

✲✲ CAUTION

Wait 90 seconds from the time the key is turned to LOCK and the negative battery cable is disconnected to begin work. This allows the SRS capacitor to discharge and prevent deployment of the air bag(s).

2. Raise and safely support the vehicle.
3. Remove the engine undercover.
4. Drain the engine coolant.

✲✲ CAUTION

Never open, service or drain the radiator or cooling system when hot; serious burns can occur from the steam and hot coolant. Also, when draining engine coolant, keep in mind that cats and dogs are attracted to ethylene glycol antifreeze and could drink any that is left in an uncovered container or in puddles on the ground. This will prove fatal in sufficient quantities. Always drain coolant into a sealable container. Coolant should be reused unless it is contaminated or is several years old.

5. Disconnect the upper radiator hose from the engine.
6. Remove the power steering drive belt.
7. Remove the air conditioning drive belt by loosening the idler pulley nut and the adjusting bolt.
8. If equipped with air conditioning, disconnect the compressor from the engine and set aside. Do not disconnect the lines from the compressor.
9. If equipped with air conditioning, disconnect the air conditioning bracket.
10. Remove the fan with the fluid coupling and fan pulleys.
11. Loosen the lockbolt, pivot bolt, and the adjusting bolt and the alternator drive belt.
12. Remove the No. 2 fan shroud by removing the two clips.
13. Disconnect the power steering pump from the engine and set aside. Do not disconnect the lines from the pump.
14. Remove the oil dipstick and the guide.
15. Remove the No. 2 timing belt cover as follows:

    a. Detach the camshaft position sensor connector from the No. 2 timing belt cover.

    b. Disconnect the four spark plug wire clamps from the No. 2 timing belt cover.

    c. Remove the six bolts and remove the timing belt cover.

16. Remove the fan bracket as follows:

    a. Remove the power steering adjusting strut by removing the nut.

    b. Remove the fan bracket by removing the bolt and nut.

17. Using SST 09213–54015 or equivalent, remove the crankshaft pulley.
18. Remove the starter wire bracket and the No. 1 timing belt cover.
19. Remove the timing belt guide.
20. Set the No. 1 cylinder at TDC of the compression stroke, as follows:

    a. Temporarily install the crankshaft pulley bolt to the crankshaft.

    b. Turn the crankshaft and align the timing marks of the crankshaft timing pulley and the oil pump body.

    c. Check that the timing marks of the camshaft timing pulleys and the No. 3 timing belt cover are aligned. If not, turn the crankshaft pulley one revolution (360°).

➡ **If reusing the timing belt, be sure that you can still read the installation marks. If not, place new installation marks on the timing belt to match the timing marks of the camshaft timing pulleys.**

21. Remove the timing belt tensioner by alternately loosening the two bolts.
22. Remove the right and left camshaft pulleys.
23. Remove the No. 2 idler pulley.
24. Using a 10mm hex wrench, remove the pivot bolt, No.1 idler pulley and the plate washer.
25. Remove the timing belt guide and remove the timing belt.
26. Remove the crankshaft timing pulley.

TOYOTA—TIMING BELTS 217

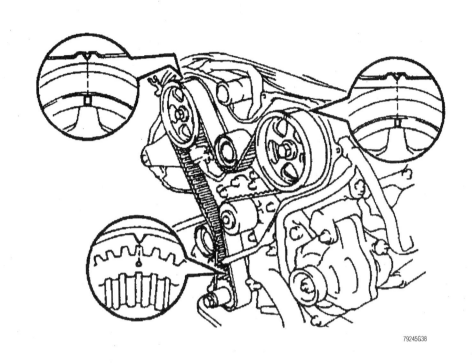

Crankshaft and camshaft timing mark locations—Toyota 4Runner 3.4L (5VZFE) engine

To install:
27. Install the crankshaft timing belt pulley, as follows:
  a. Align the timing belt pulley set key with the key groove of the timing pulley and slide on the timing pulley.
  b. Slide on the timing belt pulley with the flange side facing inward.
28. Install the plate washer and the No. 1 idler pulley with the pivot bolt and tighten it to 26 ft. lbs. (35 Nm). Check that the pulley bracket moves smoothly.
29. Install the No. 2 timing belt idler with the bolt. Tighten the bolt to 30 ft. lbs. (40 Nm). Check that the pulley bracket moves smoothly.
30. Install the left and right camshaft timing pulleys.
31. Set the No. 1 cylinder to TDC of the compression stroke, as follows:
  a. Using the crankshaft pulley bolt, turn the crankshaft and align the timing marks of the crankshaft timing pulley and the oil pump body.
  b. Using SST 09960–10010 or equivalent to turn the camshaft pulley to align the marks of the camshaft timing belt pulley and the No. 3 timing belt cover.
32. Install the timing belt, as follows:

➡ **The engine should be cold.**

  a. Face the front mark on the timing belt forward.
  b. Align the installation mark on the timing belt with the timing mark of the crankshaft timing pulley.
  c. Align the installation marks on the timing belt with the timing marks of the camshaft pulleys.
33. Install the timing belt in the following order:
- Left camshaft pulley
- No. 2 idler pulley
- Right camshaft pulley
- Water pump pulley
- Crankshaft pulley
- No. 1 idler pulley

✻✻ WARNING

If any binding is felt when adjusting the timing belt tension by turning the crankshaft, STOP turning the engine, because the pistons may be hitting the valves.

34. Set the timing belt tensioner as follows:
  a. Using a press, slowly press in the pushrod using 220–2,205 lbs. (981–9,807 N) of force.
  b. Align the holes of the pushrod and housing, pass a 1.27mm wrench through the holes to keep the setting position of the pushrod.
  c. Release the press and install the dust boot to the tensioner.
35. Install the timing belt tensioner and alternately tighten the bolts to 20 ft. lbs. (27 Nm). Using pliers, remove the 1.27mm wrench from the belt tensioner.
36. Check the valve timing, as follows:
  a. Slowly turn the crankshaft and align the timing marks of the crankshaft timing pulley and the oil pump body. Always turn the crankshaft pulley clockwise.
  b. Check that the timing marks of the right and left timing pulleys align with the timing marks of the No. 3 timing belt cover. If the marks do not align, remove the timing belt and reinstall it.
37. Install the timing belt guide with the cup side facing outward.

218 TIMING BELTS—TOYOTA, VOLKSWAGEN

38. Install the No. 1 timing belt cover and starter wire bracket. Tighten the timing belt cover fasteners to 80 inch lbs. (9 Nm).
39. Install the crankshaft pulley, as follows:
   a. Align the pulley set key with the key groove of the pulley and slide the pulley.
   b. Using SST 09213-54014 or equivalent, tighten the bolt to 184 ft. lbs. (250 Nm).
40. Install the fan bracket with the bolt and nut.
41. Install the No. 2 timing belt cover, and tighten the bolts to 80 inch lbs. (9 Nm). Install the remaining components.
42. Fill the cooling system with coolant.
43. Connect the negative battery cable.
44. Start the engine and check for leaks.
45. Check the ignition timing.

Volkswagen

1.6L DIESEL ENGINE

Some special tools are required. A flat bar, VW tool 2065A or equivalent, is used to secure the camshaft in position. A pin, VW tool 2064 or equivalent, is used to fix the pump position while the timing belt is removed. The camshaft and pump work against spring pressure and will move out of position when the timing belt is removed. It is not difficult to find substitutes but do not remove the timing belt without these tools.

➡ Do not turn the engine or camshaft with the timing belt removed. The pistons will contact the valves and cause internal engine damage.

1. Disconnect the negative battery cable and remove the accessory drive belts.
2. To remove the crankshaft accessory drive pulley, hold the center crankshaft sprocket bolt with a socket and loosen the pulley bolts.
3. The cover is now accessible. It comes off in 2 pieces, remove the upper half first. Take note of any special spacers or other hardware.
4. Temporarily reinstall the crankshaft pulley bolt and turn the crankshaft to TDC of No. 1 piston. The mark on the camshaft sprocket should be aligned with the mark on the inner timing belt cover or the edge of the cylinder head.
5. With the engine at TDC, insert the bar into the slot at the back of the camshaft. The bar rests on the cylinder head to will hold the camshaft in position.
6. Insert the pin into the injection pump drive sprocket to hold the pump in position.
7. Loosen the locknut on the tensioner pulley and turn the tensioner counterclockwise to relieve the tension on the timing belt. Slide the timing belt from the sprockets.

To install:

8. Install the new timing belt and adjust the tension so the belt can be twisted 45 degrees at a point between the camshaft and pump sprockets. Tighten the tensioner nut to 33 ft. lbs. (45 Nm).
9. Remove the holding tools.
10. Turn the engine 2 full revolutions to return to TDC of No. 1 piston. Recheck belt tension and timing mark alignment, readjust as required.

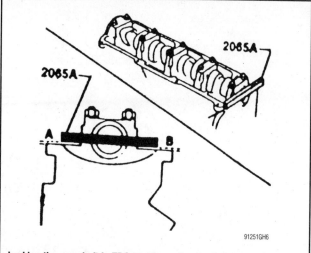

Locking the camshaft in TDC position using the Volkswagen tool—Volkswagen 1.6L Diesel engine

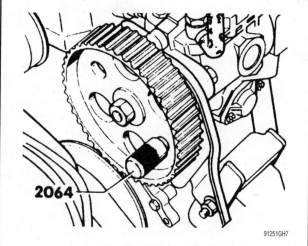

Locking the camshaft sprocket in TDC position using the Volkswagen tool—Volkswagen 1.6L Diesel engine

11. Install the belt cover, tighten the cover bolts to 87 inch lbs. (10 Nm) and accessory drive belts.
12. If the belt is too tight, there will be a growling noise that rises and falls with engine speed.

1.8L ENGINE

➡ Do not turn the engine or camshaft with the camshaft drive belt removed. The pistons will contact the valves and cause internal engine damage.

1. Disconnect the negative battery cable.
2. Remove the accessory drive belts.
3. To remove the crankshaft accessory drive pulley, hold the center crankshaft sprocket bolt with a socket and loosen the pulley bolts.

VOLKSWAGEN—TIMING BELTS

4. The cover is now accessible. It comes off in 2 pieces; remove the upper ½ first. Take note of any special spacers or other hardware.

5. Temporarily reinstall the crankshaft pulley bolt, if removed and turn the crankshaft to TDC of No. 1 piston. The mark on the camshaft sprocket should be aligned with the mark on the inner drive belt cover, if equipped, or the edge of the cylinder head.

6. On 8-valve engines, the notch on the crankshaft pulley should align with the dot on the intermediate shaft sprocket. With the distributor cap removed, the rotor should be pointing toward the No. 1 mark on the rim of the distributor housing.

7. Loosen the locknut on the tensioner pulley and turn the tensioner counterclockwise to relieve the tension on the timing belt.

8. Slide the timing belt from the sprockets.

To install:

9. Install the new timing belt and tension the belt so it can be twisted 90 degrees at the middle of its longest section, between the camshaft and intermediate sprockets.

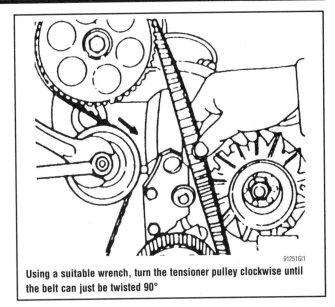
Using a suitable wrench, turn the tensioner pulley clockwise until the belt can just be twisted 90°

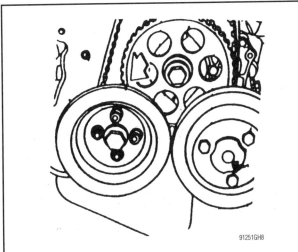
Timing marks on the crankshaft pulley and the intermediate shaft sprocket (arrow)—Volkswagen 1.8L engine

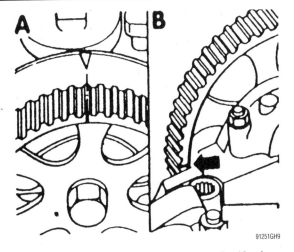
Volkswagen timing marks on camshaft sprockets—A = 16-valve engine, B = 8-valve engine

10. Recheck the alignment of the timing marks and, if correct, turn the engine 2 full revolutions to return to TDC of No. 1 piston. Recheck belt tension and timing marks. Readjust as required. Tighten the tensioner nut to 33 ft. lbs. (45 Nm).

11. Reinstall the belt cover and accessory drive belts.

12. When running the engine, there will be a growling noise that rises and falls with engine speed if the belt is too tight.

1.9L DIESEL ENGINE

Some special tools are required to perform this procedure properly. A flat bar, VW tool 2065A or equivalent, is used to secure the camshaft in position. A pin, VW tool 2064 or equivalent, is used to fix the pump position while the timing belt is removed. The camshaft and pump work against spring pressure and will move out of position when the timing belt is removed. It is not difficult to find substitutes but do not remove the timing belt without these tools.

✴✴ WARNING

Do not turn the engine or camshaft with the timing belt removed. The pistons will contact the valves and cause internal engine damage.

1. Disconnect the negative battery cable and remove the accessory drive belts, crankshaft pulley and the timing belt cover(s). Remove the camshaft cover and rubber plug at the back end of the camshaft.

2. Temporarily reinstall the crankshaft pulley bolt and turn the crankshaft to TDC of No. 1 piston. The mark on the camshaft sprocket should be aligned with the mark on the inner timing belt cover or the edge of the cylinder head.

3. With the engine at TDC, insert the bar into the slot at the back of the camshaft. The bar rests on the cylinder head to will hold the camshaft in position.

4. Insert the pin into the injection pump drive sprocket to hold the pump in position.

5. Loosen the locknut on the tensioner pulley and turn the tensioner counterclockwise to relieve the tension on the timing belt. Slide the timing belt from the sprockets.

220 TIMING BELTS—VOLKSWAGEN

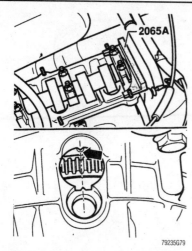

Use the VW tool (or equivalent) to lock the camshaft at TDC for timing belt replacement—Volkswagen 1.9L Diesel engines

1. Disconnect the negative battery cable and remove the accessory drive belts, crankshaft pulley and the timing belt cover(s).
2. Temporarily reinstall the crankshaft pulley bolt, if removed, and turn the crankshaft to TDC of No. 1 piston. The mark on the camshaft sprocket should be aligned with the mark on the inner drive belt cover, if equipped, or the edge of the cylinder head.
3. On 8-valve engines, the notch on the crankshaft pulley should align with the dot on the intermediate shaft sprocket. With the distributor cap removed, the rotor should be pointing toward the No. 1 mark on the rim of the distributor housing.
4. Loosen the locknut on the tensioner pulley and turn the tensioner counterclockwise to relieve the tension on the timing belt.
5. Slide the timing belt off the sprockets.

To install:

6. Install the new timing belt and tension the belt so that it can be twisted 90 degrees at the middle of its longest section, between the camshaft and intermediate sprockets.

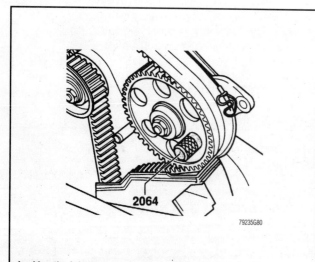

Locking the injection pump with the VW tool (or equivalent) as shown—Volkswagen 1.9L Diesel engines

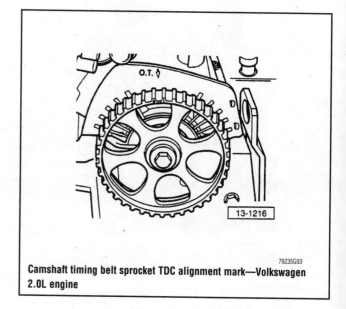

Camshaft timing belt sprocket TDC alignment mark—Volkswagen 2.0L engine

To install:

6. Install the new timing belt and adjust the tension so the belt can be twisted 45 degrees at the half-way point between the camshaft and pump sprockets. Tighten the tensioner nut to 33 ft. lbs. (45 Nm).
7. Remove the holding tools.
8. Turn the engine 2 full revolutions to return to TDC for the No. 1 cylinder. Recheck belt tension and timing mark alignment, readjust as required.
9. Install the belt cover and accessory drive belts.

➡ If the belt is too tight, there will be a growling noise that rises and falls with engine speed.

2.0L ENGINE

➡ Do not turn the engine or camshaft with the camshaft drive belt removed. The pistons will contact the valves and cause internal engine damage.

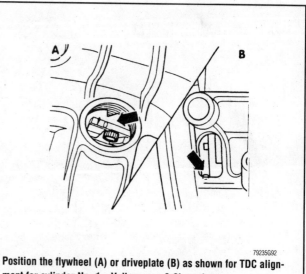

Position the flywheel (A) or driveplate (B) as shown for TDC alignment for cylinder No. 1—Volkswagen 2.0L engine

VOLKSWAGEN, VOLVO—TIMING BELTS 221

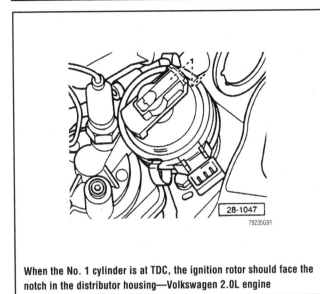

When the No. 1 cylinder is at TDC, the ignition rotor should face the notch in the distributor housing—Volkswagen 2.0L engine

7. Recheck the alignment of the timing marks, if correct, turn the engine 2 full revolutions to return to TDC of No. 1 piston. Recheck belt tension and timing marks. Readjust as required. Tighten the tensioner nut to 33 ft. lbs. (45 Nm).
8. Reinstall the belt cover and accessory drive belts.

➡ When running the engine, there will be a growling noise that rises and falls with engine speed if the belt is too tight.

Volvo

2.3L (B230F & B230FT) ENGINES

1990–94 Models

1. Disconnect the negative battery cable. Loosen the fan shroud and remove the fan. Remove the shroud.
2. Loosen the alternator, power steering pump and air conditioning compressor and remove the drive belts.
3. Remove the water pump pulley.
4. Remove the 4 retaining bolts and lift off the timing belt cover.
5. To remove the tension from the belt, loosen the nut for the tensioner and press the idler roller back. The tension spring can be locked in this position by inserting the shank end of a 3mm drill through the pusher rod.
6. Remove the 6 retaining bolts and the crankshaft pulley.
7. Remove the belt, taking care not to bend it at any sharp angles. The belt should be replaced at 45,000 mile intervals, if it becomes oil soaked or frayed or if on a vehicle that has not been operated for any length of time.

To install:

8. If the crankshaft, idler shaft or camshaft were disturbed while the belt was out, align each shaft with its corresponding index mark to assure proper valve timing and ignition timing, as follows:
   a. Rotate the crankshaft so the notch in the convex crankshaft gear belt guide aligns with the embossed mark on the front cover (12 o'clock position).

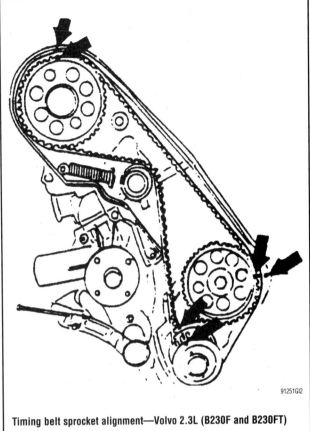

Timing belt sprocket alignment—Volvo 2.3L (B230F and B230FT) engines

   b. Rotate the idler shaft so the dot on the idler shaft drive sprocket aligns with the notch on the timing belt rear cover (4 o'clock position).
   c. Rotate the camshaft so the notch in the camshaft sprocket inner belt guide aligns with the notch in the forward edge of the valve cover (12 o'clock position).
9. Install the timing belt (don't use any sharp tools) over the sprockets and then over the tensioner roller. New belts have yellow marks. The 2 lines on the drive belt should fit toward the crankshaft marks. The next mark should then fit toward the intermediate shaft marks, etc. Loosen the tensioner nut and let the spring tension automatically take up the slack. Tighten the tensioner nut to 37 ft. lbs. (51 Nm).
10. Rotate the crankshaft 1 full revolution clockwise and make sure the timing marks still align.
11. Clean all gasket mating surfaces thoroughly. Install the timing belt cover using a new gasket.
12. Install the water pump pulley, drive belts, air conditioning compressor, power steering pump and alternator.
13. Install the fan and shroud. Install the accessory drive belts. Connect the negative battery cable. Start the engine and check for leaks.

1995–99 Models

1. Disconnect the negative battery cable.
2. Remove the drive belts.
3. Remove the cooling fan and shroud.

4. Remove the timing belt cover.
5. Remove the 6 retaining bolts and the crankshaft pulley.
6. To remove the tension from the belt, loosen the nut for the tensioner and press the idler roller back. The tension spring can be locked in this position by inserting the shank end of a 3mm drill through the pusher rod.
7. Remove the belt, taking care not to bend it at any sharp angles.

To install:

8. If the crankshaft, idler shaft or camshaft were disturbed while the belt was out, align each shaft as follows with its corresponding index mark to assure proper valve timing and ignition timing:

 a. Rotate the crankshaft so the notch in the convex crankshaft gear belt guide aligns with the embossed mark on the front cover (12 o'clock position).

 b. Rotate the idler shaft so the dot on the idler shaft drive sprocket aligns with the notch on the timing belt rear cover (4 o'clock position).

 c. Rotate the camshaft so the notch in the camshaft sprocket inner belt guide aligns with the notch in the forward edge of the valve cover (12 o'clock position).

9. Install the timing belt over the sprockets, then over the tensioner roller. Do not use any sharp tools. New belts have yellow marks. The 2 lines on the drive belt should fit toward the crankshaft marks. The next mark should fit toward the intermediate shaft marks, etc.

10. Loosen the tensioner nut and let the spring tension automatically take up the slack. Tighten the tensioner nut to 37 ft. lbs. (51 Nm).

11. Rotate the crankshaft two full revolutions clockwise and be sure the timing marks align.

12. Install the drive belts, radiator fan and shroud. Connect the negative battery cable.

2.3L (B234F) ENGINE

➡ **The B234F engine has 2 belts, one driving the camshafts and one driving the balance shafts. The camshaft belt may be removed separately. The balance shaft belt requires removal of the camshaft belt. During reassembly, the exact placement of the belts and pulleys must be observed.**

1. Remove the negative battery cable and the alternator belt.
2. Remove the radiator fan, its pulley and the fan shroud.
3. Remove the drive belts for the power steering belts and the air conditioning compressor.
4. Beginning with the top cover, remove the retaining bolts and remove the timing belt covers.
5. Turn the engine to TDC, of the compression stroke, on cylinder No. 1. Make sure the marks on the camshaft pulleys align with the marks on the backing plate and that the marking on the belt guide plate (on the crankshaft) is opposite the TDC mark on the engine block.
6. Remove the protective cap over the timing belt tensioner locknut. Loosen the locknut, compress the tensioner to release tension on the belts and re-tighten the locknut, holding the tensioner in place.
7. Remove the timing belt from the camshafts. Do not crease or fold the belt.

➡ **The camshafts and the crankshaft must not be moved when the belt is removed.**

8. Check the tensioner by spinning it counterclockwise and listening for any bearing noise within. Check also that the belt contact surface is clean and smooth. In the same fashion, check the timing belt idler pulleys. Make sure the are tightened to 18.5 ft. lbs. (25 Nm).

9. If the balance shaft belt is to be removed:

 a. Remove the balance shaft belt idler pulley from the engine.

 b. Loosen the locknut on the tensioner and remove the belt. Slide the belt under the crankshaft pulley assembly. Check the tensioner and idler wheels carefully for any sign of contamination. Check the ends of the shafts for any sign of oil leakage.

 c. Check the position of the balance shafts and the crankshaft after belt removal. The balance shaft markings on the pulleys should align with the markings on the backing plate and the crankshaft marking should still be aligned with the TDC mark on the engine block.

 d. When refitting the balance shaft belt, observe that the belt has colored dots on it. These marks assist in the critical placement of the belt. The yellow dot will align the right lower shaft, the blue dot will align on the crank and the other yellow dot will match to the upper left balance shaft.

 e. Carefully work the belt in under the crankshaft pulley. Make sure the blue dot is opposite the bottom (TDC) marking on the belt guide plate at the bottom of the crankshaft. Fit the belt around the left upper balance shaft pulley, making sure the yellow mark is opposite the mark on the pulley. Install the belt around the right lower balance shaft pulley and again check that the mark on the belt aligns with the mark on the pulley.

 f. Work the belt around the tensioner. Double check that all the markings are still aligned.

 g. Set the belt tension by inserting an Allen key into the adjusting hole in the tensioner. Turn the crankshaft carefully through a few degrees on either side of TDC to check that the belt has properly engaged the pulleys. Return the crank to the TDC position and set the adjusting hole just below the 3 o'clock position when tightening the adjusting bolt. Use the Allen

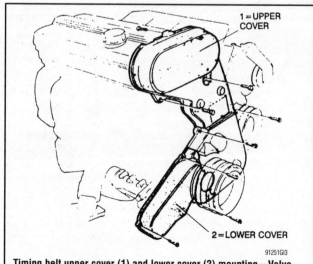

Timing belt upper cover (1) and lower cover (2) mounting—Volvo 2.3L (B234F) engine

Volvo—TIMING BELTS 223

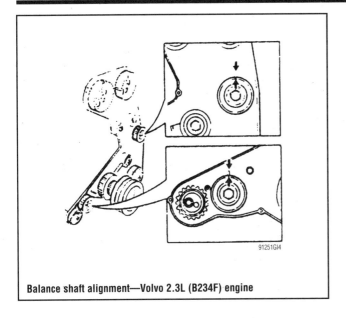

Balance shaft alignment—Volvo 2.3L (B234F) engine

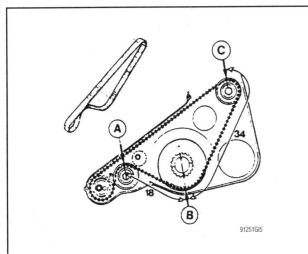

Balance shaft belt markings. There should be 18 teeth between A and B and 34 teeth between B and C—Volvo 2.3L (B234F) engine

wrench, in the adjusting hole, as a counter hold and tighten the locking bolt to 29.5 ft. lbs. (40 Nm).

  h. Use tool 998 8500 or equivalent to check the tension of the belt. Install the gauge over the position of the removed idler pulley. The tension must be 1–4 units on the scale or the belt must be readjusted.

To install:

10. Reinstall the camshaft belt by aligning the double line marking on the belt with the top marking on the belt guide plate at the top of the crankshaft. Stretch the belt around the crank pulley and place it over the tensioner and the right side idler. Place the belt on the camshaft pulleys. The single line marks on the belt should align exactly with the pulley markings. Route the belt around the oil pump drive pulley and press the belt onto the left side idler.

11. Check that all the markings align and that the engine is still positioned at TDC, of the compression stroke, for cylinder No. 1.

12. Loosen the tensioner locknut.

13. Turn the crankshaft clockwise. The camshaft pulleys should rotate 1 full turn until the marks again align with the marks on the backing plate.

➡ **The engine must not be rotated counterclockwise during this procedure.**

14. Smoothly rotate the crankshaft further clockwise until the camshaft pulley markings are 1½ teeth beyond the marks on the backing plate. Tighten the tensioner locknut.

15. Check the tension on the balance shaft belt; it should now be 3.8 units on a suitable belt tension gauge. If the tension is too low, adjust the tensioner clockwise. If the tension is too high, repeat Step 9g.

16. Check the belt guide for the balance shaft belt and make sure it is properly seated. Install the center timing belt cover, the one that covers the tensioner, the fan shroud, fan pulley and fan. Install all the drive belts and connect the battery cable.

17. Double check all installation items, paying particular attention to loose hoses or hanging wires, untightened nuts, poor routing of hoses and wires (too tight or rubbing) and tools left in the engine area.

18. Start the engine and allow it to run until the thermostat opens.

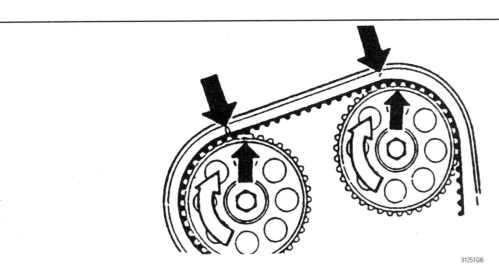

Rotate the crankshaft clockwise to position the camshaft 1½ teeth beyond the alignment marks to properly tension the belt—Volvo 2.3L (B234F) engine

224 TIMING BELTS—VOLVO

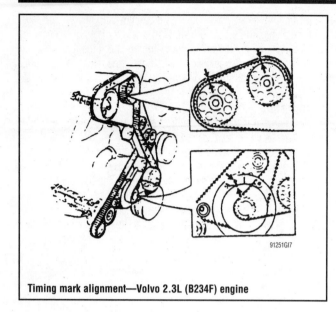

Timing mark alignment—Volvo 2.3L (B234F) engine

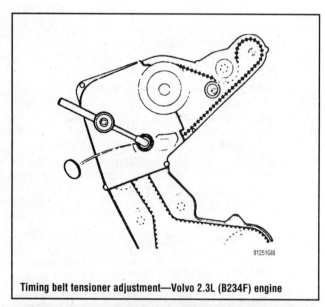

Timing belt tensioner adjustment—Volvo 2.3L (B234F) engine

✲✲ CAUTION

The upper and lower timing belt covers are still removed. The belt and pulleys are exposed and moving at high speed.

19. Shut the engine OFF and bring the motor to TDC, of the compression stroke, on cylinder No. 1.
20. Check the tension of the camshaft belt. Position the gauge between the right (exhaust) camshaft pulley and the idler. Belt tension must be 5.5 plus or minus 0.2 units on a suitable belt tension gauge. If the belt needs adjustment, remove the rubber cap over the tensioner locknut, cap is located on the timing belt cover, and loosen the locknut.
21. Insert a suitable tool between the tensioner wheel and the spring carrier pin to hold the tensioner. If the belt needs to be tightened, move the roller to adjust the tension to 6.0 units. If the belt is too tight, adjust to obtain a reading of 5.0 units on the gauge. Tighten the tensioner locknut.
22. Rotate the crankshaft so the camshaft pulleys move through 1 full revolution and recheck the tension on the camshaft belt. It should now be 5.5 plus or minus 0.2 units. Install the plastic plug over the tensioner bolt.
23. Final check the tension on the balance shaft belt by fitting the gauge and turning the tensioner clockwise. Only small movements are needed. After any needed readjustments, rotate the crankshaft clockwise through 1 full revolution and recheck the balance shaft belt. The tension should now be on the final specification of 4.9 plus or minus 0.2 units.
24. Install the idler pulley for the balance shaft belt. Reinstall the upper and lower timing belt covers.
25. Start the engine and final check performance.

2.3L (B5234) & 2.4L ENGINES

1. Disconnect the negative battery cable.
2. Remove the coolant expansion tank and place it on top of the engine.
3. Remove the spark plug cover and drive belts.
4. Remove the timing belt cover.
5. Wait five minutes after lining up marks, then install Volvo Gauge 998–8500 (or equivalent) between the exhaust camshaft and water pump. Read the gauge using a mirror, while still installed. For 23mm belts, the tension should be 2.7–4.0 units.

➡**If the belt tension is incorrect, the tensioner must be replaced.**

6. Remove the upper tensioner bolt and loosen the lower bolt, turning the tensioner to free up the pulley.
7. Remove the lower bolt and the tensioner. Remove the timing belt.

To install:
8. Turn all the pulleys listening for bearing noise. Check to see that the contact surfaces are clean and smooth. Remove the tensioner pulley lever and idler pulley, lubricate the contact surfaces and bearing with grease. If the tensioner pulley lever or idler is seized replace it.
9. Install the tensioner pulley lever and idler pulley and tighten to 18 ft. lbs. (25 Nm).
10. Compress the tensioner with Volvo Tool 999–5456 (or equivalent) and insert a 0.079 in. (2.0mm) lockpin in the piston. If the tensioner leaks, has no resistance, or will not compress, replace it. Install the tensioner and tighten to 18 ft. lbs. (25 Nm).
11. Install the timing belt in the following order:
    a. Around the crankshaft sprocket.
    b. Around the right idler pulley
    c. Around the camshaft sprockets
    d. Around the water pump
    e. Onto the tensioner pulley
12. Pull the lockpin out from the tensioner and install the upper timing cover. Turn the crankshaft two complete revolutions and check to see that the timing marks on the crankshaft and camshaft pulleys are lined up.
13. Install the timing belt covers and the fuel line clips.
14. Install the accessory belts.
15. Install the vibration damper guard and the inner fender well.

Volvo—TIMING BELTS

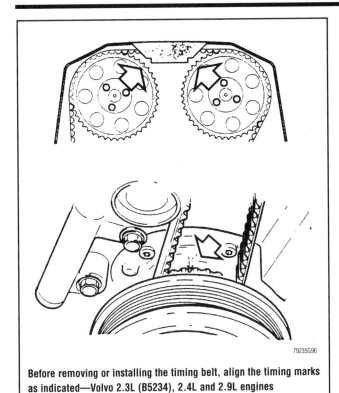

Before removing or installing the timing belt, align the timing marks as indicated—Volvo 2.3L (B5234), 2.4L and 2.9L engines

16. Install the spark plug cover and any remaining components.

2.9L (6304F) ENGINES

1990–94 Models

1. Disconnect the negative battery cable.
2. Remove the auxiliary drive belts.
3. Remove the front timing belt cover.
4. Remove the splash guard, vibration damper guard and ignition coil cover.
5. Rotate the crankshaft clockwise, until the timing marks on the camshaft pulleys and timing belt mounting plate and crankshaft pulley/oil pump housing are aligned. Remove the upper timing belt cover.
6. Check the belt tensioner. Replace the tensioner, if required.
7. Remove the tensioner upper mounting bolts. Loosen the tensioner lower mounting bolt and twist the tensioner to free the plunger. Remove the lower mounting bolt and remove the tensioner.
8. Remove the timing belt.

➡ **Do not rotate the crankshaft while the timing belt is removed.**

9. Check the tensioner and idler pulleys, as follows:
   a. Spin the pulleys and listen for bearing noise.
   b. Check that the pulley surfaces in contact with the belt are clean and smooth.
   c. Check the tensioner pulley arm and idler pulley mountings.
   d. Tighten the tensioner pulley arm to 30 ft. lbs. (40 Nm) and the idler pulley to 18 ft. lbs. (25 Nm).

To install:

10. Place the belt around the crankshaft pulley and right side idler. Place the belt over the camshaft pulleys. Position the belt around the water pump and press over tensioner pulley.
11. Insert the tensioner mounting bolts. Tighten to 18 ft. lbs. (25 Nm).
12. Remove the locking pin. Install the front timing belt cover.
13. Turn the crankshaft through 2 revolutions and check that the timing marks on the crankshaft and camshaft pulleys are correctly aligned.
14. Install the ignition coil, front timing belt cover, auxiliary drive belts, vibration damper guard and splash guard.
15. Connect the negative battery lead, start and check the engine operation.

➡ **The lever bushing, on the B234F and B6304F engines, must be greased every time the belt is replace or the tensioner pulley removed. This is necessary to help prevent seizure of the bushing, with the possible risk of incorrect belt tension. Service the bushing, using the following procedure:**

- Remove the lever mounting bolt, tensioner pulley and sleeve behind the bolt.
- Grease the surfaces of the bushing, bolt and sleeve, using part 1161246-2 or equivalent.
- Install the sleeve, tensioner pulley and lever mounting bolt.
- Tighten the bolt 30 ft. lbs. (40 Nm).

1995–99 Models

1. Disconnect the negative battery cable.
2. Remove the drive belts.
3. Remove the timing belt cover.
4. Remove the splash guard, vibration damper guard and ignition coil cover.
5. Rotate the crankshaft clockwise, until the timing marks on the camshaft pulleys and timing belt mounting plate and crankshaft pulley/oil pump housing are aligned.
6. Remove the tensioner upper mounting bolts. Loosen the tensioner lower mounting bolt and twist the tensioner to free the plunger. Remove the lower mounting bolt and remove the tensioner.
7. Remove the timing belt.

➡ **Do not rotate the crankshaft while the timing belt is removed.**

8. Check the tensioner and idler pulleys, as follows:
   a. Spin the pulleys and listen for bearing noise.
   b. Check that the pulley surfaces in contact with the belt are clean and smooth.
   c. Check the tensioner pulley arm and idler pulley mountings.
   d. Tighten the tensioner pulley arm to 30 ft. lbs. (40 Nm) and the idler pulley to 18 ft. lbs. (25 Nm).
   e. Compress the tensioner using tool 5456 or equivalent. Mount the tensioner in the tool and tighten the center nut fully. Wait until compression has taken place and insert a 2mm locking pin in the plunger.

➡ **The tensioner must be replaced if leakage is observed or the plunger offers no resistance when depressed, or cannot be depressed.**

226 TIMING BELTS—VOLVO

To install:

9. Place the belt around the crankshaft pulley and right-side idler. Place the belt over the camshaft pulleys. Position the belt around the water pump and press over the tensioner pulley.

10. Insert the tensioner mounting bolts. Tighten to 18 ft. lbs. (25 Nm).

11. Remove the locking pin from the tensioner. Install the front timing belt cover.

➡ **The lever bushing must be greased every time the belt is replace or the pulley is removed. Service the bushing, using the following procedure:**

   a. Remove the lever mounting bolt, tensioner pulley and sleeve.
   b. Grease the surfaces of the bushing, bolt and sleeve, using Volvo Part No. 1161246–2 or equivalent.
   c. Install the sleeve, tensioner pulley and lever mounting bolt.
   d. Tighten the bolt to 30 ft. lbs. (40 Nm).

12. Turn the crankshaft two revolutions and check that the timing marks on the crankshaft and camshaft pulleys are correctly aligned.

13. Install the remaining components.

14. Connect the negative battery lead, start and check the engine operation.

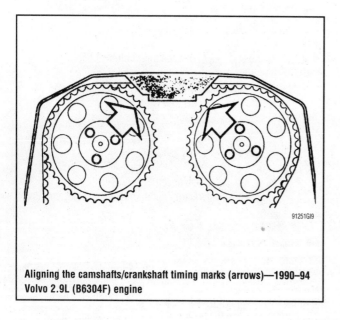

Aligning the camshafts/crankshaft timing marks (arrows)—1990–94 Volvo 2.9L (B6304F) engine

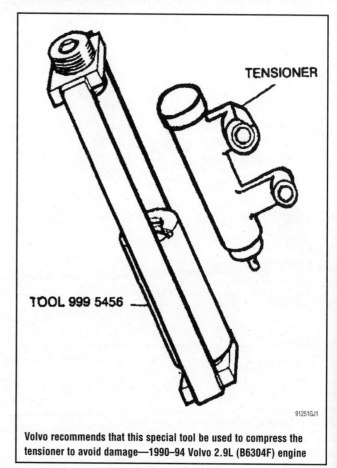

Volvo recommends that this special tool be used to compress the tensioner to avoid damage—1990–94 Volvo 2.9L (B6304F) engine

CONVERSION FACTORS

LENGTH–DISTANCE

Inches (in.)	x 25.4	= Millimeters (mm)	x .0394	= Inches
Feet (ft.)	x .305	= Meters (m)	x 3.281	= Feet
Miles	x 1.609	= Kilometers (km)	x .0621	= Miles

VOLUME

Cubic Inches (in3)	x 16.387	= Cubic Centimeters	x .061	= in3
IMP Pints (IMP pt.)	x .568	= Liters (L)	x 1.76	= IMP pt.
IMP Quarts (IMP qt.)	x 1.137	= Liters (L)	x .88	= IMP qt.
IMP Gallons (IMP gal.)	x 4.546	= Liters (L)	x .22	= IMP gal.
IMP Quarts (IMP qt.)	x 1.201	= US Quarts (US qt.)	x .833	= IMP qt.
IMP Gallons (IMP gal.)	x 1.201	= US Gallons (US gal.)	x .833	= IMP gal.
Fl. Ounces	x 29.573	= Milliliters	x .034	= Ounces
US Pints (US pt.)	x .473	= Liters (L)	x 2.113	= Pints
US Quarts (US qt.)	x .946	= Liters (L)	x 1.057	= Quarts
US Gallons (US gal.)	x 3.785	= Liters (L)	x .264	= Gallons

MASS–WEIGHT

Ounces (oz.)	x 28.35	= Grams (g)	x .035	= Ounces
Pounds (lb.)	x .454	= Kilograms (kg)	x 2.205	= Pounds

PRESSURE

Pounds Per Sq. In. (psi)	x 6.895	= Kilopascals (kPa)	x .145	= psi
Inches of Mercury (Hg)	x .4912	= psi	x 2.036	= Hg
Inches of Mercury (Hg)	x 3.377	= Kilopascals (kPa)	x .2961	= Hg
Inches of Water (H_2O)	x .07355	= Inches of Mercury	x 13.783	= H_2O
Inches of Water (H_2O)	x .03613	= psi	x 27.684	= H_2O
Inches of Water (H_2O)	x .248	= Kilopascals (kPa)	x 4.026	= H_2O

TORQUE

Pounds–Force Inches (in–lb)	x .113	= Newton Meters (N·m)	x 8.85	= in–lb
Pounds–Force Feet (ft–lb)	x 1.356	= Newton Meters (N·m)	x .738	= ft–lb

VELOCITY

Miles Per Hour (MPH)	x 1.609	= Kilometers Per Hour (KPH)	x .621	= MPH

POWER

Horsepower (Hp)	x .745	= Kilowatts	x 1.34	= Horsepower

FUEL CONSUMPTION*

Miles Per Gallon IMP (MPG)	x .354	= Kilometers Per Liter (Km/L)
Kilometers Per Liter (Km/L)	x 2.352	= IMP MPG
Miles Per Gallon US (MPG)	x .425	= Kilometers Per Liter (Km/L)
Kilometers Per Liter (Km/L)	x 2.352	= US MPG

*It is common to covert from miles per gallon (mpg) to liters/100 kilometers (1/100 km), where mpg (IMP) x 1/100 km = 282 and mpg (US) x 1/100 km = 235.

TEMPERATURE

Degree Fahrenheit (°F) = (°C x 1.8) + 32
Degree Celsius (°C) = (°F − 32) x .56

CHILTON'S...
For Professionals, From The Professionals

Hard-Cover Shop Editions

Designed for the professional technician, Chilton's Shop Manuals on automotive repair are more comprehensive and technically detailed. Information is provided in an easy-to-read format, supported by quick-reference sections as well as exploded-view illustrations, diagrams and charts. All this at the mechanic's fingertips insures fast, accurate repairs.

Chilton's Auto Repair Manual 1995-99
Part No. 7922, List price $59.95

Chilton's Import Car Repair Manual 1995-99
Part No. 7923, List price $59.95

Chilton's Truck and Van Repair Manual 1995-99
Part No. 7924, List price $59.95

Ordering Is Easy!
Call Toll-Free: 800-477-3996
Or Fax: 610-738-9370

CHILTON'S®

1020 Andrew Drive, Suite 200 • West Chester, PA 19380-4291
Phone: 610-738-9280 • 800-695-1214 • Fax: 610-738-9373

NEW!

Chilton's Professional Series has now been expanded to include quick-reference manuals for the automotive professional. These manuals provide complete coverage on repair and maintenance, adjustments, and diagnostic procedures for specific systems or components.

TIMING BELTS 1990-99
Part No. 9125, List price $34.95
A model-specific, easy-to-use guide providing general information on the replacement, inspection and installation of timing belts for virtually all domestic and import cars, trucks and vans. Includes detailed illustrations of timing mark locations and belt positioning.

ELECTRIC COOLING FANS, ACCESSORY DRIVE BELTS & WATER PUMPS 1995-99
Part No. 9126, List price $34.95
Offers information for accessory serpentine and V-belt inspection, adjustment, removal and installation, as well as model-specific drive belt routing diagrams. Coverage of electric cooling fans includes troubleshooting, removal and installation, as well as illustrations showing mounting and complete wiring schematics. Water pump removal and installation is also covered.

POWERTRAIN CODES & OXYGEN SENSORS 1990-99
Part No. 9127, List price $34.95
Provides general information on how the O2 sensor works, as well as specific testing and locations, and removal and installation. Includes complete listing of all codes and information on how to read and clear Maintenance Indicator Lamp (MIL) and OBD-I and OBD-II Diagnostic Trouble Codes (DTC).

SPECIFICATIONS AND MAINTENANCE INTERVALS 1995-99
Part No. 9128, List price $34.95
Contains factory-recommended maintenance intervals, along with detailed charts on vehicle identification, engine identification, brakes, tune-up specifications, capacities and engine rebuilding torque and valve specifications.